Biopolymer Science for Proteins and Peptides

Biopolymer Science for Proteins and Peptides

KEIJI NUMATA
Department of Material Chemistry, Kyoto University, Kyoto, Japan

Biomacromolecules Research Team, RIKEN Center for Sustainable Resource Science, Saitama, Japan

ELSEVIER

Elsevier
Radarweg 29, PO Box 211, 1000 AE Amsterdam, Netherlands
The Boulevard, Langford Lane, Kidlington, Oxford OX5 1GB, United Kingdom
50 Hampshire Street, 5th Floor, Cambridge, MA 02139, United States

British Library Cataloguing-in-Publication Data
A catalogue record for this book is available from the British Library

Library of Congress Cataloging-in-Publication Data
A catalog record for this book is available from the Library of Congress

ISBN: 978-0-12-820555-6

For Information on all Elsevier publications
visit our website at https://www.elsevier.com/books-and-journals

Publisher: Matthew Deans
Acquisitions Editor: Edward Payne
Editorial Project Manager: Ruby Smith
Production Project Manager: Prem Kumar Kaliamoorthi
Cover Designer: Mark Rogers

Typeset by MPS Limited, Chennai, India

Contents

Preface

In April 2020 my primary appointment moved from RIKEN, a national research institute in Japan, to the Department of Material Chemistry, Kyoto University. In that Japanese spring, we suffered COVID-19 and were not allowed to move around. Scientists and researchers could not go out for any experiments, including environmental tests and synchrotron measurements, and could not join any international conferences and events. Everything related to science seemed to shift into virtual and online systems. These research activities' limitations motivated me to write a textbook for the classes and lectures on biopolymer (biomacromolecules) chemistry and biomaterial science. The goal of this textbook is to introduce the basic science of structural proteins and related peptides, and also to share my structural protein research.

Biopolymers have much potential in terms of sustainability as well as their physical, chemical, and biological properties. In this textbook, the general introduction including the comparison with synthetic polymers (Chapter 1, General Introduction of Polypeptide and Protein Materials), synthesis (Chapter 2, Synthesis), structure (Chapter 3, Structure), physical property (Chapter 4, Physical Properties), biological property (Chapter 5, Biological Properties With Cells), stability (Chapter 6, Stability), examples in nature (Chapter 7, Structural Proteins in Nature), material design (Chapter 8, Biopolymer Material and Composite), and industrial application (Chapter 9, Applications as Bulk Material and Future Perspective) of structural proteins are summarized based on the many references. Chapter 10, Experimental Details, lists the experimental details that are described in the other chapters. I hope this book will be helpful for the students and researchers in polymer science, material science, biochemistry, and biology to understand the current situation of structural proteins.

This textbook introduces many examples concerning silk proteins because I have been studying them for 12 years so far. As the author of this textbook, I was supposed to include all the structural proteins. However, due to the limitation of my knowledge and time, the other proteins are not included in this textbook very much. I will expand my research activity and knowledge and then will include all types of structural proteins in the future.

I would like to express my gratitude to all of my colleagues, especially, Stern Family Professor in Engineering, Professor David L. Kaplan, Tufts University. Professor Kaplan introduced me to the silk world, which further took me to biopolymer sciences, including structural protein research. I express my gratitude to Professor Emeritus Yoshiharu Doi, Tokyo Institute of Technology and RIKEN, who supervised me when I joined the environmentally degradable polymer science fields. I am also grateful to my laboratory members who discussed and develop the structural protein-related research fields.

Finally, I sincerely thank my family, Kanna, Futaba, Daichi, and Aya, for their continuous encouragement.

Keiji Numata
Department of Material Chemistry, Kyoto University, Kyoto, Japan
February 2021

CHAPTER 1

General introduction of polypeptide and protein materials

1.1 Calcification of biological polymers

In nature, there are four types of biopolymers, namely nucleic acid, polyhydroxyalkanoate, polysaccharide, and protein. This textbook focuses on biopolymers composed of amino acids, namely, protein and polypeptides (Fig. 1.1), while the differences from other biopolymers are introduced here.

1.1.1 Nucleic acids

Nucleic acids are DNA (deoxyribonucleic acid) and RNA (ribonucleic acid), which are essential molecules encoding the biological information of all lives on Earth. Until now, the functions of all DNA sequences have not yet been clarified; recently, open reading frames have been studied widely to clarify the function of DNA that does not encode proteins. Molecular weights of nucleic acids are dependent on their biological function and roles. They are composed of nucleotides, which are the monomers composed of three structures: a 5-carbon sugar, a phosphate group, and a nitrogenous base (Fig. 1.2). RNA and DNA have ribose and deoxyribose as the sugar, respectively.

DNA is an attractive molecule as a component of nanoscale functional material, especially due to its biocompatibility, molecular recognition ability, and nanoscale controllability [1]. DNA nanotechnology has created various DNA materials such as DNA origami and DNA-based hydrogels.

1.1.2 Polyhydroxyalkanoate

Poly(hydroxyalkanoate)s (PHAs)—eco-friendly and biodegradable polymeric materials—are polyesters synthesized by a variety of bacteria as an intracellular storage material of carbon and energy [2–4]. Monomer units of PHA are shown in Fig. 1.3. Poly[(*R*)-3-hydroxybutyrate] (P(3HB)) was the first PHA isolated from *Bacillus megaterium* in the 1920s and later identified as a microbial reserve polyester [5]. PHAs are attractive as biomaterials, such as scaffolds for tissue engineering, because of their biocompatibility, processability, and wide range of mechanical properties related

Biopolymer Science for Proteins and Peptides
DOI: https://doi.org/10.1016/B978-0-12-820555-6.00002-1

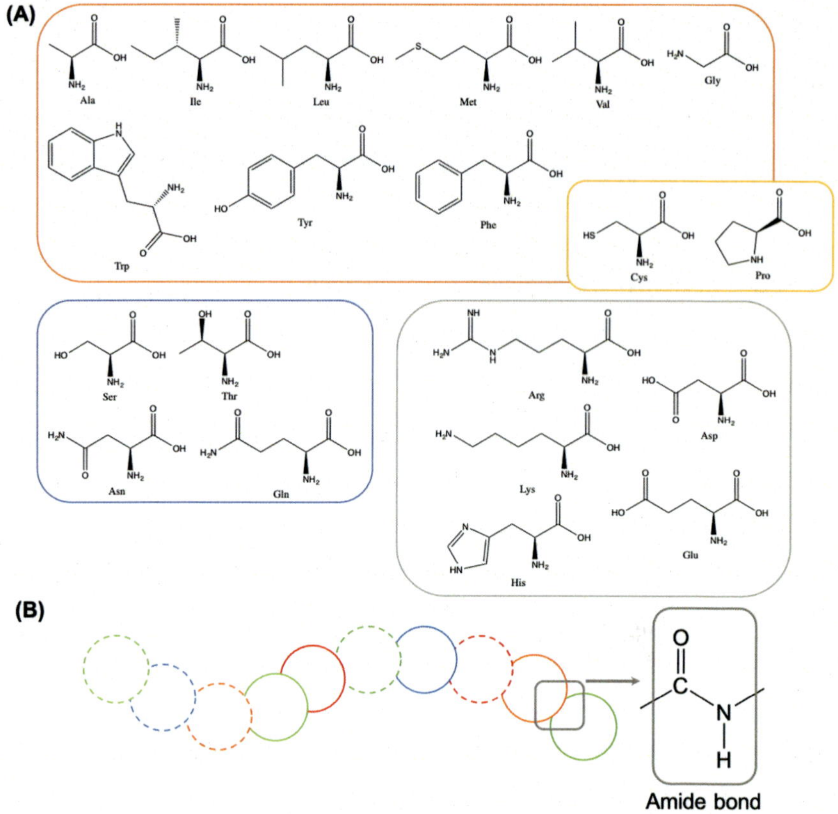

Figure 1.1 (A) Chemical structures of 20 natural L-amino acids. Depending on the biochemical properties of the amino acid, it can be classified as a hydrophobic (non-polar) amino acid, neutral amino acid, polar amino acid, or acidic amino acid. (B) Schematic illustration of a protein. A protein is a polymer (high-molecular-weight chain molecule) in which amide bonds connect amino acids.

to the chemistry of the secondary monomer units. The identification of hydroxyalkanoate units other than 3HB as constituents of microbial reserve polyesters has had a significant impact on research and commercial interests in these microbial polyesters, since the incorporation of different hydroxyalkanoate units into P(3HB) can change the mechanical, thermal, and biological properties [6–9]. Poly[(*R*)-3-hydroxybutyrate-*co*-(*R*)-3-hydroxyhexanoate] (P(3HB-*co*-3HH_x)), one of the PHAs, was also reported to show lower cytotoxicity and higher biocompatibility in comparison to P(3HB) and poly(L-lactide) (PLLA) [10]. In 1988 PHA with 4-hydroxybutyrate units (4HB), the secondary units of PHAs, which has higher

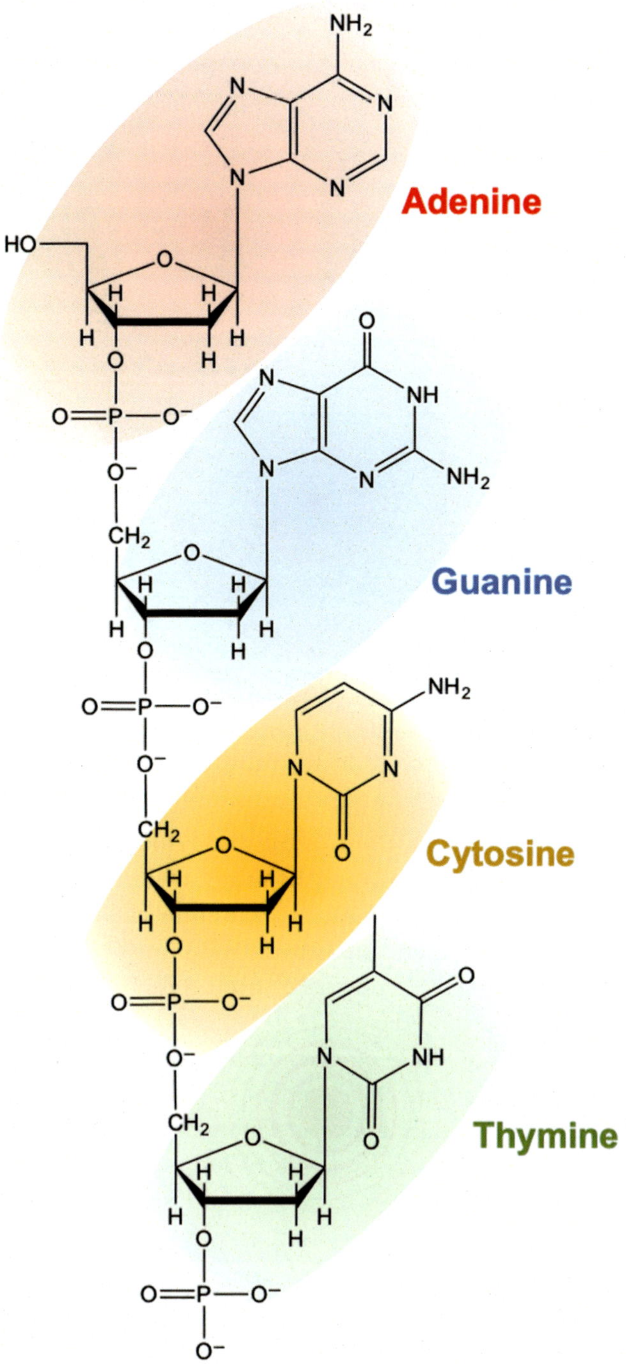

Figure 1.2 Chemical structure of DNA. DNA is composed of four types of nucleic acids that contain different nitrogen bases, namely, adenine, guanine, cytosine, and thymine.

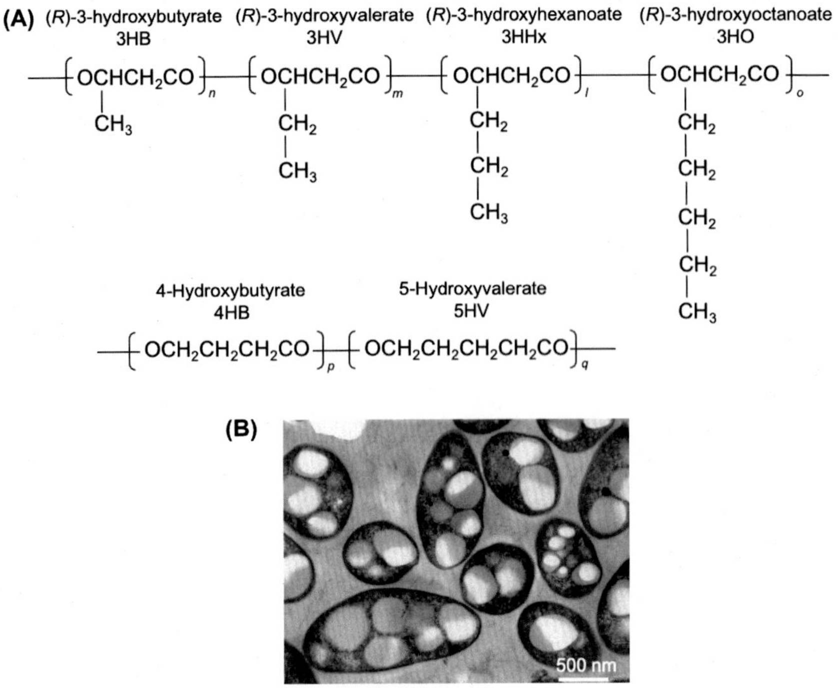

Figure 1.3 (A) Chemical structures of various PHA. (B) TEM image of PHA granules accumulated in photosynthetic bacteria.

in vivo biodegradability in comparison to the other PHAs, was synthesized via the bacterial pathways of *Ralstonia eutropha* H16 (Currently, *Cupriavidus necator* H16) [11]. PHAs containing 4HB units have been particularly attractive as biomaterials for tissue engineering and implantation.

As eco-friendly polymeric materials, PHAs have several advantages: the ability to be directly synthesized from various types of biomass, including lignin and carbon dioxide [12–16], and excellent biodegradability [17,18]. However, because of a lack of toughness, this biopolyester has limited use and application. Development of an engineered biopolymer, such as biomass-derived polyamides and protein polymers, is an emerging interest in the field of sustainable chemistry and materials engineering.

1.1.3 Polysaccharide

Polysaccharides, which are biosynthesized in woods, plants, algae, bacteria, and fungi, consist of a variety of polymers with glycosidic linkages (Fig. 1.4). Polysaccharides have useful properties for biologically derived

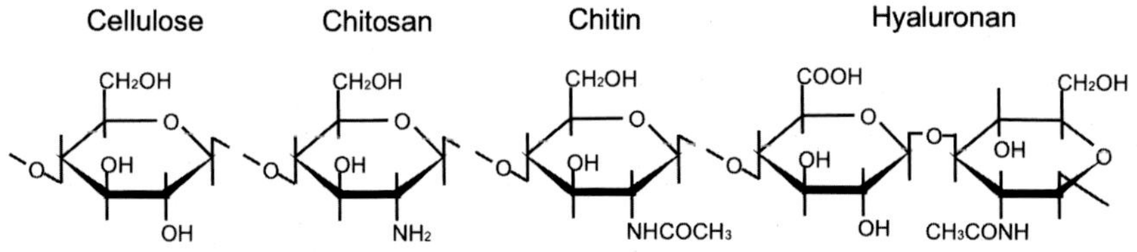

Figure 1.4 Chemical structures of monomeric units of polysaccharides.

scaffolds, such as nontoxicity, water-solubility, stability to variations in pH, and can be chemically modified to achieve various functions; the disadvantages are low mechanical properties, and poor thermal and chemical stability [19].

Cellulose, a polysaccharide composed of D-glucose via β-1,4 linkages, is an essential component of the cell wall to form the cellular structures. It is the most abundant and renewable biopolymer on Earth. Bacterial cellulose secreted from some bacteria such as *Gluconacetobacter xylinus* (*Acetobacter xylinum*) has been explored as food additives and biomedical materials. In 1920 Staudinger, a Nobel Prize laureate for Chemistry in 1953, determined the polymer structure of cellulose [12]. Furthermore, in 1991 Kobayashi chemically synthesized cellulose compound by utilizing in vitro cellulase reactions [13].

Chitosan is a linear polysaccharide composed of randomly distributed β-1,4-linked D-glucosamine (deacetylated unit) and *N*-acetyl-D-glucosamine (acetylated unit). It is made by treating the chitin shells of shrimp and other crustaceans with an alkaline substance, like sodium hydroxide. Chitosan has many commercial and possible biomedical uses. It can be used in agriculture as a seed treatment and biopesticide, helping plants to fight off fungal infections. Chitin, a biopolymer of *N*-acetylglucosamine with some glucosamine, is the main component of the cell walls of fungi, the exoskeletons of arthropods such as crustaceans and insects, the radulas of mollusks, and the beaks of cephalopods. Hyaluronan (hyaluronic acid or hyaluronate) is the main nonprotein glycosaminoglycan component of the extracellular matrix, and is composed of repeating units of glucuronic acid and *N*-acetylglucosamine [14,15]. As a component of the extracellular matrix, hyaluronan plays a vital role in cell proliferation, cell-migration, matrix assembly, and tissue development. It may be involved in the progression of some malignant tumors. These polysaccharides have great potential as biopolymer materials in various fields; however, the variation of monomeric units is relatively less compared to proteins and polypeptides.

1.1.4 Biological polyamides other than protein/polypeptide

In terms of chemical structure, protein and polypeptide are classified as biological polyamide. However, the polyamides generally considered as bioplastics are polyamide 4 (nylon 4) and polyamide 11 (nylon 11). They are widely recognized as bio-based polyamides because they are synthesized from gamma aminobutyric acid (GABA) and castor oil by decarboxylation of glutamic acid as a raw material. In general, polyamides are expected to have good mechanical properties, thermal properties, and gas barrier properties, and are expected to have physical properties different from those of polyester that is often found in bioplastic materials. Polyamide 4 has also been reported to have excellent biodegradability in natural environments and is a rare polyamide. On the other hand, it is considered that polyamide 4 has a thermal decomposition temperature close to the melting point and that the range of process temperature in injection molding and extrusion molding is narrow, which is a technical bottleneck.

In addition to nylon that a polyamide derived from these biological resources, bio-based polymers that are drawing attention in recent years are macromolecules composed of amino acids such as proteins. In the natural world there are many proteins with special mechanical properties which cannot be imitated by synthetic polymers, such as silk, keratin, and resilin. However, many unsolved points remain in the mechanism of expression of the physical properties, and understanding of the molecular level based on a hierarchical structure is required. In particular, some spider silk threads composed of silk proteins have been reported to exhibit high toughness, which cannot be achieved by synthetic polymers. This high toughness is also useful as a structural material used by human beings and attracts attention from a wide range of material fields.

1.2 Protein and polypeptides

Polypeptide and protein are biopolymers composed of amino acids as monomeric units with amide bonds. The use of polypeptide and protein as functional materials has been widely studied, resulting in the development of bioactive polypeptides and biological applications such as in tissue engineering, regenerative medicine, and drug/gene delivery systems. Polypeptides are further endowed with remarkable biological functions such as target specificity, and their degradation products can be readily metabolized. Due to those functions, various hybrid polypeptides with synthetic polymers have also been studied as functional materials. In contrast, the use of polypeptide and protein as structural materials is still limited and challenging, due to a lack of understanding of the structure–function relationship of protein-based materials containing multiple types of water molecules. The role of water molecules in thermal stability and the biological and mechanical properties of structural proteins and polypeptides have been characterized widely [16]. However, this information has not been sufficient to develop artificial protein materials with the desired thermal stability and toughness. Furthermore, one of the significant drawbacks of protein-based materials is limited synthesis methods. To synthesize polypeptides and proteins, current synthesis techniques, such as solid-phase peptide synthesis (SPPS) and recombinant DNA techniques, still have limitations in their production capacity, atom economy, and sequence regulation (see details in Chapter 2: Synthesis). Synthesis through an environment-friendly and scalable process will be needed to establish protein and polypeptide as bio-based materials.

1.2.1 Structural protein

In many cases, a structural protein has a repeated characteristic amino acid sequence and forms a higher-order structure by intermolecular and/or intramolecular hydrogen bonding. For example, silk protein has a tandem sequence of polyalanine and glycine-rich sequences. Polyalanine is a potentially crystalline motif, whereas the glycine-rich sequence plays a role in an amorphous region. This characteristic repetitive arrangement allows higher-order structures in which the microcrystals of the beta sheet structure are highly oriented along the fiber axis. However, in general, the amino acid sequence and the structure to be formed depend on the type and condition of protein, and it isn't easy to define the sequence or structure. In particular, considering nonnatural sequences created by gene recombination techniques and chemical synthesis, the definition is even more difficult and complicated. Therefore, in this review, the definition of structural protein is defined as "a protein in which a characteristic amino acid sequence or motif repeatedly exists and forms a skeleton of a living organism, cell, or material." Preferably, it is expected to form a higher-order structure by intermolecular or intramolecular interaction, leading to express physical characteristics. However, considering protein-based amorphous material, the hierarchical structure should not be included in the definition here. Some examples of structural proteins in nature are introduced in Chapter 7, Structural Proteins in Nature.

1.3 Differences and similarities to synthetic polymers

Structural protein materials are different in comparison with synthetic polymer materials from the viewpoint of physical properties and structure. Here, let me clarify the fact that structural protein has very different characteristics in terms of chemical structures and material properties. The following six points can be mentioned as characteristic and fundamental differences. Here, the relationship between the structure and physical properties will be outlined with an example of spider silk proteins.

1.3.1 Molecular weight and its distribution

When a polymer is synthesized by a chemical reaction, monomers are connected to obtain a polymer. This reaction is called a polymerization reaction, and polymer products derived from fossil resources are synthesized by this polymerization reaction. Polymeric materials that do not require this chemical reaction are limited to biologically derived substances. The polymer synthesis reaction, that is, the polymerization reaction, has been studied by many researchers. One challenge in this synthesis is to control the degree of polymerization of the polymer. At present, it is very difficult to chemically synthesize a polymer having no molecular weight distribution and having a single molecular weight. SPPS can connect monomers one by one, but it is not practical to use as a method for synthesizing bulk polymeric materials from the viewpoint of economics, time, and chemical efficiency. Proteins synthesized in vivo by ribosome through transcription and translation from DNA have exactly the same molecular weight, and have no molecular weight distribution, and become a single component (Fig. 1.5). This is the same for structural proteins, which is very different from polymer materials synthesized chemically from fossil fuels. Such unity likely contributes to the various hierarchical structures that structural proteins often form. On the other hand, there are physical properties that do not depend on the molecular weight, and it is necessary to pay attention to the correlation between the target physical properties and the molecular weight.

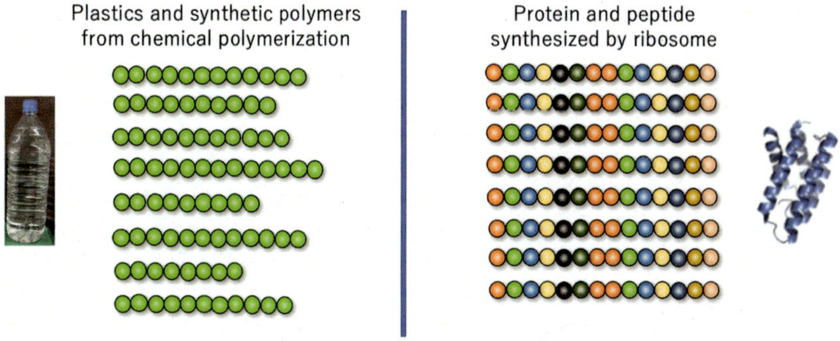

Figure 1.5 Understanding molecular weight distribution and chemical composition. Polymers such as plastics and proteins such as structural proteins exhibit molecular weight distribution based on synthetic methods.

1.3.2 Primary structure

A structural protein is a chain molecule in which amino acids are connected, and is the same as other proteins. As the primary structure, it is necessary to pay attention to changes in the side-chain structure due to posttranslational modification in addition to the chemical modifications. As mentioned above, in the case of structural proteins synthesized from ribosomes, the molecule is linear, but in the case of chemically synthesized proteins that mimic structural proteins, they can be telechelic, star-shaped, or branched in addition to linear chains. It is necessary to consider the primary structure and stereoregularity due to the inclusion of the D-form. Such a description gives a very complex impression, but natural structural proteins have a relatively simple primary structure.

Proteins are polymers in which 20 types of natural amino acids are linked by amide bonds. The natural amino acids include selenocysteine and pyrrolysine (Fig. 1.6), even though they are not contained in many structural proteins. Twenty kinds of natural amino acids and their derivatives mainly consist of structural proteins. There are also amino acids that are not directly synthesized from ribosomes, such as 3,4-dihydroxyphenylalanine (DOPA), hydroxyproline, and selenomethionine. In general, DOPA has an adhesive property due to a catechol group, and hydroxyproline has a characteristic helical structure due to a periodic hydrogen bond.

As an example, Major Ampullate Spidroin 1 (MaSp1), which is a major protein contained in spider dragline, is described here. MaSp has sequences that form relatively large three-dimensional structures of about 10–25 kDa at the N-terminus and C-terminus. Moreover, it has a

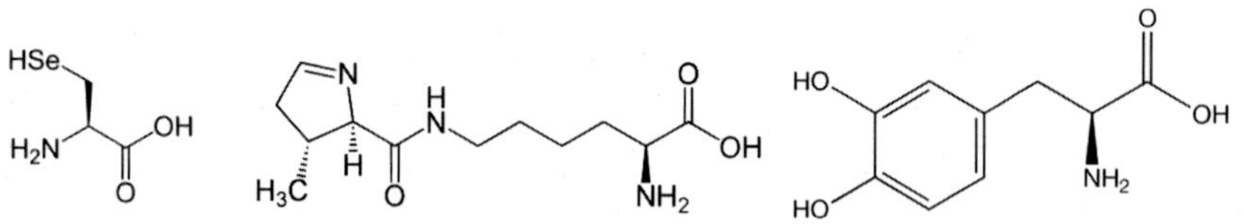

Figure 1.6 Chemical structures of selenocysteine (left), pyrrolysine (center), and DOPA (right).

repetitive sequence about 100 times between the N-terminus and the C-terminus [17]. This repetitive sequence is a glycine-rich amorphous region rich in Gly and a crystalline region containing polyalanine (6–13 alanine residue) forming a beta sheet structure [18]. This repetitive sequence is known to vary depending on spider species and proteins, but the motif where amorphous and crystalline sequences exist alternately is conserved. This trend is not limited to spiders, and silkworm silk has a motif $[(GAGA)_nG(Y/S)]$ (G = glycine, A = alanine, Y = tyrosine, and S = serine), which can be a feature of the *Bombyx mori* silk protein [20].

1.3.3 Configuration

In polymers, including proteins and polypeptides, configuration means the atoms' spatial arrangement in a polymer chain. The configurational versatility of polypeptide is a consequence of the considerable degree of rotational freedom about single bonds of the polypeptide chain. The polypeptides configuration is significantly different from the simple synthetic polymers such as polyethylene and polypropylene, due to the amide bonds and various sidechain groups. To clear the configurational differences between the synthetic polymer and polypeptide chains, tacticity, bonding mode, and branch structures need to be noted in this section.

The physical and chemical characters of polymers depend on monomeric units and the stereochemical/spatial arrangements of the atoms. The steric order of the sidechain group along the main chain is called tacticity (Fig. 1.7). In the case that all chiral centers have the same configuration, the tacticity is called isotactic, whereas if every other chiral center has the same arrangement, the tacticity is syndiotactic. On the other hand, a random arrangement of the side groups is called atactic or heterotactic. In the case of poly(L-amino acid) such as protein and peptide, all the stereochemistry is L-form, and hence protein and peptide are isotactic polymers.

Head/tail configuration is another configuration to describe the difference between synthetic polymers and proteins. In synthetic polymers,

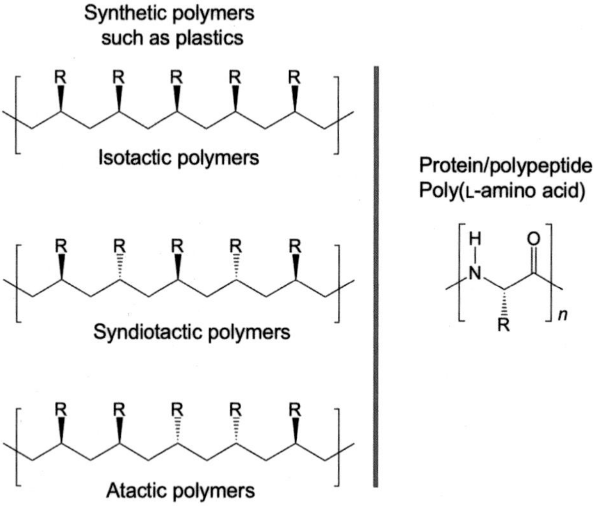

Figure 1.7 Comparison of configurations, tacticity, between synthetic polymer and protein.

some monomeric units have the molecular direction due to the asymmetric sidechain groups. The configuration of the molecular direction is described as "Head to Tail," "Head to Head," or "Tail to Tail" (Fig. 1.8). In the case of protein, the configuration is a perfect head to tail structure (see the poly(alanine) in Fig. 1.8C).

Another molecular structure is the main chain configuration structure, as shown in Fig. 1.9. Generally speaking, natural proteins and peptides are linear polymers. However, inter- and intramolecular networks are formed by posttranslational modification; for example, some cysteines are cross-linked via disulfide bonds. Also, the oxidation reaction induces dityrosine formation between two tyrosine residues. These posttranslational modifications provide network structures even in natural proteins. In the case of synthetic polymers, star-shaped, branched, and hyperbranched structures can be synthesized according to target physical and chemical properties. Recently, telechelic and branched peptides were designed and synthesized by chemical synthesis by using artificial initiator/terminator.

1.3.4 Hierarchical structure

The hierarchical structure of a protein is essential to realize its function and property. The excellent physical properties of spider silk are attributed not only to chemical structures, such as amino acid sequences, but also to

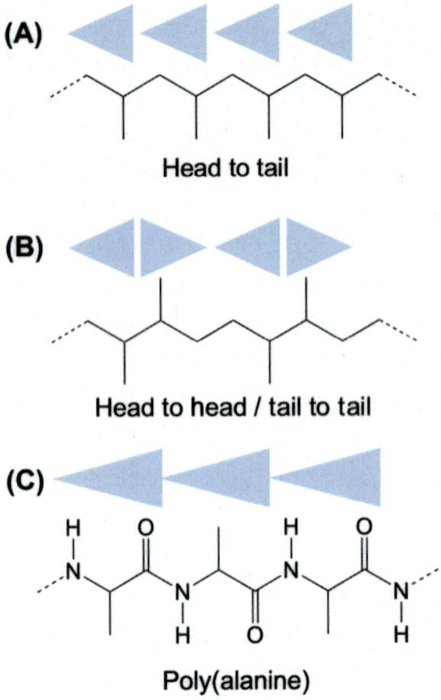

Figure 1.8 Comparison of configurations, tacticity, between synthetic polymer and protein. (A) Head to tail, (B) Head to head or tail to tail, and (C) An example of poly (alanine).

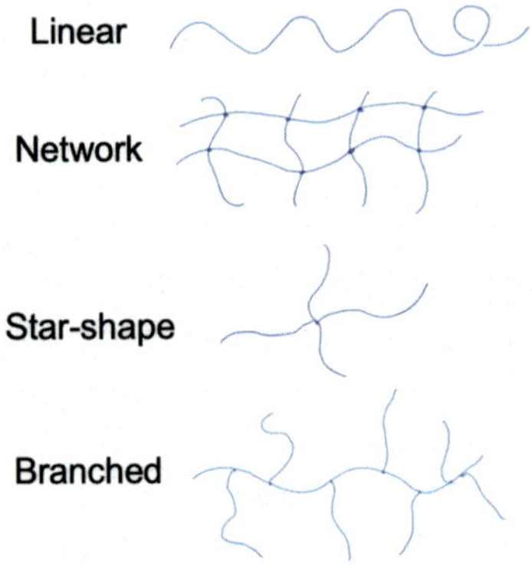

Figure 1.9 Configurations of main chain structure.

multiple hierarchical structures. The hierarchy of biological materials is discussed when architectural structures are formed across multiple scales, from the atomic- to the macroscale. In this respect, silk has a hierarchical structure similar to other complex natural materials that exhibit structural hierarchy, such as collagen, nacre, and mussels. As will be described below, using the spider dragline as an example, understanding the molecular mechanism at each level is indispensable for understanding the entire material (Fig. 1.10).

The spider dragline is surrounded by skin layers containing glycoproteins and lipids, which are thought to protect the fiber from biological degradation reactions [21]. The N-terminal and C-terminal domains of MaSp, the main protein of the silk protein part present in the skin layer, change according to pH changes and ionic environment changes that occur during the spinning process. The N-terminal domain dimerizes in response to a decrease in pH in the spinning duct and initiates conformation [22]. On the other hand, the shear force generated during spinning induces self-organization of the beta sheet crystal and orientation along the fiber axis, and realizes a hierarchical structure in the amorphous matrix

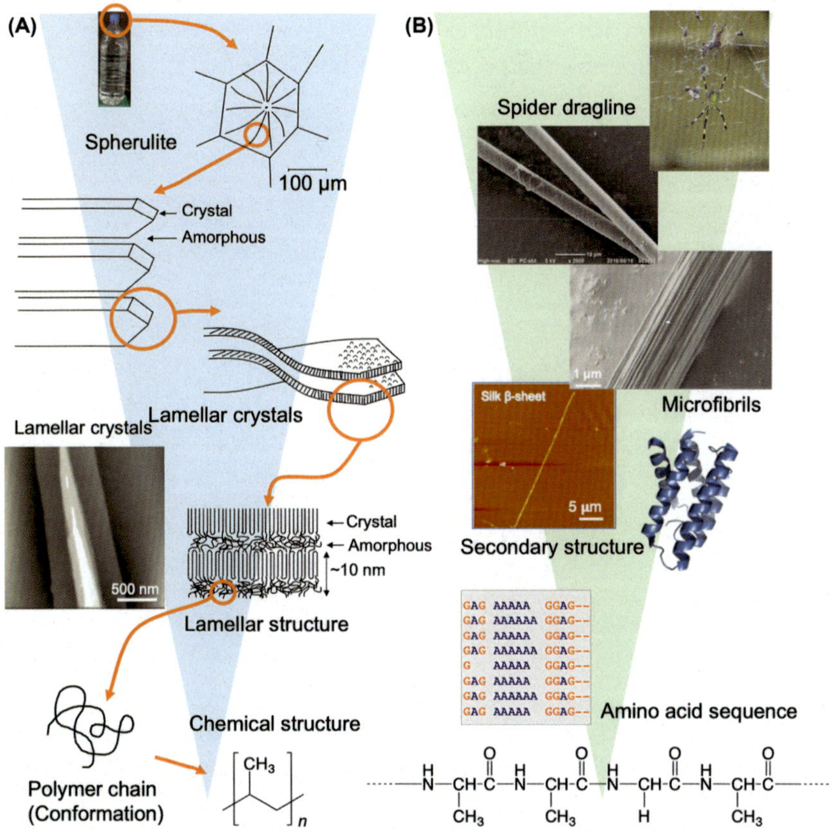

Figure 1.10 Schematic diagram of the spider silk hierarchy and structural comparison between polypropylene (A) and spider silk (B).

that cannot be imitated by an artificial process. This combination of crystal and amorphous causes the characteristic strength, elasticity, and viscoelasticity of silk fibers [23]. To realize the hierarchical structure of the spider dragline, a plurality of structural proteins is assembled in the silk protein fiber, which is a composite material composed of structural proteins. Spider draglines are usually composed of multiple proteins (such as MaSp1, MaSp2, and also smaller proteins). MaSp1 and MaSp2 are similar in overall amino acid sequence, but the amino acid motifs and compositions of the repetitive domains are obviously different; proline residues and diglutamine (QQ) motifs not found in MaSp1 can be confirmed in MaSp [24]. Differences in the MaSp1/MaSp2 ratio have been speculated to affect mechanical properties such as tensile strength and extensibility. However, the hierarchical interactions between MaSp1 and MaSp2 have not been clarified in detail. MaSp2 is believed to cause phenomena such as supercontraction in response to moisture (water molecules), which is very likely, but experimental proof has not been reported yet.

References

[1] D.Y. Yang, M.R. Hartman, T.L. Derrien, S. Hamada, D. An, K.G. Yancey, et al., DNA materials: bridging nanotechnology and biotechnology, Acc. Chem. Res. 47 (6) (2014) 1902–1911.

[2] Y. Doi, Microbial Polyesters, VCH Publishers, New York, 1990.

[3] Y. Doi, A. Steinbüchel, Biopolymers, Polyesters I and II, Wiley-VCH, Weinheim, Germany, 2001.

[4] R.W. Lenz, R.H. Marchessault, Bacterial polyesters: biosynthesis, biodegradable plastics and biotechnology, Biomacromolecules 6 (1) (2005) 1–8.

[5] M. Lemoignei, Products of dehydration and of polymerization of β-hydroxybutyric acid, Bull. Soc. Chim. Biol. 8 (1926) 770–782.

[6] S. Nakamura, Y. Doi, M. Scandola, Microbial synthesis and characterization of poly (3-hydroxybutyrate-Co-4-hydroxybutyrate), Macromolecules 25 (17) (1992) 4237–4241.

[7] S.A. Gordeyev, Y.P. Nekrasov, Processing and mechanical properties of oriented poly(beta-hydroxybutyrate) fibers, J. Mater. Sci. Lett. 18 (20) (1999) 1691–1692.

[8] K. Numata, T. Hirota, Y. Kikkawa, T. Tsuge, T. Iwata, H. Abe, et al., Enzymatic degradation processes of lamellar crystals in thin films for poly[(R)-3-hydroxybutyric acid] and its copolymers revealed by real-time atomic force microscopy, Biomacromolecules 5 (6) (2004) 2186–2194.

[9] K. Numata, Y. Kikkawa, T. Tsuge, T. Iwata, Y. Doi, H. Abe, Enzymatic degradation processes of poly[(R)-3-hydroxybutyric acid] and poly[(R)-3-hydroxybutyric acid-co-(R)-3-hydroxyvaleric acid] single crystals revealed by atomic force microscopy: effects of molecular weight and second-monomer composition on erosion rates, Biomacromolecules 6 (4) (2005) 2008–2016.

[10] K. Zhao, Y. Deng, G.Q. Chen, Effects of surface morphology on the biocompatibility of polyhydroxyalkanoates, Biochem. Eng. J. 16 (2) (2003) 115–123.

[11] Y. Doi, M. Kunioka, Y. Nakamura, K. Soga, Nuclear magnetic-resonance studies on unusual bacterial copolyesters of 3-hydroxybutyrate and 4-hydroxybutyrate, Macromolecules 21 (9) (1988) 2722–2727.
[12] H. Staudinger, Concerning polymerisation, Ber. Dtsch. Chem. Ges. 53 (1920) 1073–1085.
[13] S. Kobayashi, K. Kashiwa, T. Kawasaki, S. Shoda, Novel method for polysaccharide synthesis using an enzyme - the 1st in vitro synthesis of cellulose via a nonbiosynthetic path utilizing cellulase as catalyst, J. Am. Chem. Soc. 113 (8) (1991) 3079–3084.
[14] T.C. Laurent, I.M. Dahl, L.B. Dahl, A. Engstrom-Laurent, S. Eriksson, J.R. Fraser, et al., The catabolic fate of hyaluronic acid, Connect. Tissue Res. 15 (1-2) (1986) 33–41.
[15] J.R. Fraser, T.C. Laurent, Turnover and metabolism of hyaluronan, Ciba Found. Symp. 143 (1989) 41–53. Discussion 53-9, 281-5.
[16] K. Yazawa, K. Ishida, H. Masunaga, T. Hikima, K. Numata, Influence of water content on the beta-sheet formation, thermal stability, water removal, and mechanical properties of silk materials, Biomacromolecules 17 (3) (2016) 1057–1066.
[17] C. Holland, K. Numata, J. Rnjak-Kovacina, F.P. Seib, The biomedical use of silk: past, present, future, Adv. Healthc. Mater. 8 (1) (2019) e1800465.
[18] J. Gatesy, C. Hayashi, D. Motriuk, J. Woods, R. Lewis, Extreme diversity, conservation, and convergence of spider silk fibroin sequences, Science 291 (5513) (2001) 2603–2605.
[19] I.W. Sutherland, Polysaccharides from microorganisms, plants and animals, in: A. Steinbüchel (Ed.), Biopolymers, Wiley-VCH, Weinheim, 1996, pp. 1–19.
[20] A.D. Malay, R. Sato, K. Yazawa, H. Watanabe, N. Ifuku, H. Masunaga, et al., Relationships between physical properties and sequence in silkworm silks, Sci. Rep. 6 (2016) 27573.
[21] K. Yazawa, A.D. Malay, H. Masunaga, K. Numata, Role of skin layers on mechanical properties and supercontraction of spider dragline silk fiber, Macromol. Biosci. 19 (3) (2019) e1800220.
[22] S. Schwarze, F.U. Zwettler, C.M. Johnson, H. Neuweiler, The N-terminal domains of spider silk proteins assemble ultrafast and protected from charge screening, Nat. Commun. 4 (2013) 2815.
[23] C.J. Fu, Z.Z. Shao, V. Fritz, Animal silks: their structures, properties and artificial production, Chem. Commun. 43 (2009) 6515–6529.
[24] A.D. Malay, K. Arakawa, K. Numata, Analysis of repetitive amino acid motifs reveals the essential features of spider dragline silk proteins, PLoS One 12 (8) (2017).

Questions for this chapter

1. Draw the chemical structure of cationic amino acid(s).
2. Based on the natural morphologies of proteins, proteins are classified into fibrous and globular proteins. How would proteins be classified in terms of their natural role?
3. Show four types of major biological polymers.
4. What are structural proteins?
5. Describe the differences between protein and polypropylene in terms of chemical structures.

CHAPTER 2
Synthesis

Most of the proteins in living systems and organisms are biologically synthesized via the translation of messenger ribonucleic acid encoding genetic information by ribosomes in the central dogma. The various functionalities and properties of natural proteins are associated with their primary, secondary, tertiary, and higher-order structures. In addition to the natural proteins containing precisely regulated amino acid sequences, synthetic polypeptide materials, including homopolymers, random or block copolymers, and special polymeric architectures, such as star-shaped, branched, and telechelic polymers, have been synthesized chemically. These synthetic polypeptides have attracted intensive attention as biological polymers. The synthetic peptide/protein-based material's physical and biological characters can be designed and controlled by monomer units [amino acid and its posttranslational modification (PTM)], their sequences, and hierarchical structures.

Generally speaking, there are many ways to produce polypeptides and proteins. This chapter will detail three approaches to synthesizing artificial polypeptides: chemical, biological, and biochemical synthesis methods. In this chapter, synthesis methods for peptides and proteins are listed and introduced with their advantages and disadvantages.

2.1 Biological synthesis

The biological method mainly indicates the biosynthesis of proteins in vivo (in cell), which is processed generally by a combination of multiple biomacromolecules, such as nucleic acid (template), enzymes (catalyst), and amino acid (substrate). In the biosynthesis process, a protein's primary structure (amino acid sequence) is encoded in a gene, a sequence of deoxyribonucleic acid (DNA). A two-step biological process, namely, transcription and translation, is needed to extract the genetic information from DNA (Fig. 2.1) [1]. Transcription is the step where the genetic information encoded in DNA is transcribed into messenger ribonucleic acid (mRNA) in the form of an overlapping degenerate triplet code. The transcription process proceeds in three stages: initiation, chain elongation,

Biopolymer Science for Proteins and Peptides
DOI: https://doi.org/10.1016/B978-0-12-820555-6.00001-X

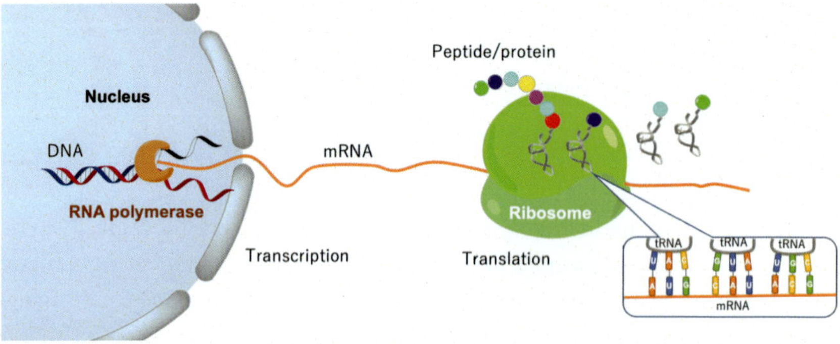

Figure 2.1 Ribosomal synthesis: transcription and translation are required to extract this information from a DNA and to synthesize peptide/protein.

and termination. Initiation occurs when the RNA polymerase binds to the promoter gene sequence on the DNA strand. Elongation begins as the RNA polymerase is guided along the template DNA unwinding the double-stranded DNA molecule and synthesizing a complementary single-stranded RNA molecule. Finally, transcription is terminated with the release of RNA polymerase from template DNA. Translation involves decoding the mRNA to specifically and sequentially link together amino acids in a growing polypeptide chain. The decoding of mRNA occurs in the ribosome, a macromolecular complex composed of nucleic acids and proteins. mRNA is read in three-nucleotide increments called codons; each codon specifies a particular amino acid in the growing polypeptide chain. Ribosomes bind to mRNA at a ribosome-binding domain just before the start codon (AUG) recognized by initiator transfer RNA (tRNA). Ribosomes proceed to the elongation phase of protein synthesis. During this stage, complexes composed of an amino acid linked to tRNA sequentially bind to the appropriate codon in mRNA by forming complementary base pairs with the tRNA anticodon. The ribosome moves from codon to codon along the mRNA. Amino acids are added one by one, translated into polypeptide sequences dictated by DNA and represented by mRNA. In the end, a release factor binds to the stop codon, terminating translation and releasing the complete polypeptide from the ribosome.

Polypeptide sequences can be designed based on the corresponding DNA sequence to be transformed into host organisms [2]. Not only natural proteins but also artificial polypeptides and recombinant proteins can be designed and tuned in terms of amino acid sequence for biosynthesis.

Tirrell et al. pioneered the biosynthesis of desired polypeptides using living organisms and demonstrated that artificial polypeptides were successfully designed and expressed by in vivo biosynthesis using microbes transformed with plasmids encoding target sequences [3,4]. The synthetic polypeptides consisting of periodic sequences were designed and encoded in plasmid DNA to transform *Escherichia coli* [5,6]. These studies are the first cases to demonstrate the potential of polypeptides and recombinant proteins as functional biomaterial.

The recombinant DNA technology has allowed the facile, large-scale production of many polypeptide sequences; however, practical use of this biological synthesis system suffers from several disadvantages. For example, expression of the target proteins is occasionally suppressed to a low level by various factors, mainly due to the cytotoxicity of the expressed proteins or the gene silencing of target DNA. Besides, polypeptides/proteins containing highly repetitive sequences like silk protein are challenging to express recombinantly due to undesirable elements in the secondary structure of the mRNA [7–10]. The multiple purification processes to isolate target polypeptides/proteins from cell lysates are also a technical issue for cost-effective production.

2.1.1 *Escherichia coli*

E. coli is the most convenient, established, and widely used protein production host. In academic and industrial research laboratories, many researchers use it in their experiments. At the laboratory level (small scale), purification procedure from *E. coli* has been established using affinity tag system(s). The synthesis of recombinant proteins is composed of two major steps (Fig. 2.2): (1) design, construction, and cloning of the genes; and (2) expression and purification of the protein/polypeptide. *E. coli* such as BL21(DE3) and JM109 have been used as host systems, resulting in yields of proteins that were inversely correlated with the size of the synthetic genes. In general, expression levels obtained from synthetic protein genes depend upon the size, charges, and hydrophobicity of the protein of interest. By optimizing the sequence and culture conditions, the productivity of recombinant proteins can be improved drastically.

After protein expression, *E. coli* cells are harvested by centrifugation. In the case one would like to obtain denatured protein or plan to perform refolding process, the resultant cell pellets are resuspended and lysed in a denaturing buffer such as urea, guanidinium chloride, or guanidinium thiocyanate. On the other hand, to maintain the native folding structure

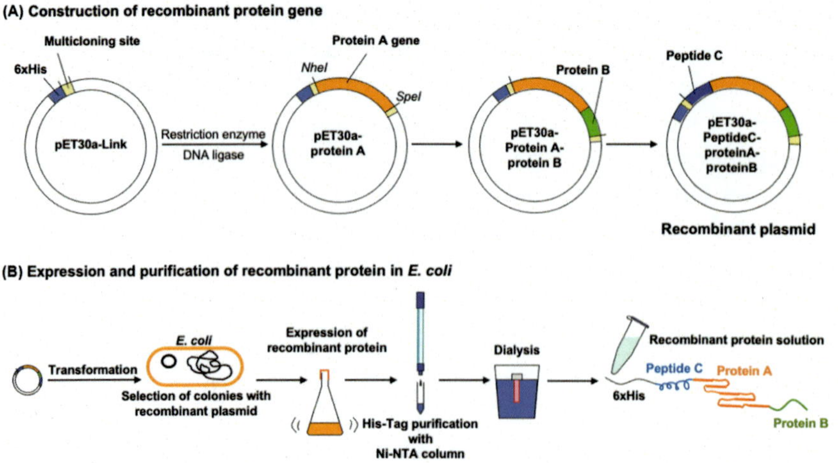

Figure 2.2 The synthesis of recombinant proteins using bacteria such as *Escherichia coli*. The experimental processes are composed of two major steps: (A) design, construction, and cloning of the genes and (B) expression and purification of the protein polymers. *Source: Adapted from K. Numata, D.L. Kaplan, Silk-based delivery systems of bioactive molecules, Adv. Drug. Deliv. Rev. 62 (15) (2010) 1497–1508 [11].*

of protein without denaturation, several lysis buffers containing surfactants, lysozyme, nuclease, and protease inhibitors can be used for the protein extraction from *E. coli* cells. For example, the author's laboratory uses the cold lysis solution containing 50 mM Tris-HCl pH 7.5, 10% glycerol, 0.1% Triton X-100, 50 mg of lysozyme, 250 units of nuclease, and ethylenediaminetetraacetic acid (EDTA)-free protease inhibitor cocktail [12]. Purification of the expressed proteins has been performed by metal affinity chromatography, such as with nickel–nitrilotriacetic acid (Ni-NTA) and $[His]_6$ systems. His-tag purification of the proteins is performed by the addition of Ni-NTA agarose resin to the supernatant (batch purification) under denaturing conditions. After washing the column with denaturing buffer at pH 6.3, the proteins are eluted with denaturing buffer at pH 4.5 or other elution buffers. Purified samples are dialyzed against acetic buffer or NH_4HCO_3 solution, and then Milli-Q water.

As an example, the biosynthesis of silk protein is introduced in this paragraph. Synthetic genes encoding dragline silk from *Trichonephila clavipes* (formerly known as *Nephila clavipes*) have been successfully constructed, cloned, and expressed [8]. The synthesis of these recombinant genes is based on the repetitive sequences found in native dragline silk genes. The methods for constructing these repeats were previously reported [8], using smaller oligonucleotide repeats and subsequent multimerization by traditional methods with restriction enzymes and ligase [13]. Thus the size and sequence of the protein generated can be controlled by the primary sequence synthesized as oligonucleotides (the building blocks). The use of synthetic gene technology to control silk protein size allows for studying relationships between sequence length/variations and structure—function. Partial complementary DNA sequences from *T. clavipes* encoding the proteins containing repetitive sequences in nature, such as partial dragline protein MaSp1 sequence [GRGGLGGQGAGAAAAAGGAGQGGYGGLGSQG], have been cloned and used to construct synthetic genes [8]. Plasmid vectors such as pET vectors (common protein synthesis vectors) have been used to place the synthetic genes under the control of either bacteriophage T5 or T7 promoters and add six histidine tags at the *N*-terminus of the recombinant protein to simplify purification by metal affinity chromatography. Many groups have synthesized many block copolymers containing silk-like sequences based on these recombinant DNA techniques. These variants include modified spider silks bioengineered to include Arg-Gly-Asp cell-binding domains to enhance cell adhesion, cell-penetrating peptide for drug/gene delivery system, and tumor-homing peptide for tumor-targeting treatments [14—17].

2.1.2 Yeast

A wide variety of yeast, including *Saccharomyces cerevisiae*, have been used extensively in the fermentation industry, and many companies prefer to use it in fermentation production. The spider silk production in yeast has a long history; for example, recombinant spider silk protein was successfully expressed in the methylotrophic yeast *Pichia pastoris* [18,19]. *P. pastoris* was reported to show relatively high yield and productivity with 300—1000 mg/L of recombinant spider silk proteins. One of the leading spider silk venture companies, Bolt Theread Inc., has been reported to use a yeast system to produce some of their products [20].

2.1.3 Transgenic plants

To prepare transgenic plants for the mass production of proteins and peptides, the nuclear genome and/or plastid genome are the target for gene modifications. Plastids are organelles found in plant cells and eukaryotic algae and play pivotal roles in photosynthesis and other metabolic processes [21]. Originally proplastids, these organelles can differentiate into green or nongreen plastids depending on the developmental stage and plant cell type [21,22]. Each type of plastid has its own characteristics, biological function, and individualized metabolic process in the cell [22]. Chloroplasts, the most common plastids found in greenish plant tissues, such as developed shoots, leaves, and stems, are responsible for photosynthesis as well as the production of food and oxygen for cells [21,22].

To synthesize and accumulate metabolites such as recombinant proteins, small molecules, and bioactive compounds in the plastid, the cellular degradation process needs to be prevented in the plastids as storage [23,24]. This organelle has become the focus of transplastomic engineering to improve economically important traits or to produce commercially valuable products in the target crop species [23–26]. The plastid genome (plastome) is a 100–200 kbp DNA molecule in the plastid, which harbors 100–120 highly conserved genes and has its own transcriptional and translational regulation system while sharing some plastid-targeting nuclear-encoded transcription factors with the nucleus [27].

Successful modification of the plastome has been achieved by particle bombardment, whereby DNA molecules are transferred across the plant cell boundary and plastid membrane using pressure (Fig. 2.3) [23,25,29,30]. Plasmid DNA can also be directly delivered into the plastid via lipid membrane integration and translocation using polyethylene glycol-mediated protoplast transformation [31]. Alternatively, plastid transformation may be accomplished by microinjecting a solution containing exogenous DNA molecules through a submicron diameter glass syringe directly into the plastids [32]. Transplastomic plants can then be established after multiple selection cycles and the subsequent regeneration of transformed plant cells or tissues [23,24]. These methods have several major drawbacks, however, as the preparation of plant material is tedious and its regeneration following transformation is time-consuming and requires high-cost experimental instruments; more importantly, these methods are only applicable for limited types and stages of plant material [23,24].

Tobacco and potato were used to produce spider silk proteins as host plants [33,34]. However, the synthesis of the full-length spider silk protein, namely, native-sized spidroins, from plants, is still challenging to date [35].

2.1.4 Transgenic mammals

In addition to plants, a transgenic goat was also used to synthesize recombinant spider silk proteins in its milk. Further, *Bombyx mori* silkworm was genetically modified to synthesize recombinant spider silk proteins [36].

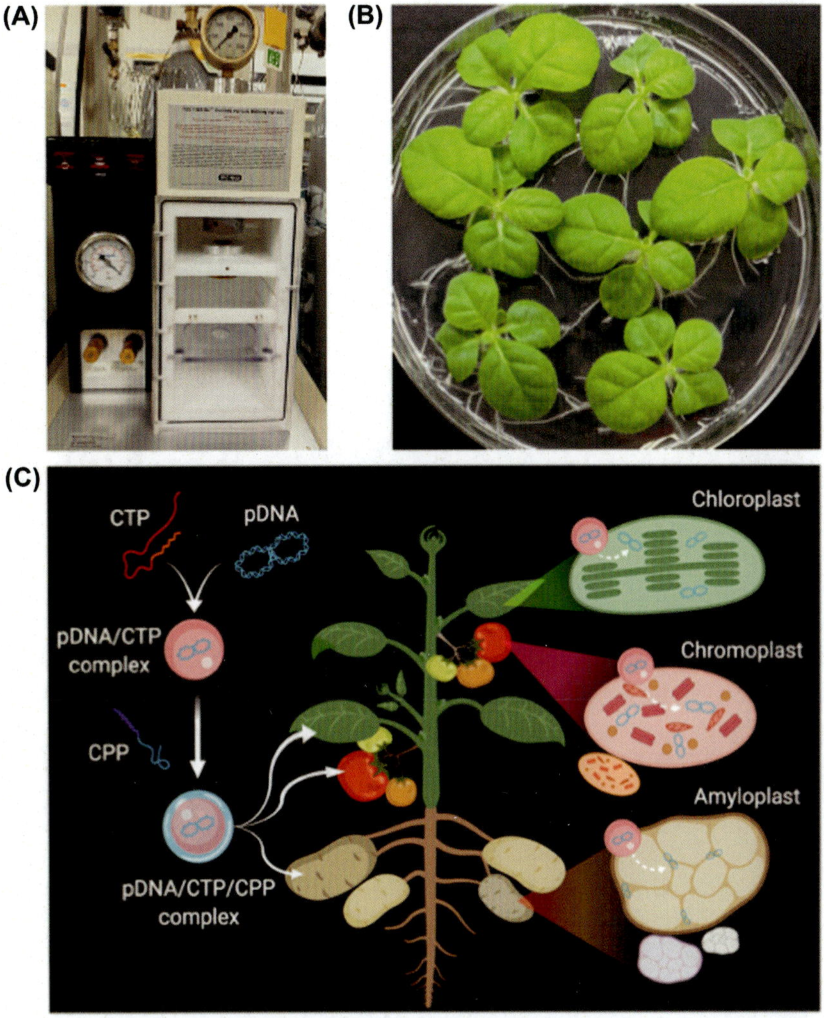

Figure 2.3 Strategy of the transgenic plants for recombinant protein productions. (A) Particle bombardment instrument. (B) Recombinant *Nicotiana tabacum* expressing spider silk protein. (C) Chloroplast transit peptide (CTP) and cell-penetrating peptide (CPP)-mediated plasmid DNA (pDNA) delivery system targeting chloroplast, chromoplast, and amyloplast in intact plants. *Source: From C. Thagun, J.A. Chuah, K. Numata, Targeted gene delivery into various plastids mediated by clustered cell-penetrating and chloroplast-targeting peptides, Adv. Sci. (Weinh.) 6 (23) (2019) 1902064 [28].*

2.1.5 Photosynthetic bacteria

Biosynthesis of high-value compounds, including proteins and peptides in photosynthetic organisms, is one of the potential methods to reduce costs and to contribute a sustainable system because they can utilize sunlight energy and carbon dioxide (CO_2) in the air for growth. As introduced in the previous section 2.1.3, cyanobacteria, algae, and plants possess two photosystems, namely, photosystem I and II, and extract electrons from water and produce oxygen [37], while the anoxygenic photosynthetic bacterium has a single photosystem, either type I or type II photosynthetic reaction center, and extracts electrons from organic compounds, sulfur compounds, and hydrogen. These anoxygenic type I and type II reaction centers are structurally and functionally similar to oxygenic photosystems, and hence photochemical reaction and electron transport in photosynthetic reaction centers have been made using anoxygenic photosynthetic bacteria because of their simple structure [38].

Purple photosynthetic bacteria, which are typical anoxygenic photosynthetic bacteria, are classified into two types, namely, purple sulfur and purple nonsulfur bacteria. Purple sulfur bacteria use sulfide and hydrogen as an electron donor, whereas purple nonsulfur bacteria utilize organic compounds [39]. Purple photosynthetic bacteria are widely distributed in the aquatic environment. Some species of purple nonsulfur bacteria are known to have nitrogen (N_2) fixation ability [40]. This means that they can use N_2 in the air as a nitrogen source for their growth and bioproduction (Fig. 2.4). However, the exact nitrogen fixation ability of purple nonsulfur bacteria has not been evaluated despite the contribution to the

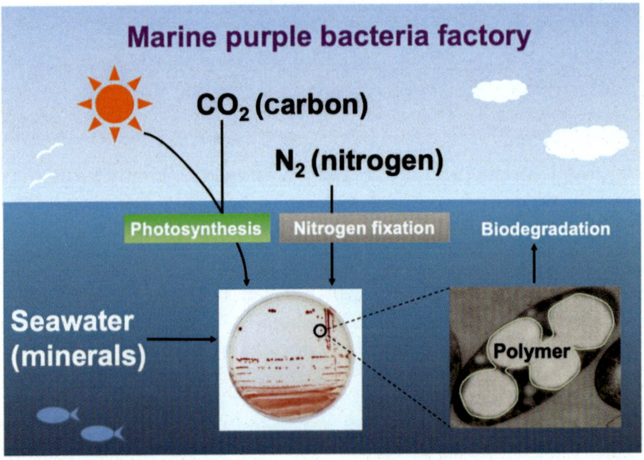

Figure 2.4 Marine purple photosynthetic bacteria show great potential as a platform to synthesize structural proteins and other biopolymers [41].

nitrogen flux in aquatic environments. The bioproduction by photosynthetic purple bacteria has been reviewed in the literature [41].

Furthermore, the utilization of marine organisms has several potential advantages for large-scale commercial production. Sterilized seawater can be used as a culture medium instead of a synthetic medium. Moreover, the high salt concentration of seawater can inhibit biological contamination during the cultivation. Considering these advantages, marine purple photosynthetic bacteria would be an ideal host organism for microbial production.

As described above, photosynthetic bacteria are attractive biological hosts, which is a "sustainable cell factory" to capture and fix CO_2 and N_2 directly [41]. Their biopolymer productivity and transformation methods have been studied [42–47]. Silk proteins were successfully synthesized by using recombinant nonsulfur photosynthetic bacteria in the marine environment. The critical point is that through photosynthesis and the nitrogen fixation process, by utilizing low-cost and abundant renewable resources such as light (energy), CO_2 (carbon source), and N_2 (nitrogen source), structural protein can be synthesized under marine conditions. The efficiency of the fixation and conversion of CO_2 and N_2 to amino acids and proteins is not quantified but shown in the metabolic pathways (Fig. 2.5). This bacterial platform is still on the way to being established but is very attractive for the sustainable production of biopolymers.

2.1.6 Posttranslational modification

In addition to amino acid sequences, modification of the sidechain, namely, PTM, is another crucial factor in determining protein's chemical (primary) structures. PTM generally refers to the enzymatic modification of amino acids within proteins immediately following ribosomal protein biosynthesis in cells. PTMs form an integral part of polypeptide chains folding correctly into their final protein conformation. In this section, various PTMs with various silk proteins are introduced and summarized.

PTMs within silk have generally been studied using alternating collision-induced dissociation and electron transfer dissociation mass spectrometry [49], and more recently solid-state nuclear magnetic resonance spectroscopy [50]. Interestingly, although many PTMs have been demonstrated within spider silk, little is known about their exact role and structural impact [51]. PTMs have been shown both within MaSp1 and MaSp2 of spider major ampullate (MA) silk, which displays notable and

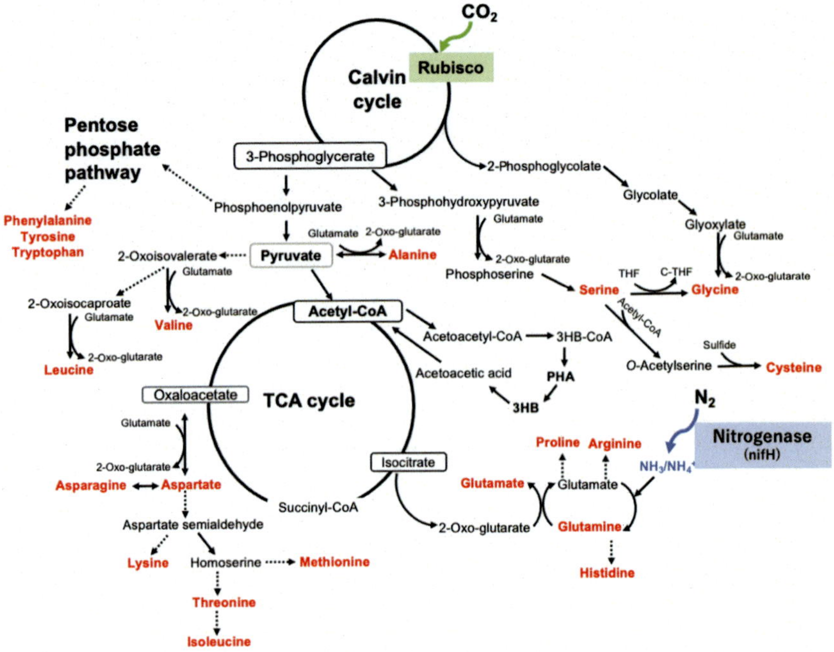

Figure 2.5 Metabolic pathways from CO_2 and N_2 to amino acids in the typical photosynthetic bacteria [48].

regular phosphorylation of the serine and tyrosine within the repetitive region (Fig. 2.6) [52,53]. The position and enhancement of negative charge associated with the phosphorylation have known impacts on the formation of helical structures in other proteins and show similar patterns in spider silk with a molecular model of MaSp1 incorporating phosphorylated serine and tyrosine showing increased coil formation, with the greater formation of alpha-helices and a reduction of the amount of 3_{10} helices than in previous models [54].

The presence of the oxidized residue dityrosine has also been confirmed within spider silk proteins, MaSp1 and MaSp2 [52,53]. However, the exact abundance or structural role of dityrosine within spider silk has not been clarified to date. Dityrosine is known to also form in small amounts within Tussah silk, forming cross-links in the crystalline region [55]. Although, in this case, dityrosine is not considered to influence on the physicochemical properties of the Tussar silk, the utility of dityrosine cross-linkage will likely prove to be an essential tool in de novo development of future biomaterials [56].

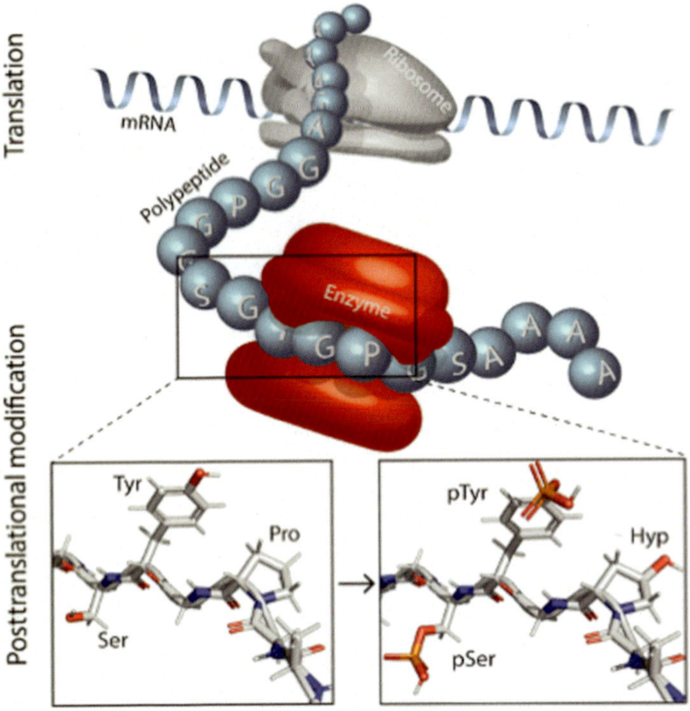

Figure 2.6 The process of posttranslational modification from messenger RNA to the final posttranslationally modified polypeptide displaying three major forms of modification found within spider silk, phosophoserine, phosophotyrosine, and hydroxyproline formation.

A hydroxylated form of tyrosine and an intermediate in the formation of dityrosine, 3,4-dihydroxyphenylalnine has also been shown to be used in marine silk from caddisfly larvae and sandcastle worms, utilized as an adhesive sight used to bind dissimilar materials under aqueous conditions [57–59]. A further example of PTMs being necessary for adhesion within silk can be observed in the gumfoot threads found in *Latrodectus* species. The glycosylation of residues within the MA silk appears to play an important role in the adhesion of gumfoot glue to the gumfoot lines that form the capture threads within these species [60].

The most recent PTM to be confirmed within silk is hydroxyproline, which has been demonstrated within both the MA and flagelliform silk of spiders [49,50]. The hydrogen bonding site added through hydroxylation of proline within the GPGXX (MaSp2) and GPGGX (Flag) motifs is thought to add to the mechanoelastic properties of these fibers. This demonstrates the importance of detailed proteomic understanding of silk, pivotal when modeling silk structure necessary for a detailed understanding of silk's structure–function relationship.

Ultimately our understanding of the silk proteome is still in its infancy compared to our current genomic, transcriptomic and structural knowledge. The influence PTMs have on protein assembly and then resultant protein structure may prove a pivotal factor in understanding the structure–function relationship within spider silk. Moreover, this is a critical consideration in the production and functionalization of recombinant or synthetic silks with comparable material prosperities to natural spider silk [56].

2.2 Chemical synthesis

The chemical synthesis methods include solid-phase peptide synthesis (SPPS), which was developed by Merrifield [61], and ring-opening polymerization of amino acid *N*-carboxyanhydrides (NCA) [62]. The chemistry related to peptide synthesis is well reviewed in the literature [2,63]. In the SPPS method, polypeptides are synthesized on insoluble solid supports to propagate the peptide chain in a stepwise manner, enabling precise control of the amino acid sequence of polypeptides. On the other hand, the polymerization of amino acid NCA derivatives can provide polypeptides of higher molecular weight; however, this method is only available for the synthesis of homopolymers and random/block copolymers. These chemical synthetic methods employ the polymerization of amino acid derivatives, which are chemically synthesized in advance with organic solvents. In spite of their beneficial advantages, a recent trend to avoid petroleum-derived chemicals for a future sustainable society pushes us to pursue an environmentally benign alternative method of polypeptide synthesis.

2.2.1 Solid-phase and liquid-phase methods

The solid-phase method is one of the most common techniques used to synthesize relatively short polypeptides and peptides of up to about 50 amino acids [64,65]. In SPPS reaction (Fig. 2.7), the *N*-terminal amino acid is attached to a solid matrix with the carboxyl group and the amine group is protected. The amine group undergoes a deprotection step revealing an *N*-terminal amine. This is followed by a coupling reaction between the activated carboxyl group of the next amino acid and the amine group of the immobilized residue. Side chains of several amino

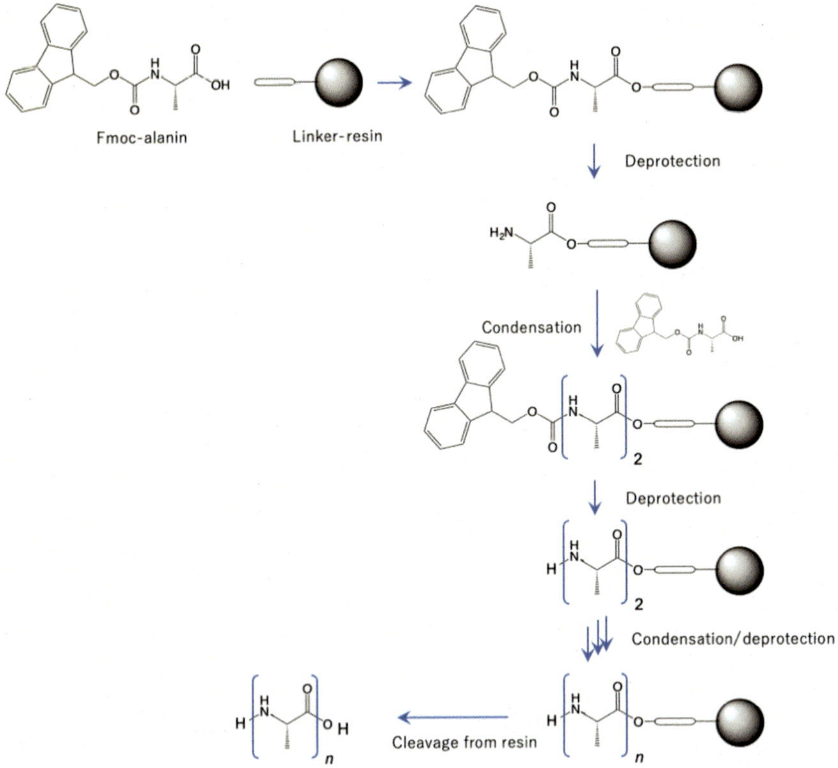

Figure 2.7 Chemical reaction scheme of solid-phase synthesis [2,66]. *Fmoc,* Fluorenylmethyloxycarbonyl group.

acids contain functional groups, which may interfere with the formation of amide bonds and must be protected. This process may continue through iterative cycles until the polypeptide has reached its desired chain length. The polypeptide is then cleaved off the resin and purified [67–69]. SPPS has enabled the synthesis of oligopeptides, ~40–50-amino acid residues. However, limitations in chemical coupling efficiency have made it impractical to synthesize longer polypeptides with reasonable yields. In general, the method is not scalable, and it is considered unsuitable as a technique for synthesizing proteins and polypeptides in bulk scale.

The reaction in a solvent without using a solid surface is called a liquid-phase method (Fig. 2.8), and the basic mechanism is similar to the solid-phase synthesis [70]. The liquid-phase method is relatively suitable for large-scale synthesis. In both cases, the aggregating peptides

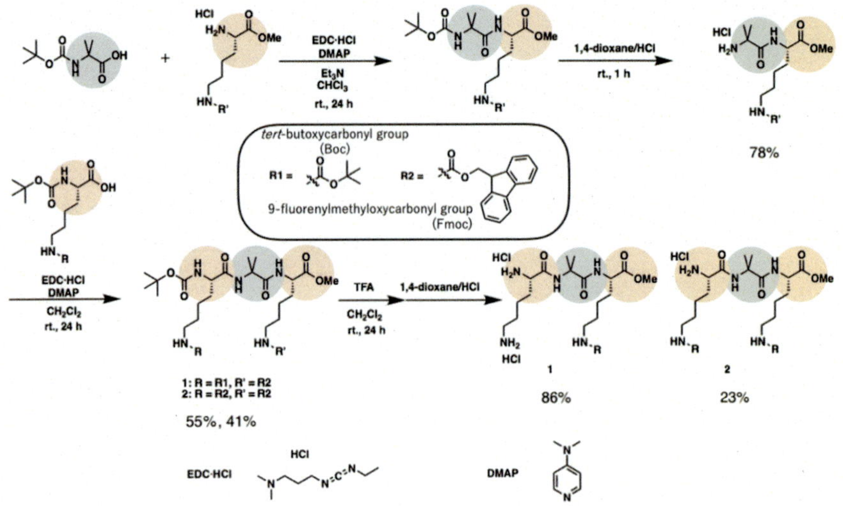

Figure 2.8 Chemical reaction scheme of liquid-phase synthesis.

which preferentially form beta-sheet structures are not so easily synthesized because of the peptide chain aggregation during the polymerization.

2.2.2 α-Amino acid-*N*-carboxyanhydride (NCA) ring-opening polymerization

In addition to the synthesis of short peptides by stepwise condensation, a successful example of solution synthesis is the ring-opening polymerization of amino acid NCAs [62]. NCA has high electrophilicity, and can efficiently obtain a polypeptide by living anion ring-opening polymerization using an amine as an initiator [71] (Fig. 2.9). It is possible to synthesize various polypeptides by devising an initiator amine [62], while this technique can be used to synthesize polypeptides composed of one or a few amino acid residue(s) rather than sequentially controlled polypeptides. The controllable nature of anionic ring-opening polymerization allows the propagating terminal to further extend the peptide chain, resulting in the formation of block copolypeptides. The ring-opening polymerization of NCAs has also been combined with the conventional anionic polymerization of other synthetic monomers; thus polymer–polypeptide conjugates can be obtained [72]. Until now, the reason why the NCA method is not used industrially has been mentioned as being that a highly toxic reagent such as phosgene or triphosgene is necessary for the preparation of NCA. In recent years, phosgene-free synthesis of NCA was reported and

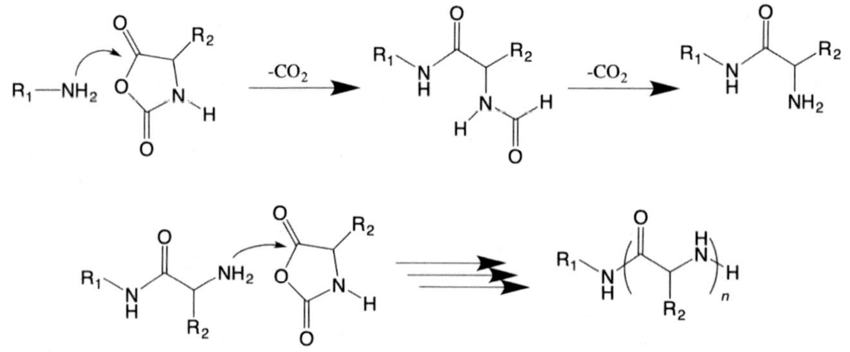

Figure 2.9 Chemical reaction scheme of NCA ring-opening polymerization for peptide synthesis [2,66].

used to synthesize polypeptides [73,74]. However, sequence control is still technically difficult, and it is not easy to mimic a protein sequence.

2.3 Biochemical synthesis

Several methods using biochemical approaches to synthesize peptides and proteins have been recently reported and studied widely. Generally, biochemical methods for peptide synthesis are relatively newer than chemical and biological methods [2,66]. Most of the biochemical methods introduced here use an enzyme as a catalyst to synthesize or modify peptides and proteins.

2.3.1 Native chemical ligation

In the solid-phase method and the liquid-phase method described above, the synthesis efficiency becomes lowered as the molecular weight increases. To compensate for this technical problem of the chemical synthesis, native chemical ligation is considered as a strong complementary reaction [75,76]. With native chemical ligation, it is possible to obtain a linked polypeptide by mixing a polypeptide having a thioester (TE) at the *C*-terminus and a polypeptide having a cysteine at the *N*-terminus, which are synthesized by other methods including solid/liquid-phase methods. Thus polypeptides and proteins with high molecular weights (more than 200 amino acid residues) can be synthesized by combining chemical synthesis (solid phase, liquid phase, and others) and native chemical ligation [77].

Native chemical ligation allows the combination of two unprotected peptide segments by the reaction of a *C*-terminal TE with an *N*-terminal cysteine peptide in a two-step process (Fig. 2.10). The first step is a transthioesterification where the thiolate group of peptide 2 attacks the *C*-terminal TE of peptide 1 under mild reaction conditions (for example, aqueous buffer, pH 7.0, 20°C), resulting in a TE intermediate. This intermediate rearranges by an intramolecular $S \rightarrow N$ acyl shift that results in a peptide bond at the ligation site [75]. In recent years, systems other than cysteine have been reported, and versatility is increasing.

Native chemical ligation has been employed for the synthesis of biomaterials such as type I collagen. The length of type I collagen far exceeds the length which can be synthesized by SPPS [78]. Recombinant expression of this protein is limited due to the high number of hydroxyproline residues [79]. Paramonov et al. were able to overcome these limitations by employing a native chemical ligation strategy [80]. Low-molecular-weight (Pro-Hyp-Gly)$_n$ repeats bearing an *N*-terminal Cys residue and a *C*-terminal TE were prepared by solid-phase synthesis. Following native chemical ligation polymeric material with $M_w = 28{,}000$, $M_n = 12{,}000$, and polydispersity index (PDI) $= 2.3$, were observed. Circular dichroism studies demonstrated that the secondary structure of the ligated collagen was in agreement with the native collagen. Transmission electron microscopy (TEM) studies of the ligated collagen revealed a dense network of fibers with diameters in the nanometer range and lengths in microns, indicating that the collagen was self-assembling in a fashion similar to natural collagen.

Native chemical ligation has been a useful tool in the development of these biomaterials. However, to further expand this technique for the development of higher-molecular-weight polypeptides significant hurdles must be addressed. High-molecular-weight polypeptides would require iterative ligations with peptides bearing both an *N*-terminal cysteine and a *C*-terminal TE. Under typical native chemical ligation conditions the expected product would be the cyclic peptide preventing the formation of larger products [81].

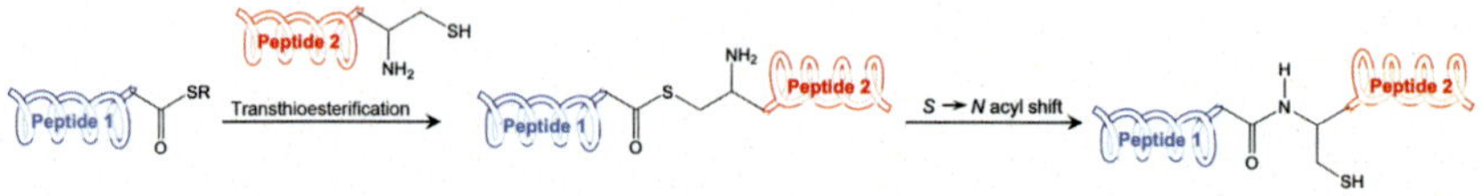

Figure 2.10 Native chemical ligation, the sulfur atom of the *N*-terminal Cys residue of peptide 2 attacks the *C*-terminal thiol group of peptide 1 producing a thioester intermediate that rearranges to yield a peptide bond [2,66].

2.3.2 Nonribosomal peptide synthetase

As described in the biological synthesis (Section 2.1), proteins are generally synthesized by ribosome in cells. However, an exceptional synthesis is nonribosomal peptide synthesis by nonribosomal peptide synthetases (NRPSs). NRPSs are multimodular complexes of enzymes found in lower organisms that assemble secondary metabolites such as polypeptides, polyketides, and fatty acids [82] (Fig. 2.11). Each NRPS module is subdivided into four catalytic domains, namely, adenylation domain (A-domain), peptidyl carrier protein (PCP-domain), condensation domain (C-domain), and TE-domain [82]. Initiation reaction occurs in A- and PCP-domains. The A-domain recognizes specific amino acids and activates its carboxyl group in an adenosine triphosphate (ATP)-dependent manner. The A-domain has broad

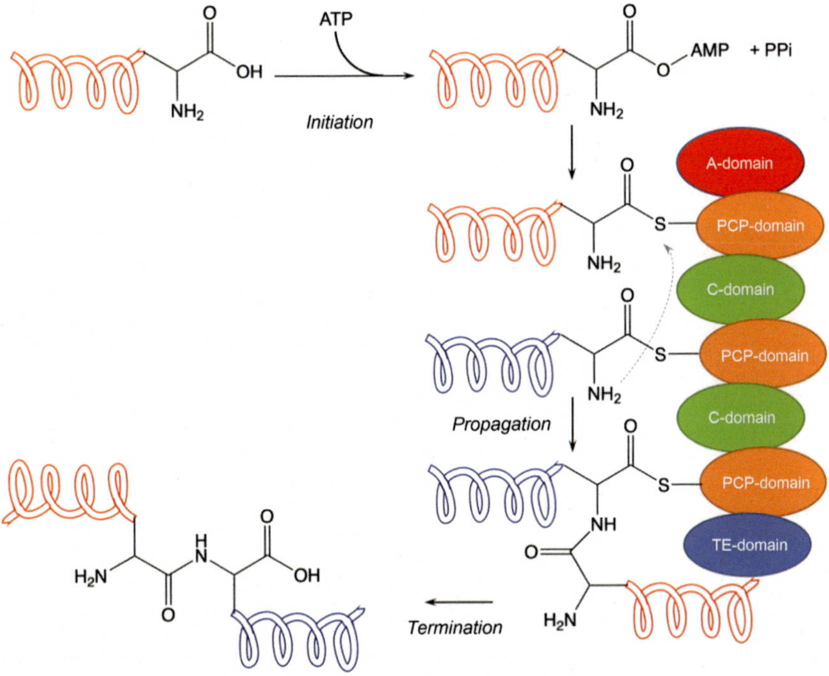

Figure 2.11 General scheme of nonribosomal peptide synthesis. Each NRPS module incorporates one amino acid into the growing peptide chain. The modules are composed of several domains: adenylation domain (*red*) is responsible for substrate selectivity, peptidyl carrier protein domain (*orange*) and condensation domain (*green*) work synergistically to form the peptide bond, and thioester domain (*blue*) which terminates the reaction, resulting in either a linear or cyclic polypeptide. The anion $P_2O_7^{4-}$ is abbreviated as PP_i, standing for inorganic pyrophosphate. It is formed by the hydrolysis of ATP into AMP (adenosine monophosphate) [2,66].

substrate recognition and allows for the facile incorporation of modified amino acids [83,84]. The activated amino acids are transferred to the PCP-domain where they are covalently tethered to the 4′-phosphopantetheinyl cofactor [85]. Propagation takes place at the C-domain, where the amino acids are linked via a condensation reaction. Termination reaction occurs at the TE-domain through either a hydrolysis or a cyclization reaction, resulting in a linear or cyclic polypeptide, respectively [86].

NRPSs have made significant advances in the synthesis of cyclic polypeptides and the ability to introduce nonnatural amino acids will assist in the development of scaffolds for biomaterials. However, NRPSs produce low-molecular-weight polypeptides and require large enzymatic complexes, which are difficult to use in the large-scale production of polypeptides.

2.3.3 Amino acid ligase

Biological synthesis of polypeptides also occurs independent of genetic information. Several enzymes, including folylpolyglutamate synthetase, poly-γ-glutamate synthetase, and D-alanine:D-alanine ligase (DDL), mediate polypeptide synthesis [87–89]. These synthetases either produce non-α-linked amino acids or use nonnatural amino acids as their substrates [90–92]. The reaction mechanism of these synthetases requires an ATP-dependent ligation of a carboxyl of one amino acid with an amino- or imino group of a second. These enzymes all belong to the carboxylate-amine/thiol ligase superfamily [93]. This superfamily is identified by structural motifs corresponding to the phosphate-binding loop and an Mg^{2+}-binding site located within the ATP-binding domain [93].

Cell-free biochemical studies suggest the existence of dipeptide synthetases, which form α-peptide linkages between L-amino acids [94,95]. Tabata et al. using in silico screening identified the L-amino acid ligase gene (*YwfE*) in *Bacillus subtilis* [96]. Recombinant YwfE (also reported as BacD, in the literature) protein demonstrated α-dipeptides synthesis activity, using from unprotected L-amino acids as the substrate. YwfE contains an ATP-binding domain, however this domain showed no sequence homology with aminoacyl-tRNA synthetase or A-domains of NRPSs. The crystal structure of YwfE was refined to 2.5 Å and is superimposable on DDL from *E. coli* (PDB ID: 2DLN) [97]. YwfE is divided into three domains: an *N*-terminal domain, a central domain, and a *C*-terminal domain. The *N*-terminal and central domains are structurally similar to DDL while the *C*-terminal domain is 100 residues longer and contains additional antiparallel beta-sheets [97]. Critical differences were observed in the active site cavity of YwfE when compared to DDL: binding mode of dipeptide moiety, and the size and electrostatic potential of the active site. These differences contribute to the substrate preference for L-amino acids and play a critical role in the stabilization of the tetrahedral intermediate state as proposed for DDL mechanism [97,98]. Future protein engineering studies will be greatly enhanced with the structure of YwfE. However, the specificity of YwfE is currently limited to nonbulky, neutral residues at the *N*-terminus and bulky, neutral residues at the *C*-terminus.

Using YwfE as a template sequence new L-amino acid ligases (L-AAL) have been identified in silico (Table 2.1) [99–101]. RizB from *B. subtilis* NBRC3134 mediated the synthesis of branch-chained L-amino acids and L-methionine homopolypeptides [102–106]. RizB also synthesized heteropeptides with high specificity at the *N*-terminal and relaxed specificity toward the *C*-terminal [99]. RizB polypeptide synthesis is similar to other L-AAL, however RizB uses amino acid monomers as well as oligopeptides as their substrates, while other L-AALs use only amino acids monomers as their substrate. In a subsequent study, RizB was used as a template sequence to identify additional L-AAL that synthesized high-molecular-weight polypeptides [100]. spr0969 and BAD_1200 from *Streptococcus pneumoniae* and *Bifidobacterium adolescentis*, respectively, were identified as RizB homologs. spr0969 showed a modest improvement in polypeptide chain length. spr0969 polymerization of Val resulted in six repeat units while RizB polymerization showed only four repeats. BAD_1200 was more promiscuous than the other L-AAL investigated. The activity toward branched amino acids was lower than RizB but BAD_1200 also polymerized aromatic amino acids [100].

Ribosomal protein S6 from *E. coli* undergoes a unique posttranslation modification where up to six glutamic acid residues are ligated to the *C*-terminus [107–109]. RimK, also a member of carboxylate-amine/thiol ligase superfamily, mediated this PTM [109]. In vitro analysis of RimK synthesis resulted in 46-mer (maximum length) of α-poly(L-glutamic acid) at pH 9.0, 30°C. The maximum chain length was pH dependent. Furthermore, RimK demonstrated strict substrate specificity for Glu [101].

The data reported in these manuscripts demonstrates the enzymatic synthesis of homo- and heteropolypeptides. Data mining has expanded both the diversity and chain length of the resultant polypeptides. Recently, the crystal structure YwfE has been reported which will assist in the development of protein design studies. Engineering the reaction media for enhanced solubility of the polypeptides may influence the molecular weight.

Table 2.1 Amino acid ligases' homologs, substrate specificity, and products [2,66].

Ligase	Species	Preference	Length	Composition	Reference
YwfE	*Bacillus subtilis*	Neutral	Dimer	Heteropolypeptide	[96]
RizB	*B. subtilis* NBRC3134	Branched	Oligomer	Heteropolypeptide	[99]
spr0969	*Streptococcus pneumoniae*	Branched	Oligomer	Heteropolypeptide	[100]
BAD_1200	*Bifidobacterium adolescentis*	Aromatic	Oligomer	Heteropolypeptide	[100]
RimK	*Escherichia coli*	Glutamic acid	Oligomer	Homopolypeptide	[101]

2.3.4 Chemoenzymatic polymerization

Chemoenzymatic polymerization is another promising candidate for a green synthetic method for polypeptide and protein synthesis (Fig. 2.12). In contrast to biological and biochemical syntheses with either ligase or ribosome, the enzymes used for chemoenzymatic polymerization of amino acid monomers are proteases, which originally cleave the peptide bonds in peptide/protein. The protease-catalyzed hydrolysis of peptide/protein is a reversible reaction; hence, the reverse reaction of the hydrolysis can propagate peptide chains under optimized reaction conditions (high monomer concentration, higher pH, etc.). Protease-catalyzed polymerization has several advantages over chemical and biological synthesis methods. Amino acids as monomers, generally in the form of an ester, are polymerized in the presence of a protease in an aqueous buffer solution, resulting in the formation of polypeptides with only small alcoholic by-products. Thus enzymatic polymerization is considered to be an eco-friendly synthetic method with high atom economy compared to chemical condensation methods.

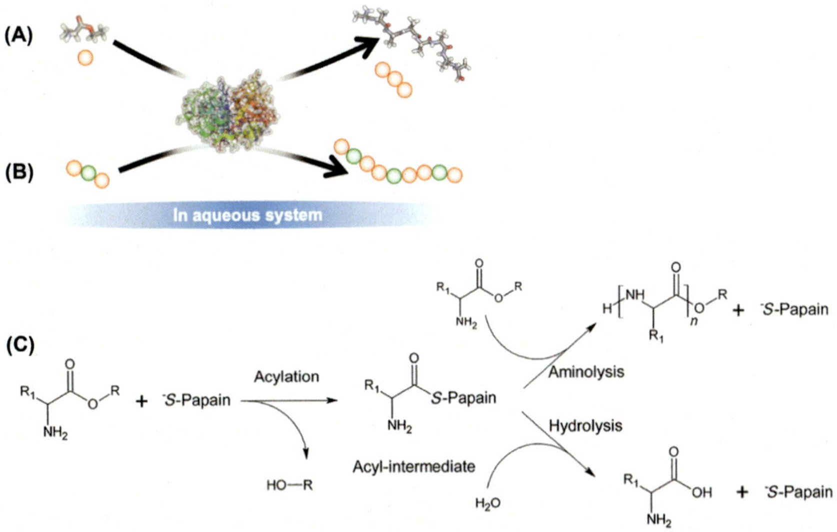

Figure 2.12 Schematic illustration of chemoenzymatic polymerization of alanine ester mediated by papain (A) and tripeptide (B). (C) Reaction scheme of the chemoenzymatic peptide synthesis mediated by papain. In the first step, acylation, formation of the acyl-intermediate enzyme occurs after release of a leaving molecule from the substrate. In the acyl-intermediate stage, both aminolysis and hydrolysis reactions may happen, depending on nucleophiles present in the reaction. *Source: Adapted from J. Gimenez-Dejoz, K. Tsuchiya, K. Numata, Insights into the stereospecificity in papain-mediated chemoenzymatic polymerization from quantum mechanics/molecular mechanics simulations, ACS Chem. Biol. 14 (6) (2019) 1280–1292 [110].*

In chemoenzymatic polymerization, the substrate specificity of proteases rationally recognizes an appropriate chiral isomer, generally the L-form, and reacting groups attached to an α-carbon. Therefore enzymatic polymerization proceeds in a stereoselective and regioselective manner. The amino acid containing the reactive sidechain can be polymerized without protection of the reactive side groups. The enzymatic polymerization reaction is a readily scalable science as the reaction conditions are simply mixing of amino acid monomers and proteases in mild conditions.

Enzymatic amide formation has been continually developed to synthesize small peptide compounds since the 1950s [111]. To date, many comprehensive studies on enzymatic peptide synthesis have not only elucidated the mechanisms of the involved reactions but also synthesized novel functional polypeptide materials. Since protease-catalyzed aminolysis generates amide bonds, an enzymatic polymerization by protease can be applied to not only polypeptides but also other synthetic polyamides. Indeed, there are some reports on enzymatic copolymerization of unnatural and natural amino acids [112–114]. The copolymers of 2-aminoisobutyric acid (Aib) and nylon units with natural amino acids show the significant changes in secondary structure and thermal property [115,116]. Further, unnatural amino acids, branched structures, and blocky sequences can be easily introduced into the molecular design of peptides [115,117–119]. The current drawback of chemoenzymatic polymerization is the difficulty in controlling molecular weight and sequence.

As described before, proteases catalyze the hydrolysis of peptide bonds and their aminolysis. The aminolysis by proteases results in the formation of a peptide bond between two amino acids. According to the long research history of enzyme-mediated polymerization, the chemoenzymatic polymerization of amino acids is classified into two reaction mechanisms [120–127], that is, thermodynamically and kinetically controlled syntheses.

2.3.4.1 Thermodynamic control

Thermodynamically controlled synthesis of peptides focuses on an equilibrium between the hydrolysis and aminolysis reactions. As shown in Fig. 2.13, protease-catalyzed peptide bond formation can be divided into two equilibrium reactions. One reaction is the equilibrium between the

$$\text{R-COO}^- + \text{H}_3\text{N}^+\text{-R'} \underset{}{\overset{K_{ion}}{\rightleftharpoons}} \text{R-COOH} + \text{H}_2\text{N-R'} \overset{K_{trans}}{\rightleftharpoons} \text{R-CO-NH-R'}$$

Figure 2.13 Equilibrium in thermodynamically controlled peptide synthesis [2,66].

ionization and deionization of carboxylic acid/amine of amino acid, and the other is the equilibrium between hydrolysis and aminolysis. These two steps need to be controlled appropriately to form peptide bonds from amino acids with free carboxylic acid and amine groups. In the first step, noncharged substrates need to be formed by the deionization of carboxylic acid and amine groups. Subsequently, the aminolysis reaction occurs via the formation of an acyl intermediate involved with proteases. In Fig. 2.13, the first equilibrium tends to drive to the left side. This effect can be explained by the equilibrium constant and the ΔG of the reaction [128]. In the cases using two unprotected amino acids, the equilibrium constant of the first step is $10^{-7.5}$ and the ΔG is approximately $10\ \text{kcal} \cdot \text{mol}^{-1}$, which means that the equilibrium favors not the nonionic but rather the ionic reactants [128]. The equilibrium constant of the second step is $10^{3.7}$, and its ΔG is about $-5\ \text{kcal} \cdot \text{mol}^{-1}$. Therefore the equilibrium constant of the whole reaction is $10^{-3.8}$, and its ΔG is approximately $5\ \text{kcal} \cdot \text{mol}^{-1}$. This value indicates that the equilibrium reaction proceeds toward an endothermic reverse reaction. Significant disincentives to forming peptide bonds in this step are a slow reaction speed and a low yield because of the low equilibrium constant and positive ΔG. When the aminolysis reaction in homopolymerization proceeds, the equilibrium constant of the reaction becomes lower and the ΔG of the reaction becomes higher. Hence, it is not easy to obtain the polypeptide by iterative aminolysis reactions. Thus the thermodynamically controlled synthesis is a method to control the thermodynamic equilibrium constants of these two elementary reactions. To increase the equilibrium constant of aminolysis, we need to improve the control of either the first or the second reaction.

In the first step, deionization of an ionized substrate occurs by proton transfer from ammonium to carboxylate groups at a pH above its isoelectric point. However, amino acids exist as zwitterions at pH 5–9, at which most proteases show high catalytic activity. In this pH range, the neutral form of amino acids is a minor substrate because the equilibrium constant of the first step is much lower than 1. In contrast, the equilibrium constant of the reaction between *N*-protected and *C*-protected amino acids is $10^{-4.0}$. This value is still lower than 1, but it is much higher than that of protection-free amino acids. Therefore combining *N*- and *C*-protected amino acids is favorable for driving the equilibrium toward the neutral forms.

The thermodynamically controlled synthesis is limited to the synthesis of oligopeptides via a coupling between two reactants. It is not favorable for the synthesis of polypeptides, which generally result from multistep reactions. The major factors that can control the first equilibrium reaction include pH and ionic strength. For example, the addition of water-miscible organic solvents induces a reduction in ionic strength, thereby decreasing the frequency of the proton transfer from ammonium to carboxylate groups. By using the *N*- or *C*-protected amino acids as substituents and organic/water-miscible solution as solvents, the equilibrium constant can increase. The equilibrium constant can move to the right side, namely, peptide synthesis. Homandberg et al. examined the equilibrium constant of the reaction between benzyloxycarbonyl-protected tryptophan (Z-Trp) and glycinamide (Gly-NH_2) in water, 60 v/v% triethylene glycol solution, or 85 v/v% 1,4-butanediol solution [129]. The equilibrium constant in 85 v/v% 1,4-butanediol solution showed the largest value. The reaction rate and yields can be controlled by the content of organic solvent and water.

When products precipitate from the solution, the concentration of products dissolved in the solution becomes lower and the equilibrium shifts toward the right side in Fig. 2.13. In this way, peptide formation proceeds by removing products from a system. In other words, by using a biphasic condition, products can be removed from water media. The reaction occurs around the surface between water and organic layers. A nonpolar organic solvent, such as ethyl acetate, chloroform, and toluene, is used as the organic layer. Casselles and Halling reported protease-catalyzed peptide synthesis in water/organic two-phase systems [130]. The reaction rate of Z-Phe-Phe-OMe synthesis with a silica-supported thermolysin in a reaction mixture composed predominantly of ethyl acetate was as fast as that with thermolysin in the corresponding high-water emulsion system. In this case, the reaction proceeds as follows: substrates dissolved in the organic layer diffused to water media in which proteases exist and reacted to afford products. Substrates are continuously supplied from the organic layer to water media, whereas products are removed from water to the organic layer. This removal of the peptide product drives the equilibrium toward aminolysis and prevents hydrolysis of the product. However, the reaction rate is very slow because of the slow diffusion of the substrate, and the reaction suffers from denaturing of proteases at the surface.

Another strategy to enhance the polymerization efficiency is the removal of water molecules from the system. Hydrolysis of peptides is induced by the nucleophilic attack of water molecules (H_2O). Reducing the amount of H_2O or the activity of H_2O is a useful method to prevent hydrolysis. An organic solvent suppresses the activity of H_2O. However, proteases do not maintain their structures and functions in an organic solvent. Therefore water-miscible organic solvents (such as ethanol, dimethylsulfoxide, *N*,*N*-dimethylformamide, dioxane, acetone, and acetonitrile) are used with 2%–5% water as the equilibrium condition [129,131,132].

2.3.4.2 Kinetic control

The kinetic control synthesis is the other way for the chemoenzymatic polymerization of amino acids. Based on the synthetic scheme of kinetically controlled peptide synthesis (Fig. 2.14) [133], an ester group is generally introduced at a carboxyl end and used as a mildly activated acyl donor. The ester donor rapidly reacts at the catalytic center in proteases and an acyl-protease intermediate forms. Subsequently, the acyl-protease intermediate is nucleophilically attacked by nucleophiles, that is, H_2O or an amino group of another amino acid substrate. Either hydrolysis or aminolysis occurs by deacylation with H_2O or free amino groups. Hydrolysis and aminolysis are not reversible reactions but rather cooperative reactions in kinetically controlled peptide synthesis. Therefore the degree of the nucleophilic attack affects the rate of hydrolysis/aminolysis. Thermodynamic hydrolysis proceeds with proteases, peptides, and water molecules. Nucleophilic hydrolysis and aminolysis are much faster than the equilibrium reaction. In the initial stage of the reaction, the concentration of peptide product rapidly increases. In contrast, in the latter half of the reaction, the concentration of peptide product reaches a plateau. Thus the aminolysis reaction rapidly occurs under kinetically controlled peptide synthesis.

Several computational studies have investigated the hydrolysis reaction of peptide bonds by cysteine proteases, including papain [134–140]. Recent quantum mechanics/molecular mechanics (QM/MM) molecular

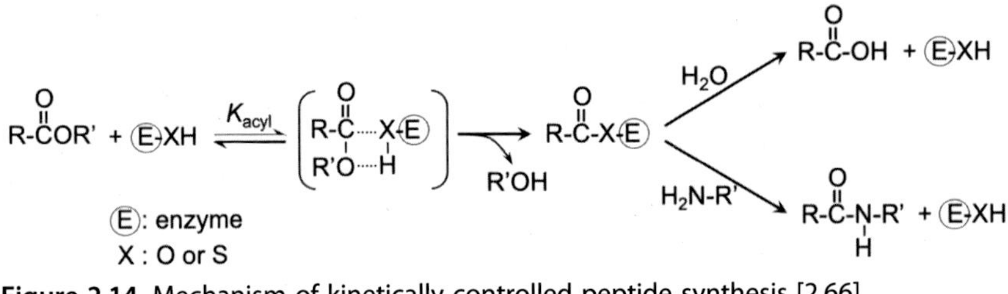

Figure 2.14 Mechanism of kinetically controlled peptide synthesis [2,66].

dynamics simulations tried to demonstrate how the chemoenzymatic polymerization is accomplished via papain-catalyzed aminolysis at the atomic level of detail [110,141]. The papain-mediated chemoenzymatic peptide synthesis reactions using L- and D-Ala-OEt stereoisomers as substrates was investigated to determine the enzyme stereospecificity and its molecular mechanism. The papain-catalyzed acylation and aminolysis reactions of L- and D-Ala-OEt determine their free energy landscapes. In particular, we employed the adaptively biased molecular dynamics (ABMD) method [142,143], based on metadynamics [144], to enhance the sampling efficiency of the simulations, together with the first-principle density functional theory framework. The experimental and computational results reveal that papain can recognize both L- and D-Ala-OEt stereoisomers in the acylation step, forming an acyl-intermediate enzyme with a similar energy barrier (Fig. 2.15), and demonstrate that the different reactivity of the enzyme toward the Ala-OEt stereoisomers arises from the aminolysis step of the polymerization reaction (Fig. 2.16).

Both serine and cysteine proteases can be used for kinetically controlled synthesis to generate peptide bonds. The hydroxyl group of serine and the thiol group of cysteine play a critical role in forming the acyl-protease intermediate. Fig. 2.17 demonstrates molecular behaviors around

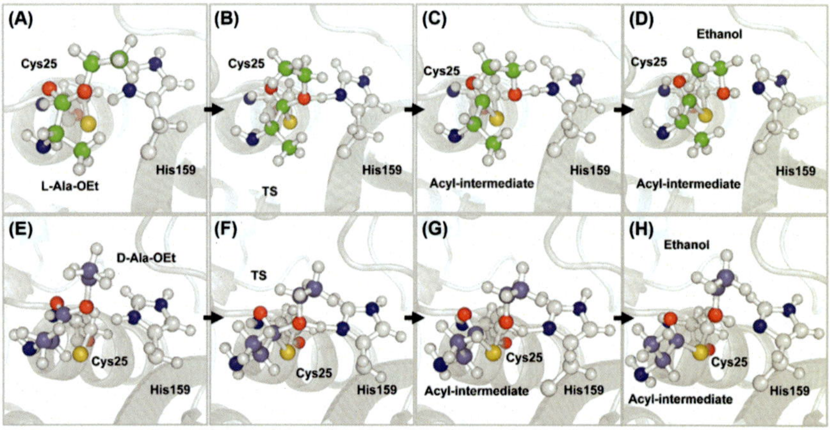

Figure 2.15 Representative snapshots of the mechanism of the acylation reaction of L-Ala-OEt (A–D) and D-Ala-OEt (E–H) extracted from the lowest free energy pathway calculated with the QM/MM ABMD simulations. Papain is represented as a gray cartoon, while atoms QM-treated are illustrated using ball-and-stick representations. Carbon atoms for L- and D-Ala-OEt are displayed in green and blue color, respectively [110]. *TS,* Transition state.

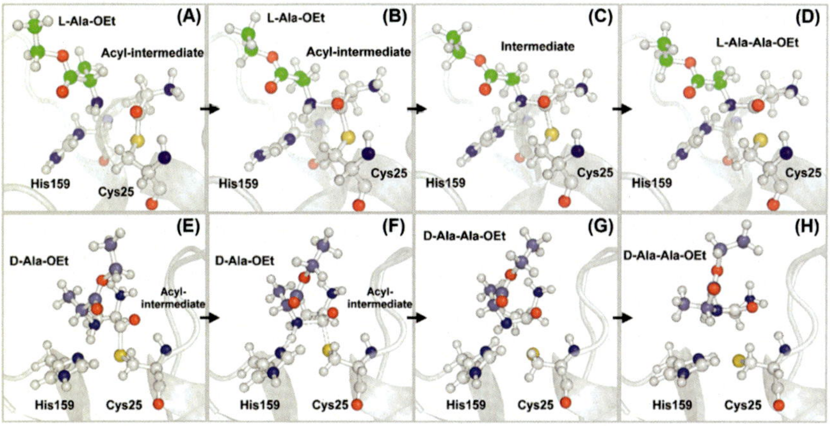

Figure 2.16 Representative snapshots of the mechanism of the aminolysis reaction of L-Ala-OEt (A–D) and D-Ala-OEt (E–H) extracted from the lowest free energy pathway calculated with the QM/MM ABMD simulations. Papain is represented as a gray cartoon, while atoms QM-treated are illustrated using ball-and-stick representations. Carbon atoms for L- and D-Ala-OEt are displayed in green and blue color, respectively [110].

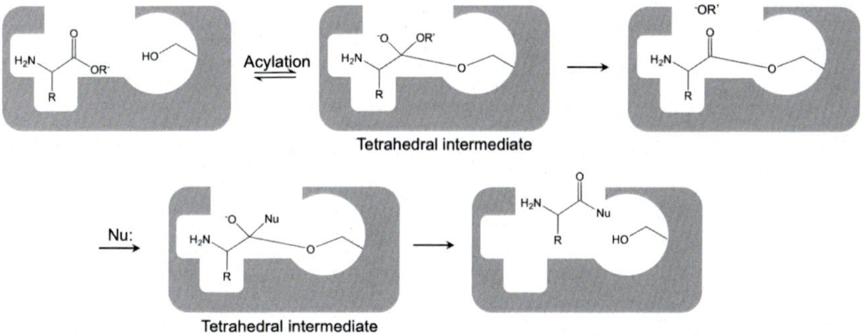

Figure 2.17 The behavior of acyl donors and nucleophiles at the active site of serine proteases [2,66].

the active site of serine proteases. When carboxylic ester approaches the catalytic center, a tetrahedral intermediate is formed with a C—O covalent bond between the ester and the catalytic serine residue. The acyl-protease intermediate is generated by releasing an alcohol corresponding to the ester groups of the starting substrates (Fig. 2.15). Subsequently, the tetrahedral intermediate is formed again by the nucleophilic attack of nucleophiles. Finally, an amide bond is formed through the equilibrium between the tetrahedral intermediate and a product (Fig. 2.16).

As described in the previous section 2.3.4, many factors affect the rate of aminolysis. The concentration of amine nucleophilic groups, pH, and temperature affect the yield of peptide products. The rate of nucleophilic attack by amines to form acyl-enzyme intermediates is enhanced by raising the concentration of amine substrates (monomer). The amine nucleophilic group exists in its reactive neutral form at high pH. A suitable pH is in a range of 5–9 for the enzymatic reaction in the thermodynamic control method. In contrast, the kinetic control method needs a higher pH condition to keep monomers active for aminolysis [145]. Besides, higher temperature induces a lower reaction rate of aminolysis. Based on the thermal characteristics of aminolysis by proteases, frozen aqueous synthesis has been developed by many research groups [146–148]. In frozen aqueous conditions, namely, −20°C to 5°C, the hydrolysis rate is decreased, and the yield of the resultant peptide is improved.

In contrast to the thermodynamic control method, the biphasic solution system is inappropriate for kinetically controlled synthesis [149–152]. Proteases exist in an aqueous phase, whereas amino acid ester substrates exist in both phases. Therefore the concentration of acyl-protease intermediates is low and hardly controlled in this system. On the other hand, the activity of H_2O toward the acyl-protease is reduced in a miscible water/organic solvent system, which suppresses the competing hydrolysis reaction. Kisee et al. [153] and Viswanathan et al. [154] reported a miscible water/organic (methanol, ethanol, tetrahydrofuran, and so on) solvent system. The yield of resulting peptides was dependent on the content of water and organic solvent. Thus the water/organic solvent system is a highly potential reaction system. However, it is noted that organic solvent has the potential to cause protease denaturation and dysfunction. To avoid the denaturation and dysfunction of proteases, the enzyme immobilization on solid substrates could be the easiest solution [155].

2.3.4.3 Specificity of proteases

The substrate specificity of proteases affects the rate of aminolysis as well as hydrolysis. An acyl ester donor is recognized as a substrate at the active sites of proteases, and acyl-proteases are formed. Therefore the affinity of acyl esters to proteases is most important in forming acyl-protease intermediates. Also, the affinity of amine nucleophilic groups to the catalytic site is dependent on the types of amino acids. Fastrez and Fersht reported a comparison of the deacylation rate of Acyl-Phe-chymotrypsin [156]. The aminolysis rate with glycinamide (Gly-NH_2) was 11 times faster than the hydrolysis rate with H_2O, while the aminolysis rate with L-alanine amide (Ala-NH_2) was 44 times faster than the hydrolysis rate with H_2O. Thus the affinity of Ala-NH_2 to chymotrypsin is higher than that of Gly-NH_2 to chymotrypsin. The substrate specificity of proteases can be evaluated by the partition constant p, which can be calculated from the reaction rates of hydrolysis and aminolysis. Sehellenberger et al. reported on the partition constant of nucleophilic amino acids using maleyl-(3-carboxyacryloyl)-Phe-chymotrypsin as an acyl-protease [157]. The order of nucleophilicity toward the protease was Arg-NH_2 > Leu-NH_2 > Val-NH_2 > Ala-NH_2 > Gly-NH_2 > H_2O. The yields and degree of polymerization of the polypeptides depend on selecting proteases and amino acids because the affinity of amino acids to proteases affected the aminolysis rate.

Proteases have some subsites that align on both sides of the catalytic site, contributing to various substrate specificity [158,159]. These subsites determine the cleavage point of hydrolysis and the reaction rate of aminolysis. Schechter and Berger reported the subsite of papain, one of the major proteases for chemoenzymatic polymerization [158]. The size of the binding site of papain was estimated using the oligo(L-Ala) with or without D-Ala; the existence of D-Ala reduces the rate of hydrolysis. Based on this experimental result, the size of the papain subsite was estimated at over 25 Å. In the case of papain, there are four subsites (S_1—S_4) located on one side and three subsites (S_1'—S_3') located on the other (Fig. 2.18), and each site can accommodate one amino acid

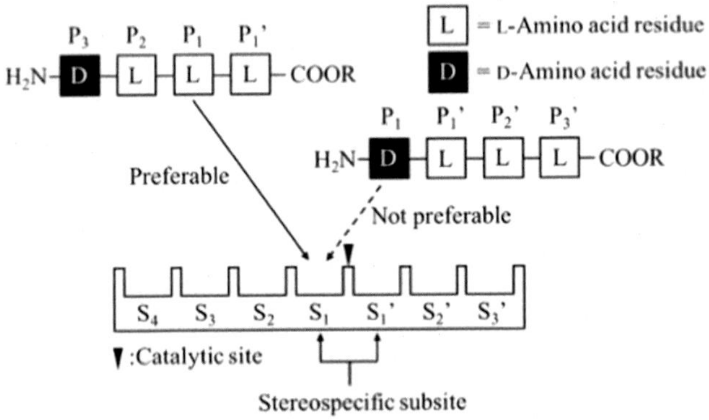

Figure 2.18 Schematic illustration of the catalytic site (*black triangle*) and subsites (denoted as S_n and S_n') in the papain active site. Each subsite can accommodate an amino acid residue in polypeptide substrates (denoted as P_n and P_n'), and subsites S_1 and S_1' strictly regulate the stereospecificity of oligoalanine substrates. Peptide bonds next to the L-alanine residue are preferred for hydrolytic cleavage compared to the bonds next to the D-alanine residue [2,66,158].

residue of a substrate (P_1–P_4 and P_1'–P_3'). Peptide bonds far from the D-Ala residue were labile to papain-catalyzed hydrolysis, whereas peptide bonds next to the D-Ala residue (P_1 and P_1' position) hardly reacted. This experimental result indicates that the chirality of amino acid substrates is most importantly recognized in S_1 and S_1'. On the other hand, D-residues are acceptable to occupying the subsites away from the catalytic center. Furthermore, they concluded that active site S_2 tends to bind the hydrophobic amino acids, such as Phe, Leu, and Val. In the case of enzymatic polymerization, this enzymatic specificity influences on the reaction efficiency.

2.3.4.4 Copolymerization

By chemoenzymatic polymerization, we can copolymerize nonamino acid molecules with natural amino acids. In addition to linear polypeptides, chemoenzymatic copolymerization realizes various polypeptide architectures with unique shapes such as star-shaped and telechelic-type polypeptides [118,119,160,161]. Besides, some proteases, in particular papain and proteinase K, exhibit a relatively broad substrate specificity, which allows unnatural amino acid derivatives to be involved in chemoenzymatic polymerization. For example, ω-aminoalkanoates, monomer units for synthetic polyamides such as nylon, were successfully copolymerized with diethyl L-glutamate by papain-catalyzed chemoenzymatic polymerization [116,162]. However, the resulting polymer showed relatively low nylon contents even using an excess amount of ω-aminoalkanoates in feed, and contaminated with the homopolymer of glutamate monomer (Fig. 2.19). Further, a

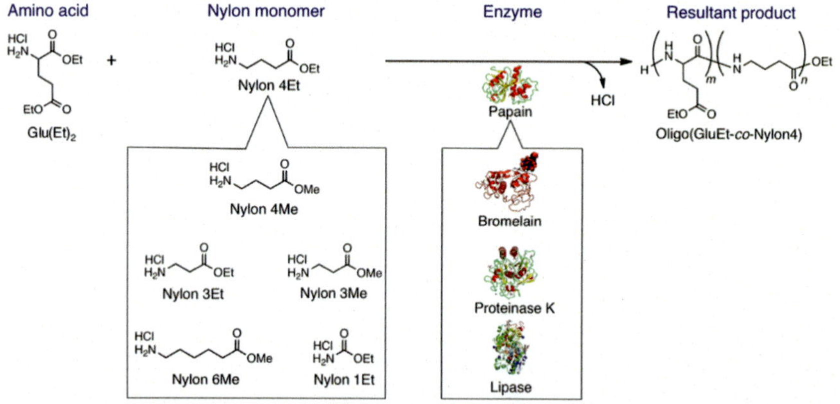

Figure 2.19 Chemoenzymatic copolymerization of amino acids and nylon units by various enzymes. In this case, papain was reported to demonstrate the polymerization acitivity of nylon 4 and nylon 6 [162].

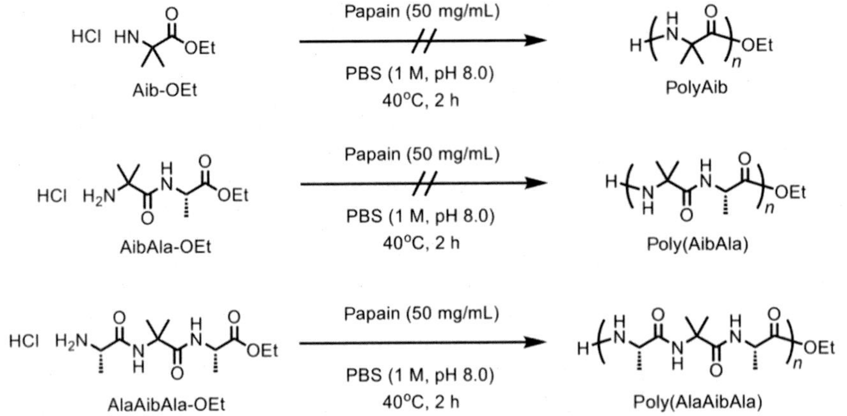

Figure 2.20 Chemoenzymatic polymerization of Aib-containing monomers [115]. *PBS,* Phosphate buffer.

papain-catalyzed reaction with mild aqueous conditions was reported to synthesize oligo(LeuEt-*co*-nylon) containing approximately 14 mol% nylon units. The oligo(LeuEt-*co*-nylon) exhibited a melting point, and the nylon unit prevented the growth of peptide crystals consisting of Leu units [116]. The presence of the nylon monomer units most likely changed the density of the intermolecular hydrogen bonds, inducing the melting transition. Another example is the copolymerization with unnatural amino acids [115]. Polypeptides containing Aib units as an unnatural amino acid residue were synthesized by papain-catalyzed chemoenzymatic polymerization of a tripeptide ethyl ester L-Ala-Aib-L-Ala-OEt in an aqueous medium (Fig. 2.20). The Aib-containing polypeptide adopted an alpha-helix conformation in both the solid and solution phases, which was induced by the periodic Aib residue.

2.4 Future perspectives of the syntheses

Amide bonds are the key and essential chemical linkages in proteins and polypeptides. In 2007 the American Chemical Society of Green Chemistry Institute named amide bond formation as a top challenge for organic chemistry [163]. Since then, several new amide bond synthesis reactions have been developed. Recently, chemoenzymatic synthesis and native chemical ligation have advanced the more traditional polypeptide production routes. Despite the successes outlined above, these techniques have been modest in their production of new biomaterials. In the near future, the further development of amide bond formation, as well as its application to polymerization reactions, will lead to the development of the chemistry related to polypeptides and proteins. In addition to chemistry, we need to consider biological and biochemical methods and combine them to develop the next-generation biomaterials and production process.

References

[1] P.J. Baker, K. Numata, Polymerization of peptide polymers for biomaterial applications, in: F. Yilmaz (Ed.), Polymer Science, 2013, pp. 229–246.

[2] K. Tsuchiya, Y. Miyagi, T. Miyamoto, P.G. Gudeangadi, K. Numata, Synthesis of polypeptides, in: S. Kobayashi, H. Uyama, J.-i Kadokawa (Eds.), Enzymatic Polymerization towards Green Polymer Chemistry, Springer, Singapore, 2019, pp. 233–265.

[3] M.J. Dougherty, M.T. Kreichi, H.S. Creel, T.L. Mason, M.J. Fournier, D.A. Tirrell, Genetic synthesis of chain-folded periodic polypeptides, Faseb J. 6 (1) (1992). A265-A265.

[4] K.P. Mcgrath, D.A. Tirrell, M. Kawai, T.L. Mason, M.J. Fournier, Chemical and biosynthetic approaches to the production of novel polypeptide materials, Biotechnol. Progr. 6 (3) (1990) 188–192.

[5] K.P. McGrath, M.J. Fournier, T.L. Mason, D.A. Tirrell, Genetically directed syntheses of new polymeric materials. Expression of artificial genes encoding proteins with repeating-(AlaGly)3ProGluGly-elements, J. Am. Chem. Soc. 114 (2) (1992) 727–733.

[6] G. Zhang, M.J. Fournier, T.L. Mason, D.A. Tirrell, Biological synthesis of monodisperse derivatives of poly(.alpha.,L-glutamic acid): model rodlike polymers, Macromolecules 25 (13) (1992) 3601–3603.

[7] S. Arcidiacono, C. Mello, D. Kaplan, S. Cheley, H. Bayley, Purification and characterization of recombinant spider silk expressed in Escherichia coli, Appl. Microbiol. Biotechnol. 49 (1) (1998) 31–38.

[8] J.T. Prince, K.P. McGrath, C.M. DiGirolamo, D.L. Kaplan, Construction, cloning, and expression of synthetic genes encoding spider dragline silk, Biochemistry 34 (34) (1995) 10879–10885.

[9] S. Winkler, S. Szela, P. Avtges, R. Valluzzi, D.A. Kirschner, D. Kaplan, Designing recombinant spider silk proteins to control assembly, Int. J. Biol. Macromol. 24 (2-3) (1999) 265–270.

[10] A. Rising, M. Widhe, J. Johansson, M. Hedhammar, Spider silk proteins: recent advances in recombinant production, structure-function relationships and biomedical applications, Cell Mol. Life Sci. 68 (2) (2011) 169–184.

[11] K. Numata, D.L. Kaplan, Silk-based delivery systems of bioactive molecules, Adv. Drug. Deliv. Rev. 62 (15) (2010) 1497–1508.

[12] A.D. Malay, T. Suzuki, T. Katashima, N. Kono, K. Arakawa, K. Numata, Spider silk self-assembly via modular liquid-liquid phase separation and nanofibrillation, Sci. Adv. 6 (45) (2020).

[13] K. Numata, B. Subramanian, H.A. Currie, D.L. Kaplan, Bioengineered silk protein-based gene delivery systems, Biomaterials 30 (29) (2009) 5775–5784.

[14] K. Numata, J. Hamasaki, B. Subramanian, D.L. Kaplan, Gene delivery mediated by recombinant silk proteins containing cationic and cell binding motifs, J. Control. Release 146 (1) (2010) 136–143.

[15] K. Numata, M.R. Reagan, R.H. Goldstein, M. Rosenblatt, D.L. Kaplan, Spider silk-based gene carriers for tumor cell-specific delivery, Bioconjug Chem. 22 (8) (2011) 1605–1610.
[16] K. Numata, A.J. Mieszawska-Czajkowska, L.A. Kvenvold, D.L. Kaplan, Silk-based nanocomplexes with tumor-homing peptides for tumor-specific gene delivery, Macromol. Biosci. 12 (1) (2012) 75–82.
[17] K. Numata, D.L. Kaplan, Silk-based gene carriers with cell membrane destabilizing peptides, Biomacromolecules 11 (11) (2010) 3189–3195.
[18] S.R. Fahnestock, L.A. Bedzyk, Production of synthetic spider dragline silk protein in Pichia pastoris, Appl. Microbiol. Biotechnol. 47 (1) (1997) 33–39.
[19] R. Jansson, C.H. Lau, T. Ishida, M. Ramstrom, M. Sandgren, M. Hedhammar, Functionalized silk assembled from a recombinant spider silk fusion protein (Z-4RepCT) produced in the methylotrophic yeast Pichia pastoris, Biotechnol. J. 11 (5) (2016) 687–699.
[20] D.N. Breslauer, Recombinant protein polymers: a coming wave of personal care ingredients, ACS Biomater. Sci. Eng. 6 (11) (2020) 5980–5986.
[21] K.A. Pyke, Plastid division and development, Plant Cell 11 (4) (1999) 549–556.
[22] W. Sakamoto, S.Y. Miyagishima, P. Jarvis, Chloroplast biogenesis: control of plastid development, protein import, division and inheritance, Arabidopsis Book 6 (2008) e0110.
[23] P. Maliga, Plastid transformation in higher plants, Annu. Rev. Plant. Biol. 55 (2004) 289–313.
[24] M. Adem, D. Beyene, T. Feyissa, Recent achievements obtained by chloroplast transformation, Plant Methods 13 (2017) 30.
[25] S. Ruf, M. Hermann, I.J. Berger, H. Carrer, R. Bock, Stable genetic transformation of tomato plastids and expression of a foreign protein in fruit, Nat. Biotechnol. 19 (9) (2001) 870–875.
[26] M.S. Khan, P. Maliga, Fluorescent antibiotic resistance marker for tracking plastid transformation in higher plants, Nat. Biotechnol. 17 (9) (1999) 910–915.
[27] S. Kahlau, R. Bock, Plastid transcriptomics and translatomics of tomato fruit development and chloroplast-to-chromoplast differentiation: chromoplast gene expression largely serves the production of a single protein, Plant Cell 20 (4) (2008) 856–874.
[28] C. Thagun, J.A. Chuah, K. Numata, Targeted gene delivery into various plastids mediated by clustered cell-penetrating and chloroplast-targeting peptides, Adv. Sci. (Weinh.) 6 (23) (2019) 1902064.
[29] J.M. Hibberd, P.J. Linley, M.S. Khan, J.C. Gray, Transient expression of green fluorescent protein in various plastid types following microprojectile bombardment, Plant J. 16 (5) (1998) 627–632.
[30] C.L. Langbecker, G.N. Ye, D.L. Broyles, L.L. Duggan, C.W. Xu, P.T. Hajdukiewicz, et al., High-frequency transformation of undeveloped plastids in tobacco suspension cells, Plant Physiol. 135 (1) (2004) 39–46.
[31] B. Sporlein, M. Streubel, G. Dahlfeld, P. Westhoff, H.U. Koop, PEG-mediated plastid transformation: a new system for transient gene expression assays in chloroplasts, Theor. Appl. Genet. 82 (6) (1991) 717–722.
[32] M. Knoblauch, J.M. Hibberd, J.C. Gray, A.J. van Bel, A galinstan expansion femtosyringe for microinjection of eukaryotic organelles and prokaryotes, Nat. Biotechnol. 17 (9) (1999) 906–909.
[33] J.J. Yang, L.A. Barr, S.R. Fahnestock, Z.B. Liu, High yield recombinant silk-like protein production in transgenic plants through protein targeting, Transgenic Res. 14 (3) (2005) 313–324.
[34] J. Scheller, K.H. Guhrs, F. Grosse, U. Conrad, Production of spider silk proteins in tobacco and potato, Nat. Biotechnol. 19 (6) (2001) 573–577.

[35] V. Hauptmann, N. Weichert, M. Rakhimova, U. Conrad, Spider silks from plants—a challenge to create native-sized spidroins, Biotechnol. J. 8 (10) (2013) 1183–1192.

[36] J. Xu, Q.L. Dong, Y. Yu, B.L. Niu, D.F. Ji, M.W. Li, et al., Mass spider silk production through targeted gene replacement in Bombyx mori, Proc. Natl. Acad. Sci. U S A 115 (35) (2018) 8757–8762.

[37] W.W. Fischer, J. Hemp, J.E. Johnson, Evolution of oxygenic photosynthesis, Annu. Rev. Earth Planet Sci. 44 (2016) 647–683.

[38] W. Hillier, G.T. Babcock, Photosynthetic reaction centers, Plant Physiol. 125 (1) (2001) 33–37.

[39] M.T. Madigan, D.O. Jun, An overview of purple bacteria: systematics, physiology, and habitats. Advances in Photosynthesis and Respiration series in: C.N. Hunter, et al. (Eds.), The Purple Phototrophic Bacteria, 2009, pp. 1–15.

[40] J.B. McKinlay, C.S. Harwood, Photobiological production of hydrogen gas as a biofuel, Curr. Opin. Biotech. 21 (3) (2010) 244–251.

[41] M. Higuchi-Takeuchi, K. Numata, Marine purple photosynthetic bacteria as sustainable microbial production hosts, Front. Bioeng. Biotech. 7 (2019) 258. Available from: https://doi.org/10.3389/fbioe.2019.00258.

[42] M. Higuchi-Takeuchi, K. Morisaki, K. Numata, Method for the facile transformation of marine purple photosynthetic bacteria using chemically competent cells, MicrobiologyOpen 9 (1) (2020) e00953.

[43] M. Higuchi-Takeuchi, K. Numata, Acetate-inducing metabolic states enhance polyhydroxyalkanoate production in marine purple non-sulfur bacteria under aerobic conditions, Front. Bioeng. Biotech. 7 (2019) 118. Available from: https://doi.org/10.3389/fbioe.2019.00118.

[44] C.P. Foong, M. Higuchi-Takeuchi, K. Numata, Optimal iron concentrations for growth-associated polyhydroxyalkanoate biosynthesis in the marine photosynthetic purple bacterium Rhodovulum sulfidophilum under photoheterotrophic condition, PLoS One 14 (4) (2019) e0212654.

[45] M. Higuchi-Takeuchi, Y. Motoda, T. Kigawa, K. Numata, Class I polyhydroxyalkanoate synthase from the purple photosynthetic bacterium Rhodovulum sulfidophilum predominantly exists as a functional dimer in the absence of a substrate, ACS Omega 2 (8) (2017) 5071–5078.

[46] M. Higuchi-Takeuchi, K. Morisaki, K. Toyooka, K. Numata, Synthesis of high-molecular-weight polyhydroxyalkanoates by marine photosynthetic purple bacteria, PLoS One 11 (8) (2016) e0160981.

[47] M. Higuchi-Takeuchi, K. Morisaki, K. Numata, A screening method for the isolation of polyhydroxyalkanoate-producing purple non-sulfur photosynthetic bacteria from natural seawater, Front. Microbiol. 7 (2016) 1509.

[48] K. Numata, How to define and study structural proteins as biopolymer materials, Polym. J. 52 (9) (2020) 1043–1056.

[49] J. dos Santos-Pinto, H. Arcuri, F. Esteves, M. Palma, G. Lubec, Spider silk proteome provides insight into the structural characterization of Nephila clavipes flagelliform spidroin, Sci. Rep. 8 (2018) 14674.

[50] H. Craig, S. Blamires, M. Sani, M. Kasumovic, A. Rawal, J. Hook, DNP NMR spectroscopy reveals new structures, residues and interactions in wild spider silks, Chem. Commun. 55 (32) (2019) 4687–4690.

[51] A. Whaite, T. Wang, J. Macdnald, S. Cummins, Major ampullate silk gland transcriptomes and fibre proteomes of the golden orb-weavers, Nephila plumipes and Nephila pilipes (Araneae: Nephilidae), PLoS One 13 (10) (2018) e0204243.

[52] J. dos Santos-Pinto, G. Lamprecht, W. Chen, S. Heo, J. Hardy, H. Priewalder, et al., Structure and post-translational modifications of the web silk protein spidroin-1 from Nephila spiders, J. Proteom. 105 (2014) 174–185.

[53] J. dos Santos-Pinto, H. Arcuri, G. Lubec, M. Palma, Structural characterization of the major ampullate silk spidroin-2 protein produced by the spider Nephila clavipes, Biochim. Biophy. Acta 1864 (10) (2016) 1444–1454.
[54] J. dos Santos-Pinto, H. Arcuri, H. Priewalder, H. Salles, M. Palma, G. Lubec, Structural model for the spider silk protein spidroin-1, J. Proteome Res. 14 (9) (2015) 3859–3870.
[55] D. Raven, C. Earland, M. Little, Occurrence of dityrosine in Tussah silk fibroin and keratin, Biochim. Biophys. Acta 251 (1) (1971) 96–99.
[56] B. Partlow, M. Applegate, F. Omenetto, D. Kaplan, Dityrosine cross-linking in designing biomaterials, ACS Biomater. Sci. Eng. 2 (12) (2016) 2108–2121.
[57] R. Jensen, D. Morse, The bioadhesive of *Phragmatopoma californica* tubes: a silk-like cement containing L-DOPA, J. Comp. Physiol. B 158 (3) (1988) 317–324.
[58] R. Stewart, C. Wang, H. Shao, Complex coacervates as a foundation for synthetic underwater adhesives, Adv. Colloid Interface Sci. 167 (1-2) (2011) 85–93.
[59] B. Partlow, M. Bagheri, J. Harden, D. Kaplan, Tyrosine templating in the self-assembly and crystallization of silk fibroin, Biomacromolecules 17 (11) (2016) 3570–3579.
[60] D. Jain, C. Zhang, L. Cool, T. Blackledge, C. Wesdemiotis, T. Miyoshi, et al., Composition and function of spider glues maintained during the evolution of cob-webs, Biomacromolecules 16 (10) (2015) 3373–3380.
[61] B. Merrifield, Solid phase synthesis, Science 232 (4748) (1986) 341–347.
[62] T.J. Deming, Synthetic polypeptides for biomedical applications, Prog. Polym. Sci. 32 (8–9) (2007) 858–875.
[63] S. Kobayashi, A. Makino, Enzymatic polymer synthesis: an opportunity for green polymer chemistry, Chem. Rev. 109 (11) (2009) 5288–5353.
[64] G.B. Fields, R.L. Noble, Solid phase peptide synthesis utilizing 9-fluorenylmethoxycarbonyl amino acids, Int. J. Pept. Protein Res. 35 (3) (1990) 161–214.
[65] R.B. Merrifield, Solid-phase peptide synthesis, Adv. Enzymol. Relat. Areas Mol. Biol. 32 (1969) 221–296.
[66] P.J. Baker, K. Numata, Chemoenzymatic synthesis of poly(L-alanine) in aqueous environment, Biomacromolecules 13 (4) (2012) 947–951.
[67] R. Arlinghaus, J. Shaefer, R. Schweet, Mechanism of peptide bond formation in polypeptide synthesis, Proc. Natl. Acad. Sci. U S A 51 (1964) 1291–1299.
[68] C.D. Chang, J. Meienhofer, Solid-phase peptide synthesis using mild base cleavage of N alpha-fluorenylmethyloxycarbonylamino acids, exemplified by a synthesis of dihydrosomatostatin, Int. J. Pept. Protein Res. 11 (3) (1978) 246–249.
[69] R.B. Merrifield, Solid-phase peptide synthesis. 3. An improved synthesis of bradykinin, Biochemistry 3 (1964) 1385–1390.
[70] E. Bayer, M. Mutter, Liquid-phase synthesis of peptides, Nature 237 (5357) (1972) 512–513.
[71] M. Goodman, J. Hutchison, Mechanisms of polymerization of N-unsubstituted N-carboxyanhdrides, J. Am. Chem. Soc. 88 (15) (1966) 3627–3630.
[72] H.R. Marsden, A. Kros, Polymer-peptide block copolymers—an overview and assessment of synthesis methods, Macromol. Biosci. 9 (10) (2009) 939–951.
[73] K. Koga, A. Sudo, T. Endo, Revolutionary phosgene-free synthesis of alpha-amino acid N-carboxyanhydrides using diphenyl carbonate based on activation of alpha-amino acids by converting into imidazolium salts, J. Polym. Sci. A Polym. Chem. 48 (19) (2010) 4351–4355.
[74] S. Yamada, K. Koga, A. Sudo, M. Goto, T. Endo, Phosgene-free synthesis of poly-peptides: useful synthesis for hydrophobic polypeptides through polycondensation of

activated urethane derivatives of -amino acids, J. Polym. Sci. A Polym. Chem. 51 (17) (2013) 3726–3731.
[75] P.E. Dawson, T.W. Muir, I. Clarklewis, S.B.H. Kent, Synthesis of proteins by native chemical ligation, Science 266 (5186) (1994) 776–779.
[76] E.C.B. Johnson, S.B.H. Kent, Insights into the mechanism and catalysis of the native chemical ligation reaction, J. Am. Chem. Soc. 128 (20) (2006) 6640–6646.
[77] E.C.B. Johnson, T. Durek, S.B.N. Kent, Total chemical synthesis, folding, and assay of a small protein on a water-compatible solid support, Angew. Chem. Int. Ed. 45 (20) (2006) 3283–3287.
[78] A.K. Tickler, J.D. Wade, Overview of solid phase synthesis of "difficult peptide" sequences, Curr. Protoc. Protein Sci. 50 (1) (2007). Unit 18.8.1–18.8.6.
[79] J. Myllyharju, Recombinant collagen trimers from insect cells and yeast, Methods Mol. Biol. 522 (2009) 51–62.
[80] S.E. Paramonov, V. Gauba, J.D. Hartgerink, Synthesis of collagen-like peptide polymers by native chemical ligation, Macromolecules 38 (18) (2005) 7555–7561.
[81] J.P. Tam, J. Xu, K.D. Eom, Methods and strategies of peptide ligation, Biopolymers 60 (3) (2001) 194–205.
[82] F. Kopp, M.A. Marahiel, Macrocyclization strategies in polyketide and nonribosomal peptide biosynthesis, Nat. Prod. Rep. 24 (4) (2007) 735–749.
[83] E. Conti, T. Stachelhaus, M.A. Marahiel, P. Brick, Structural basis for the activation of phenylalanine in the non-ribosomal biosynthesis of gramicidin S, EMBO J. 16 (14) (1997) 4174–4183.
[84] S. Lautru, G.L. Challis, Substrate recognition by nonribosomal peptide synthetase multi-enzymes, Microbiology 150 (Pt 6) (2004) 1629–1636.
[85] A. Koglin, C.T. Walsh, Structural insights into nonribosomal peptide enzymatic assembly lines, Nat. Prod. Rep. 26 (8) (2009) 987–1000.
[86] S.A. Samel, B. Wagner, M.A. Marahiel, L.O. Essen, The thioesterase domain of the fengycin biosynthesis cluster: a structural base for the macrocyclization of a nonribosomal lipopeptide, J. Mol. Biol. 359 (4) (2006) 876–889.
[87] S.J. Ritari, W. Sakami, C.W. Black, J. Rzepka, The determination of folylpolyglutamate synthetase, Anal. Biochem. 63 (1) (1975) 118–129.
[88] K.G. Scrimgeour, Biosynthesis of polyglutamates of folates, Biochem. Cell Biol. 64 (7) (1986) 667–674.
[89] C.V. Carpenter, F.C. Neuhaus, Enzymatic synthesis of D-alanyl-D-alanine. Two binding modes for product on D-alanine: D-alanine ligase (ADP), Biochemistry 11 (14) (1972) 2594–2598.
[90] S. Evers, B. Casadewall, M. Charles, S. Dutka-Malen, M. Galimand, P. Courvalin, Evolution of structure and substrate specificity in D-alanine:D-alanine ligases and related enzymes, J. Mol. Evol. 42 (6) (1996) 706–712.
[91] Y. Urushibata, S. Tokuyama, Y. Tahara, Characterization of the Bacillus subtilis ywsC gene, involved in gamma-polyglutamic acid production, J. Bacteriol. 184 (2) (2002) 337–343.
[92] X. Sun, A.L. Bognar, E.N. Baker, C.A. Smith, Structural homologies with ATP- and folate-binding enzymes in the crystal structure of folylpolyglutamate synthetase, Proc. Natl. Acad. Sci. U S A 95 (12) (1998) 6647–6652.
[93] M.Y. Galperin, E.V. Koonin, A diverse superfamily of enzymes with ATP-dependent carboxylate-amine/thiol ligase activity, Protein Sci. 6 (12) (1997) 2639–2643.
[94] M. Sakajoh, N.A. Solomon, A.L. Demain, Cell-free synthesis of the dipeptide antibiotic bacilysin, J. Ind. Microbiol. 2 (4) (1987) 201–208.
[95] H. Ueda, Y. Yoshihara, N. Fukushima, H. Shiomi, A. Nakamura, H. Takagi, Kyotorphin (tyrosine-arginine) synthetase in rat-brain synaptosomes, J. Biol. Chem. 262 (17) (1987) 8165–8173.

[96] K. Tabata, H. Ikeda, S. Hashimoto, ywfE in Bacillus subtilis codes for a novel enzyme, L-amino acid ligase, J. Bacteriol. 187 (15) (2005) 5195–5202.
[97] Y. Shomura, E. Hinokuchi, H. Ikeda, A. Senoo, Y. Takahashi, J.I. Saito, et al., Structural and enzymatic characterization of BacD, an L-amino acid dipeptide ligase from Bacillus subtilis, Protein Sci. 21 (5) (2012) 707–712.
[98] C. Fan, P.C. Moews, C.T. Walsh, J.R. Knox, Vancomycin resistance: structure of D-alanine:D-alanine ligase at 2.3 A resolution, Science 266 (5184) (1994) 439–443.
[99] K. Kino, T. Arai, D. Tateiwa, A novel L-amino acid ligase from Bacillus subtilis NBRC3134 catalyzed oligopeptide synthesis, Biosci. Biotechnol. Biochem. 74 (1) (2010) 129–134.
[100] T. Arai, K. Kino, New L-amino acid ligases catalyzing oligopeptide synthesis from various microorganisms, Biosci. Biotechnol. Biochem. 74 (8) (2010) 1572–1577.
[101] K. Kino, T. Arai, Y. Arimura, Poly-alpha-glutamic acid synthesis using a novel catalytic activity of RimK from Escherichia coli K-12, Appl. Env. Microbiol. 77 (6) (2011) 2019–2025.
[102] A. Senoo, K. Tabata, Y. Yonetani, M. Yagasaki, Identification of novel L-amino acid alpha-ligases through Hidden Markov Model-based profile analysis, Biosci. Biotechnol. Biochem. 74 (2) (2010) 415–418.
[103] K. Kino, Y. Nakazawa, M. Yagasaki, Dipeptide synthesis by L-amino acid ligase from Ralstonia solanacearum, Biochem. Biophys. Res. Commun. 371 (3) (2008) 536–540.
[104] K. Kino, A. Noguchi, T. Arai, M. Yagasaki, Identification and characterization of a novel L-amino acid ligase from Photorhabdus luminescens subsp. laumondii TT01, J. Biosci. Bioeng. 110 (1) (2010) 39–41.
[105] T. Arai, K. Kino, A novel L-amino acid ligase is encoded by a gene in the phaseolotoxin biosynthetic gene cluster from Pseudomonas syringae pv. phaseolicola 1448A, Biosci. Biotechnol. Biochem. 72 (11) (2008) 3048–3050.
[106] K. Kino, A. Noguchi, Y. Nakazawa, M. Yagasaki, A novel L-amino acid ligase from bacillus licheniformis, J. Biosci. Bioeng. 106 (3) (2008) 313–315.
[107] B. Kade, E.R. Dabbs, B. Wittmannliebold, Protein-chemical studies on Escherichia-Coli mutants with altered ribosomal protein-s6 and protein-S7, FEBS Lett. 121 (2) (1980) 313–316.
[108] S. Reeh, S. Pedersen, Post-translational modification of Escherichia-Coli ribosomal-protein S6, Mol. Gen. Genet. 173 (2) (1979) 183–187.
[109] W.K. Kang, T. Icho, S. Isono, M. Kitakawa, K. Isono, Characterization of the gene rimK responsible for the addition of glutamic acid residues to the C-terminus of ribosomal protein S6 in Escherichia coli K12, Mol. Gen. Genet. 217 (2-3) (1989) 281–288.
[110] J. Gimenez-Dejoz, K. Tsuchiya, K. Numata, Insights into the stereospecificity in papain-mediated chemoenzymatic polymerization from quantum mechanics/molecular mechanics simulations, ACS Chem. Biol. 14 (6) (2019) 1280–1292.
[111] F. Bordusa, Proteases in organic synthesis, Chem. Rev. 102 (12) (2002) 4817–4867.
[112] K. Yazawa, K. Numata, Recent advances in chemoenzymatic peptide syntheses, Molecules 19 (9) (2014) 13755–13774.
[113] K. Numata, Poly(amino acid)s/polypeptides as potential functional and structural materials, Polym. J. 47 (8) (2015) 537–545.
[114] K. Tsuchiya, K. Numata, Chemical synthesis of multiblock copolypeptides inspired by spider dragline silk proteins, ACS Macro Lett. 6 (2) (2017) 103–106.
[115] K. Tsuchiya, K. Numata, Chemoenzymatic synthesis of polypeptides containing the unnatural amino acid 2-aminoisobutyric acid, Chem. Commun. (Camb.) 53 (53) (2017) 7318–7321.

[116] K. Yazawa, J. Gimenez-Dejoz, H. Masunaga, T. Hikima, K. Numata, Chemoenzymatic synthesis of a peptide containing nylon monomer units for thermally processable peptide material application, Polym. Chem. 8 (29) (2017) 4172–4176.
[117] K. Tsuchiya, N. Kurokawa, J. Gimenez-Dejoz, P.G. Gudeangadi, H. Masunaga, K. Numata, Periodic introduction of aromatic units in polypeptides via chemoenzymatic polymerization to yield specific secondary structures with high thermal stability, Polym. J. 51 (12) (2019) 1287–1298.
[118] J.M. Ageitos, P.J. Baker, M. Sugahara, K. Numata, Proteinase K-catalyzed synthesis of linear and star oligo(L-phenylalanine) conjugates, Biomacromolecules 14 (10) (2013) 3635–3642.
[119] J. Fagerland, A. Finne-Wistrand, K. Numata, Short one-pot chemo-enzymatic synthesis of L-lysine and L-alanine diblock co-oligopeptides, Biomacromolecules 15 (3) (2014) 735–743.
[120] H.D. Jakubke, P. Kuhl, A. Könnecke, Basic principles of protease-catalyzed peptide bond formation, Angew. Chem. Int. Ed. Engl. 24 (2) (1985) 85–93.
[121] K. Morihara, Using proteases in peptide synthesis, Trends Biotechnol. 5 (6) (1987) 164–170.
[122] J. Bongers, E.P. Heimer, Recent applications of enzymatic peptide synthesis, Peptides 15 (1) (1994) 183–193.
[123] D. Kumar, T.C. Bhalla, Microbial proteases in peptide synthesis: approaches and applications, Appl. Microbiol. Biotechnol. 68 (6) (2005) 726–736.
[124] F. Guzmán, S. Barberis, A. Illanes, Peptide synthesis: chemical or enzymatic, Electron. J. Biotechnol. 10 (2) (2007) 279–314.
[125] M. Yagasaki, S.-i Hashimoto, Synthesis and application of dipeptides; current status and perspectives, Appl. Microbiol. Biotechnol. 81 (1) (2008) 13–22.
[126] F. Chen, F. Zhang, A. Wang, H. Li, Q. Wang, Z. Zeng, et al., Recent progress in the chemo-enzymatic peptide synthesis, Afr. J. Pharm. Pharmacol. 4 (10) (2010) 721–730.
[127] A.M. Białkowska, K. Morawski, T. Florczak, Extremophilic proteases as novel and efficient tools in short peptide synthesis, J. Ind. Microbiol. Biotechnol. 44 (9) (2017) 1325–1342.
[128] F.H. Carpenter, The free energy change in hydrolytic reactions: the non-ionized compound convention, J. Am. Chem. Soc. 82 (5) (1960) 1111–1122.
[129] G.A. Homandberg, J.A. Mattis, M. Laskowski, Synthesis of peptide bonds by proteinases. Addition of organic cosolvents shifts peptide bond equilibria toward synthesis, Biochemistry 17 (24) (1978) 5220–5227.
[130] J.M. Cassells, P.J. Halling, Low-water organic two-phase systems and problems affecting it, Biotechnol. Bioeng. 33 (1989) 1489–1494.
[131] P.J. Halling, Thermodynamic predictions for biocatalysis in nonconventional media: theory, tests, and recommendations for experimental design and analysis, Enzyme Microb. Technol. 16 (3) (1994) 178–206.
[132] B. Deschrevel, J.C. Vincent, C. Ripoll, M. Thellier, Thermodynamic parameters monitoring the equilibrium shift of enzyme-catalyzed hydrolysis/synthesis reactions in favor of synthesis in mixtures of water and organic solvent, Biotechnol. Bioeng. 81 (2) (2003) 167–177.
[133] V. Schellenberger, H.D.D. Jakubke, Protease-catalyzed kinetically controlled peptide synthesis, Angew. Chem. Int. Ed. Engl. 30 (11) (1991) 1437–1449.
[134] M. Ŝtrajbl, J. Florián, A. Warshel, Ab initio evaluation of the free energy surfaces for the general base/acid catalyzed thiolysis of formamide and the hydrolysis of methyl thiolformate: a reference solution reaction for studies of cysteine proteases, J. Phys. Chem. B 105 (2001) 4471–4484.

[135] S. Gul, S. Hussain, M.P. Thomas, M. Resmini, C.S. Verma, E.W. Thomas, et al., Generation of nucleophilic character in the Cys25/His159 ion pair of papain involves Trp177 but not Asp158, Biochemistry 47 (2008) 2025–2035.
[136] M. Mladenovic, R.F. Fink, W. Thiel, T. Schirmeister, B. Engels, On the origin of the stabilization of the zwitterionic resting state of cysteine proteases: a theoretical study, J. Am. Chem. Soc. 130 (2008) 8696–8705.
[137] D. Wei, X. Huang, M. Tang, C.-G. Zhan, Reaction pathway and free energy profile for papain-catalysed hydrolysis of N-acetyl-phe-gly 4-nitroanilide, Biochemistry 52 (2013) 5145–5154.
[138] A. Paasche, A. Zipper, S. Schäfer, J. Ziebuhr, T. Schirmeister, B. Engels, Evidence for substrate binding-induced zwitterion formation in the catalytic Cys-His dyad of the SARS-CoV main protease, Biochemistry 53 (2014) 5930–5946.
[139] K. Arafet, S. Ferrer, V. Moliner, First quantum mechanics/molecular mechanics studies of the inhibition mechanism of cruzain by peptidyl halomethyl ketones, Biochemistry 54 (2015) 3381–3391.
[140] A. Fekete, I. Komáromi, Modeling the archetype cysteine protease reaction using dispersion corrected density functional methods in ONIOM-type hybrid QM/MM calculations; the proteolytic reaction of papain, Phys. Chem. Chem. Phys. 18 (2016) 32847–32861.
[141] J. Gimenez-Dejoz, K. Tsuchiya, A. Tateishi, Y. Motoda, T. Kigawa, Y. Asano, et al., Computational study on the polymerization reaction of d-aminopeptidase for the synthesis of d-peptides, RSC Adv. 10 (30) (2020) 17582–17592.
[142] V. Babin, C. Roland, C. Sagui, Adaptively biased molecular dynamics for free energy calculations, J. Chem. Phys. 128 (2008) 134101–134107.
[143] V. Babin, C. Sagui, Conformational free energies of methyl-α-L-iduronic and methyl-β-D-glucuronic acids in water, J. Chem. Phys. 132 (10) (2010).
[144] A. Laio, M. Parrinello, Escaping free-energy minima, Proc. Natl. Acad. Sci. 99 (2002) 12562.
[145] U. Christensen, Effects of pH on carboxypeptidase-Y-catalyzed hydrolysis and aminolysis reactions, Eur. J. Biochem. 220 (1) (1994) 149–153.
[146] M. Hansler, H.D. Jakubke, Reverse action of hydrolases in frozen aqueous solutions, Amino Acids 11 (1996) 379–395.
[147] Å. Jönsson, E. Wehtje, P. Adlercreutz, Low reaction temperature increases the selectivity in an enzymatic reaction due to substrate solvation effects, Biotechnol. Lett. 19 (1) (1997) 85–88.
[148] A. Narai-Kanayama, T. Hanaishi, K. Aso, α-Chymotrypsin-catalyzed synthesis of poly-l-cysteine in a frozen aqueous solution, J. Biotechnol. 157 (3) (2012) 428–436.
[149] P. Kuhl, A. Konnecke, G. Doring, A. Konnecke, H. Jakubke, Enzyme-catalyzed peptide synthesis in biphasic aqueous-organic systems, Tetrahedron Lett. 21 (1980) 895–896.
[150] D.K. Eggers, H.W. Blanch, J.M. Prausnitz, Extractive catalysis: solvent effects on equilibria of enzymatic reactions in two-phase systems, Enzyme Microb. Technol. 11 (2) (1989) 84–89.
[151] H. Gaertner, A. Puigserver, Kinetics and specificity of serine proteases in peptide synthesis catalyzed in organic solvents, Eur. J. Biochem. 181 (1) (1989) 207–213.
[152] A. Nadim, I.B. Stoineva, B. Galunsky, V. Kasche, D.D. Petkov, Mass transfer induced interchange of the kinetic and thermodynamic control of enzymic peptide synthesis in biphasic water-organic systems, Biotechnol. Technol. 6 (6) (1992) 539–544.
[153] H. Kisee, K. Fujimoto, H. Noritomi, Enzymatic reactions in aqueous-organic media. VI. Peptide synthesis by α-chymotrypsin in hydrophilic organic solvents, J. Biotechnol. 8 (4) (1988) 279–290.

[154] K. Viswanathan, R. Omorebokhae, G. Li, R.A. Gross, Protease-catalyzed oligomerization of hydrophobic amino acid ethyl esters in homogeneous reaction media using L-phenylalanine as a model system, Biomacromolecules 11 (8) (2010) 2152–2160.
[155] P.J. Baker, S.V. Patwardhan, K. Numata, Synthesis of homopolypeptides by aminolysis mediated by proteases encapsulated in silica nanospheres, Macromol. Biosci. 14 (11) (2014) 1619–1626.
[156] J. Fastrez, A.R. Fersht, Demonstration of the acyl-enzyme mechanism for the hydrolysis of peptides and anilides by chymotrypsin, Biochemistry 12 (11) (1973) 2025–2034.
[157] V. Schellenberger, H.-D. Jakubke, A spectrophotometric assay for the characterization of the S′ subsite specificity of α-chymotrypsin, Biochim. Biophys. Acta 869 (1) (1986) 54–60.
[158] I. Schechter, A. Berger, On the size of the active site in proteases. I. Papain, Biochem. Biophys. Res. Commun. 27 (2) (1967) 157–162.
[159] A. Berger, I. Schechter, Mapping the active site of papain with the aid of peptide substrates and inhibitors, Philos. Trans. R. Soc. B Biol. Sci. 257 (813) (1970) 249–264.
[160] J.M. Ageitos, K. Yazawa, A. Tateishi, K. Tsuchiya, K. Numata, The benzyl ester group of amino acid monomers enhances substrate affinity and broadens the substrate specificity of the enzyme catalyst in chemoenzymatic copolymerization, Biomacromolecules 17 (1) (2016) 314–323.
[161] K. Tsuchiya, K. Numata, Chemoenzymatic synthesis of polypeptides for use as functional and structural materials, Macromol. Biosci. 17 (11) (2017).
[162] K. Yazawa, K. Numata, Papain-catalyzed synthesis of polyglutamate containing a nylon monomer unit, Polymers (Basel) 8 (5) (2016).
[163] D.J.C. Constable, P.J. Dunn, J.D. Hayler, G.R. Humphrey, J.L. Leazer, R.J. Linderman, et al., Key green chemistry research areas—a perspective from pharmaceutical manufacturers, Green. Chem. 9 (5) (2007) 411–420.

Questions for this chapter

1. What is the major advantage and disadvantage of the solid-phase peptide synthesis?
2. How do you synthesize a polypeptide, if you need to control the amino acid sequence?
3. How do you synthesize an artificial spider silk protein? Describe your synthesis strategy.
4. What kind of enzyme can be used for peptide synthesis?
5. Some proteinases are used for peptide synthesis, even though proteases cleave amide bonds of peptides. Explain the mechanism underlying the protease-mediated polymerization.

CHAPTER 3

Structure

The structure of protein material contains a wide variety of structural information, at the angstrom level to macroscale structures. The hierarchical structures of protein and polypeptide achieve and realize their physical, chemical, and biological functions/properties. The difference between protein and synthetic polymer was already introduced in Chapter 1, General Introduction of Polypeptide and Protein Materials, and hence more detail of the structures of natural and artificial proteins and polypeptides will be explained in this chapter.

3.1 Primary structure/chemical structure

Protein and polypeptide are poly(L-amino acid) and hence are composed of amino acid monomers with amide bonds. Generally, 20 types of natural amino acids are the components used to build up proteins (Fig. 3.1). In the case of proteins and polypeptides, the chemical, physical, biological, and structural properties are influenced by amino acid sequences, even though the relationship between amino acid sequences and the properties is not direct. This section shows 20 types of natural amino acids and their roles in proteins, especially structural proteins.

3.1.1 Amino acids

3.1.1.1 Glycine

Glycine is the simplest amino acid with only a single hydrogen atom as a side chain. Due to this simplicity, glycine is the only achiral proteinogenic amino acid. Glycine fits into structural/random or hydrophilic/hydrophobic environments, due to its minimal side chain. Because of this property, oligoglycine and its derivatives are used for linker sequences. Further, glycine can be integrated into the helical structure, sheet structure, and random coil.

3.1.1.2 Alanine

Alanine is the second simplest amino acid containing a methyl group side chain. L-Alanine is found in a wide variety of proteins, whereas D-alanine

Biopolymer Science for Proteins and Peptides
DOI: https://doi.org/10.1016/B978-0-12-820555-6.00005-7

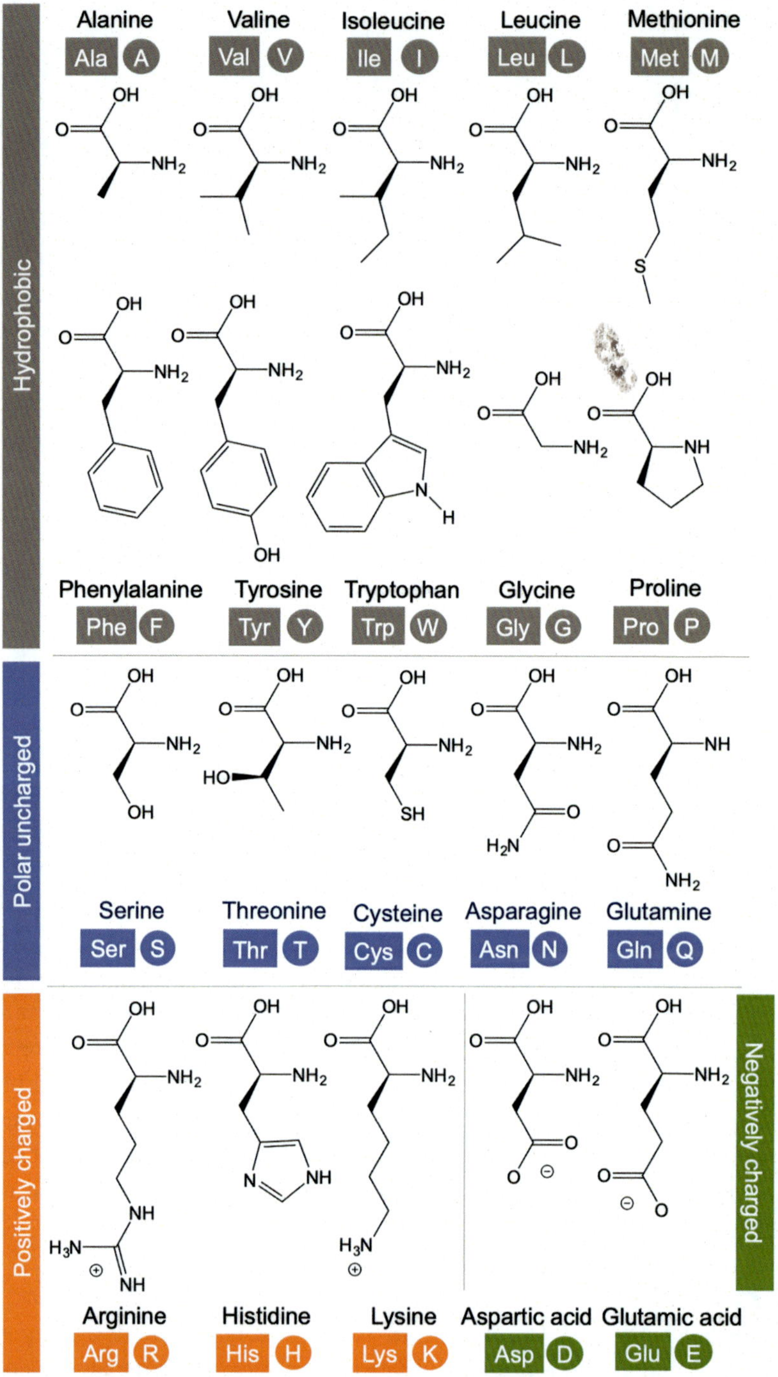

Figure 3.1 A list of 20 types of natural amino acids.

also can be found in some bacterial cell walls, the tissues of crustaceans, and mollusks. Alanine has just a methyl group as a side chain, and thus does not contribute to the function of the globular proteins such as enzymes. However, poly- and oligoalanine sequences in structural proteins are very critical for forming beta-sheet structures to stabilize the crystalline region and hierarchical structure. For example, poly(alanine) plays a critical role in spider silk to achieve its excellent mechanical properties.

3.1.1.3 Valine, threonine, isoleucine

Valine, threonine, and isoleucine are beta-branched amino acids. Valine contains an isopropyl group as a side chain, resulting in a nonpolar aliphatic amino acid. Threonine has a hydroxyl group on the beta-carbon of valine. Isoleucine has a beta-branched structure which contains one more carbon than valine. Due to the hydrophobic side groups, valine, threonine, and isoleucine prefer to form a beta-strand and also a beta-sheet, which is one of the most important secondary structures for the formation of stable structures.

3.1.1.4 Tyrosine, phenylalanine, tryptophan

Tyrosine, phenylalanine, and tryptophan have a large aromatic side-chain group and prefer to form beta-strand and beta-sheet structures. Phenylalanine shows a relatively high crystallization behavior. Tyrosine has a hydroxyl group, and hence is more hydrophilic than phenylalanine, even though tyrosine is classified as one of the hydrophobic amino acids. The chemical and physical properties of tyrosine depend on the posttranslational modification (PTM) of the hydroxyl group, namely, phosphorylation, sulfation, and oxidation. Phosphorylation of the hydroxyl group can change the biological properties of tyrosine. The phosphorylated form, phosphotyrosine, is considered to be one of the key in signal transduction and regulation of enzymatic activity. Tyrosine sulfation occurs in most eukaryotes [1].

Similar to tyrosine conversion to 3,4-dihydroxy-L-phenylalanine (L-DOPA) as a hormone by the enzyme tyrosine hydroxylase, tyrosine in protein is also converted to L-DOPA. The blue mussel (*Mytilus edulis*) foot protein 5 (Mefp-5) is an adhesive protein that is composed of the mussel adhesive plaque and consists mainly of glycine, L-lysine, and DOPA [2,3]. Its adhesion is mainly achieved by the redox-chemistry of DOPA [4,5]. The effects of the sequence of mussel proteins have also been investigated using random copolypeptides containing DOPA [6,7],

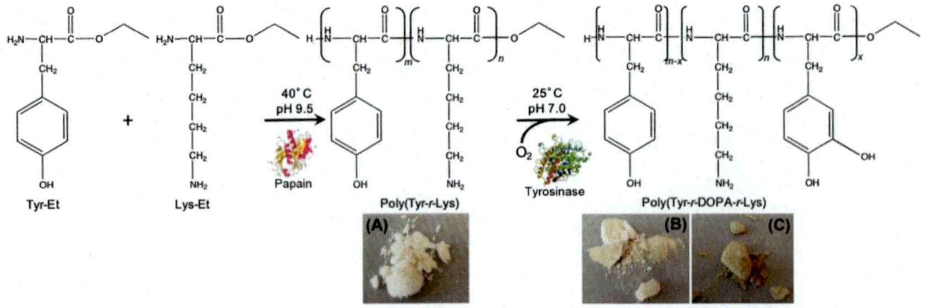

Figure 3.2 Reaction schemes for the enzymatic synthesis of poly(L-tyrosine-*r*-3,4-dihydroxy-L-phenylalanine-*r*-L-lysine) [P(Tyr-DOPA-Lys)], via poly(L-tyrosine-*r*-L-lysine) [P(Tyr-Lys)], from L-tyrosine ethyl ester (Tyr-Et) and L-lysine ethyl ester (Lys-Et). The images show the reaction intermediate (A) P(75%Tyr-25%Lys), and products (B) P (50%Tyr-25%Lys-25%DOPA), and (C) P(45%DOPA-30%Tyr-25%Lys) [9].

suggesting that DOPA is the main factor underlying the adhesive function. In addition to DOPA, Mefp-5 is composed of L-lysine (approximately 20 mol%), which is necessary to form a network of structures between catechol and amine groups [4,8]. From these reports, it is suggested that DOPA and lysine are essential for biomimetic adhesive design. Adhesive peptides containing DOPA and L-lysine via two enzymatic reactions, namely, chemoenzymatic polymerization of L-tyrosine and L-lysine by *Papaya* peptidase I (papain) as well as the enzymatic conversion of tyrosine to DOPA by tyrosinase (Fig. 3.2).

3.1.1.5 Methionine, cysteine

Methionine and cysteine are sulfur-containing amino acids. Methionine contains an S-methyl thioether side chain, whereas cysteine has the same structure as serine, but with one of its oxygen atoms replaced by sulfur. In the case of selenocysteine, its oxygen atom is replaced with selenium. Mostly, methionine residues do not have a catalytic role [10], whereas the thiol group of cysteine has a catalytic role in enzymes. The disulfide bond between two cysteine residues is important and essential to form dimerization and build up a higher order structure. The methionine's codon AUG is also the most common start codon in translation.

3.1.1.6 Proline

Proline contains a side-chain pyrrolidine and hence is the only proteinogenic amino acid containing a secondary amine. The exceptional chemical structure of proline provides conformational rigidity and bending character along the main chain compared to other natural amino acids. This conformational rigidity of proline influences the secondary structure of proteins/polypeptides around a proline residue. Proline is considered to induce beta-turns, whereas proline plays a role to disrupt the secondary structure elements such as helical and sheet structures. Hence, proline often exists at the edge of an alpha-helix and beta-strands. Hydroxyproline is converted from proline via hydroxylation, for example, by prolyl hydroxylase. The hydroxylation of proline stabilizes the triple helices of collagen. Like collagen, prolines and/or hydroxyprolines in a row can create a polyproline helix.

Proline is an important indicator to classify spider silk, which is one of the structural proteins. Spider dragline silk is mainly composed of at least two spidroin components, the major ampullate spidroins 1 and 2 (MaSp1 and MaSp2) [11]. Variations in MaSp1/MaSp2 ratios, as found across phylogenetic groups and between individuals of the same species, are thought to play an essential role in modulating dragline silk's mechanical properties, including elongation and strength. MaSp1 and MaSp2 sequences are generally differentiated based on conserved proline residues in the repetitive domain (RD), often in the context of iterated GPGXX runs, in addition to diglutamine (QQ) motifs [12]. In MaSp2, the proline-rich sections of the Gly-rich regions are thought to adopt beta-turn conformations that contribute to the high mobility of the polypeptide chains [13]. Notably, a correlation has been found between the abundance of proline residues in MaSp2 RDs and the degree of elasticity and supercontraction of the corresponding dragline fiber [13–15].

3.1.1.7 Serine

Serine contains a hydroxymethyl group as a side chain, classifying it as a polar amino acid. Serine plays an important role in the catalytic function of many enzymes, including serine proteases like chymotrypsin, trypsin, and subtilisin. Serine side chains are often hydrogen bonded via hydroxyl group; one of the resultant structures is ST (ST means serine and threonine) turns and ST motifs, often at the beginning of alpha-helices. Serine is also essential for side-chain modification of polypeptides and proteins such as glycosylation and phosphorylation; its side chain can undergo *O*-linked glycosylation, which is biologically and functionally important. More details are introduced in Section 3.1.2 on PTM.

3.1.1.8 Asparagine

Asparagine contains a side chain of carboxamide. Asparagine residues are often found near the beginning of alpha-helices as asx turns (asx motifs, which play similar roles to ST motifs) and in other turn motifs, or beta-sheets, because the side chain can form hydrogen bonding interactions with the peptide backbone. Asparagine is a key reaction site for *N*-linked glycosylation, which is a modification of the protein chain with the addition of carbohydrate chains. A reaction between asparagine and reducing sugars or other sources of carbonyls produces acrylamide, which can be found in heated foods.

3.1.1.9 Glutamine

Glutamine contains a side chain similar to that of glutamic acid, except that the carboxylic acid group is replaced with an amide. A continuous sequence of glutamine can be found in sticky and water-absorbable proteins. For example, the mutual occurrence of conserved proline and QQ motifs was also found to be a prominent feature of pyriform spidroin sequences (constituents of web attachment disks), albeit within different sequence contexts (Fig. 3.3) [16,17]. Moreover, the high-molecular-weight subunit of glutenin from wheat, responsible for the strength and elasticity of bread dough, also features a highly repetitive central domain that is rich in proline residues and QQ motifs, reminiscent of MaSp2. The glutenin repeats are predicted to adopt flexible beta-spiral conformations [18], and structural studies suggest that the prolines provide molecular

Major ampullate silk protein 2 *Argiope bruennichi*

PYGPGAAAAAGAAAGGYGPGAGQQGPGGAGQQGPYGPGSGQQGPGGAGQQGPGGAGQGGPSGRGPYGPG
AAAAAAAAGGYGPGAGQQGPGGAGQQGPGGAGQQGPGGAGQGGPGGQGPYGPAAAAAAAAAGGYGPGAG
QQGPGGAGQQGPGGAGQQGPGGAGQQGPGGQGPYGPGAAAAAAAAGGYGPGAGQQGPGGAGQQGPGGQG
PYGPGAAAAAAAAAGGYGPGAGQQGPGGAGQQGPGGAGQQGPGGAGQQGPGGAGQQGPGGAGQGGPSGRG
PYGPGAAAAAAAAASGYGPRAGQQGPGGAGQQGPGGAGQQGPGGQGPYGPGAAAAAAAAAGGYGPGAGQQG
PGGAGQQGPGGQGPYGPGAAAAAAAAAGGYGPGAGQQGPGGAGQQGPGGAGQQGPGGAGQQGPGGAGQQG
PGGQGPYGPGAAAAAAAAAAGGYGPGAGQQGPGGAGQQGPYGPGSGQQGPGGAGQQGPGGAGQGGPSGRG

High-molecular-weight glutenin *Aegilops searsii*

QQVSYYPGQASPQRPGQGQQSGQGQQGYYPTSPQQPGQRQQPEQGQPGYYPTSPQQPGQLQQPAQGQQG
QQPEQGQPGYYPTSLQLQPGQLQQPAQGQQGQQAGQGQQGQQPGQGQQLRQGQQGQQSGQGQQPEQGQQ
GQQLGQGQQGYYATSLQQSGQGQPGYYPTSLQQPGQGQSEYYPTSLQQPGQGQQPGQLQQPAQGQQPGQ
GQQGQQSGQGQQGQQPVQGQQGQQPGQGQQGQQPGQGHQPGQGQPGYYPTSPQQSGQGQPRYYPTSSQQ
PGQSQQPGQGQQGQQLGQGQQAQQPGQGQQPGQGQPGYYPTSPQQSGQGQPGYYLTSPQQPGQGQQPGQ
LQQSAQGQEGQQPGQGQQGQQSGQGQQGQQPVQGQQGQQPGQGQQGQQPGQGQPGYYPTSSHQSGQGQQ
PGQWQQPGQGQPGYYPTSPLQPGQGQPGYDPTSPQQPGQGQQPTQGQQGQQPGQGQRGQQPGQGQQGQQ

Figure 3.3 Amino acid sequences of major ampullate silk protein 2 from *Argiope bruennichi* and glutenin from *Aegilops searsii* [15].

chain mobility. At the same time, the glutamine residues participate in an extensive network of intermolecular hydrogen bonds [19].

3.1.1.10 Arginine

Arginine's side chain consists of a three-carbon aliphatic straight-chain ending in a guanidinium group, and hence arginine's side chain is amphipathic. The guanidinium group at the side chain shows a pK_a of 12.48, and is therefore always protonated and positively charged at physiological pH. The hydrophilic head group interacts with the polar environment, including cell membranes, and then oligoargine such as nona-arginine is used for the delivery system of bioactive molecules into various cells. Arginine-related PTM is citrullination and methylation.

3.1.1.11 Histidine

Histidine contains an imidazole side chain, which is partially protonated and positively charged at physiological pH. The protonated form of the imidazole side chain has a pK_a of approximately 6.0, indicating the imidazole ring is mostly protonated (positively charged) below a pH of 6. The positive charge in the imidazole ring is equally distributed between both nitrogen atoms, whereas one of the two protons is lost at pH 6. Due to the charges, histidines play a critical role in gene/drug delivery systems. The accumulation of histidine residues inside gene/drug delivery vesicles is considered to induce a proton sponge effect, enhancing their osmolarity leading to escape from the endosomes. Furthermore, histidine-based carriers are often less cytotoxic to other cationic amino acid residues-based ones.

3.1.1.12 Lysine

Lysine has a four-carbon side chain together with an amine group at the end, making lysine a basic, charged, and aliphatic amino acid. Lysine can also contribute to protein stability as its ε-amino group often participates in hydrogen bonding, salt bridges, and covalent interactions. In collagen, lysine functions as a cross-linker to stabilize the collagen hierarchical structure. Multiple lysine residues are essential for dimerization of the *N*-terminal domain (NTD) of spider silk via salt bridges. Mussel foot proteins show adhesion properties involving covalent interactions through lysine side chains.

3.1.1.13 Aspartic acid

Aspartic acid has an acidic side chain, which is negatively charged under biological conditions. The acidic side chain reacts with other amino acids and proteins. Interestingly, in addition to the L-form, the D-form of aspartic acid exists commonly in mammals. In proteins, the side chains are often hydrogen bonded to form asx turns (asx motifs), similar to asparagine. Poly(aspartic acid), a polymer of aspartic acid, is developed as bio-based and biodegradable superabsorbent polymers.

3.1.1.14 Glutamic acid

Similar to aspartic acid, glutamic acid has an acidic side chain, which is negatively charged under biological conditions, that is similar to that of glutamine except that the amine group exists instead of the carboxylic acid group. Glutamic acid (glutamate) is the precursor for the synthesis of the inhibitory gamma-aminobutyric acid (GABA). GABA is also a starting material for the synthesis of polyamide, nylon 4.

3.1.2 Posttranslational modification

From the viewpoint of chemical structure, PTMs cannot be ignored. PTM generally refers to enzymatic modifications of amino acids within proteins following protein biosynthesis, as described in Section 2.1.6. PTM is integral to the correct folding of polypeptide chains into their final protein conformation, and also are considered to have a major impact on the function of the protein. In spider silk, PTMs have been shown within both MaSp1 and MaSp2 of spider major ampullate silk, which displays notable and regular phosphorylation of serine and tyrosine within the repeat region [20,21]. The position and enhancement of the negative charge associated with phosphorylation impacts the formation of helical structures in other proteins. Spider silk also shows phosphorylated serine and tyrosine with grater coil, alpha-helices formations, and a reduction in the amount of 3_{10} helices [22]. The presence of the oxidized residue dityrosine has also been confirmed within MaSp1 and MaSp2 [20,21]. A hydroxylated form of tyrosine and an intermediate in the formation of dityrosine, 3,4-dihydroxyphenylalnine (DOPA), and glycosylation was also reported [23–25]. Hydroxyproline has been demonstrated within both the major ampullate and flagelliform silk of spiders [26,27].

3.2 Secondary structure

Proteins, including structural proteins, form characteristic secondary structures (Fig. 3.4). A beta-sheet structure, a helix (helical structure), a beta-turn, and the like, are typical secondary structures. A state with high mobility is formed as a random coil without forming a structure. These secondary structures can be predicted from amino acid sequences. However, they change depending on environmental conditions such as pH, ionic strength, and temperature, and the predictions and actual measurements do not always match.

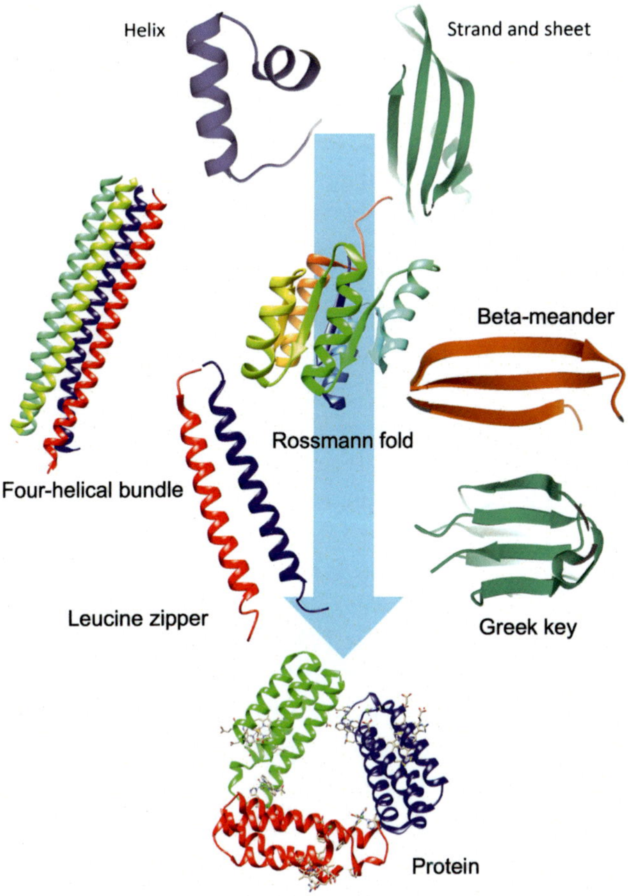

Figure 3.4 Hierarchy of secondary structure, supersecondary structure, and motif of protein.

A supersecondary structure is an intermediate between secondary and tertiary structures, and also a combination of multiple secondary structures into a motif found in widely variant proteins. Beta-sheets are rolled into a tube, resulting in a beta-barrel. Helix—loop—helix is the structure where two alpha-helices are connected via a beta-turn. Leucine zipper and zinc zipper are found in transcription factors that interact with DNA. Another typical supersecondary structure is rope-like structures (coiled coils) where the alpha-helices twist around each other.

Taking MaSp, the main silk protein derived from spider dragline as an example, you can understand its complexity. As described in the primary structure section, the *N*-terminal and *C*-terminal, which are both ends, form a three-dimensional structure in which a plurality of helical structures are gathered. The *Euprosthenops australis* MaSp, which has many reports, has an *N*-terminal structure of about 14 kDa, and five helical structures are bundled, and the *N*-terminal structure is dimerized to form threads [28,29]. This is expected to be similar in other spiders. On the other hand, repetitive sequences occupying most of the molecular weight have not been studied in detail beyond that the main secondary structure is a beta-sheet. Even if silk is crystallized as much as possible, its crystallinity is less than 70% [30]. It is difficult to crystallize high-molecular-weight proteins such as silk. In the stage before the yarn is formed, the soluble silk protein exists in a metastable state. At that time, it has been reported that the RD has a random coil as a temporary secondary structure, but also forms a less stable structure, including polyproline type II helix (PPII helix) [31]. In addition, chaotropic ions such as sodium ions suppress the intramolecular and intermolecular interactions of silk proteins and dissolve silk proteins, while kosmotropic ions such as phosphate ions promote hydrogen bond formation. However, it is suggested that the formation of the beta-sheet structure is promoted [32]. When forming a yarn, the formation of a beta-sheet structure, which is an essential structural change, occurs in a polyalanine sequence followed by an alanine residue, and significantly affects the mechanical properties of silk fibers, particularly strength [33—35]. However, it should be noted that not all polyalanine crystallizes [36,37].

3.3 Crystal structure and crystalline region

Protein and polypeptide are not thermally crystalline polymers like synthetic plastics. However, some structural proteins are semicrystalline, and their crystalline region play an important role in forming hierarchical structures as well as realizing excellent physical properties. The crystalline component of proteins consists of beta-sheet structures. Here is an example, the structural importance of beta-sheet crystals in silk fibroin.

3.3.1 Beta-sheet crystals in silk

The antiparallel beta-sheet structure of silk fibroin is one of the typical beta-sheet structures, as introduced in many biochemistry textbooks. The hierarchical structures of silks vary with the silk type (type of fibers and host organisms) and determine their outstanding physical properties [38,39]. Among the most critical components of the hierarchical structures of silk are the beta-sheet crystals (assembly of beta-sheet structures) since they play an important role in determining silk elongation, stiffness, strength, and toughness [40–42]. Aligned beta-sheet crystals contribute to silk's stiffness (modulus) and form cross-links between the beta-sheet domains embedded in a semiamorphous matrix consisting of less orderly structures, namely, random coils and beta-turns [41,42]. Based on the amino acid sequences and structures of the cocoon silk of the *Bombyx mori* silkworm and the dragline silk of the *Trichonephila clavata* (formerly known as *Nephila clavata*) spider, the beta-sheet structure is predominant in both fiber types [41,43,44]. The strongest spider dragline silk was reported to be up to 60% beta-sheet domains [45], but the general amount of beta-sheet structure is ranged from 15% to 25%. In contrast to dragline, capture silk contains 54.4 mol% glycine and 16.6 mol% proline, resulting in predominantly beta-spiral structures that resemble molecular springs [46,47]. Capture silk does not contain poly(alanine) sequences; however, the beta-sheet structure was found to contribute to the strength of captured silk fibers based on Raman spectromicroscopy measurement [48]. Thus the crystallization of silk molecules is a critical molecular transition that dictates the physical, biological, and functional properties of silk.

In addition to nuclear magnetic resonance (NMR) [42,49–52], transmission electron microscopy [53], and Raman and infrared spectroscopies [41,54–56], synchrotron radiation X-ray scattering [44,57–67] has been used to characterize the structure of silk molecules and, in particular, those of beta-sheet structures under extrusion conditions. The molecular structures of the silk fibers of silkworm and spider during deformation were investigated by several groups [56,68–72]. Lefèvre et al. investigated a single fiber of silkworm cocoon silk during stretching deformation by

Raman spectroscopy. They suggested decreases in protein-chain alignment, including beta-sheet crystals, due to the reorganization of the amorphous phase [73]. X-ray and neutron scattering measurements of a bundle of silk fibers provided that the amorphous phase of silk plays a critical role in exhibiting the toughness during stretching deformation [69,70]. Using single filaments of silk with Raman spectroscopy, Sirichaisit et al. successfully detected the structural change in spider dragline silk during approximately 25% stretching deformation [56,71].

3.3.2 Molecular packing of beta-sheet crystal

The poly(alanine) sequences in the primary structure of the silk proteins form antiparallel beta-sheets and are the main components of the crystalline region of silk fibroin. The amino acid sequences that form the beta-sheet are 7–9-mer alanine sequences for *Nephila clavipes* dragline silk and GAGAGS for *B. mori* [74], whereas other silkworm silk species use polyalanine sequences to form the beta-sheet structure [75]. Asakura et al. investigated the effect of the number of alanine residues on the secondary structure of silk, including the formation of antiparallel beta-sheets, parallel beta-sheets, and beta-turns, by using solid-state NMR spectroscopy and X-ray crystallography. Their results revealed that short poly(alanine) sequences (6-mer or shorter) assemble to form a packed rectangular arrangement, whereas longer poly(alanine) sequences pack in a staggered arrangement [37].

The beta-sheet is the most fundamental secondary structure in silk materials. The predominant beta-sheet structure plays a crucial role in stabilizing silk materials via physical cross-links, as the beta-sheet behaves as a cross-linking point. Crystal structures of silk beta-sheets have been characterized by wide-angle X-ray analysis. The crystal structure of *B. mori* silk fiber has a unit cell with the space group $P2_1$-C_2^2 [60]. The crystal lattice of the *B. mori* silk fiber reported by Marsh et al. had unit cell dimensions of $a = 9.40$ Å, $b = 9.20$ Å, and c (fiber axis) = 6.97 Å, while Takahashi et al. reported cell dimensions of $a = 9.38$ Å, $b = 9.49$ Å, and c (fiber axis) = 6.98 Å [66].

Antheraea yamamai contains motifs composed of long poly(alanine) sequences (10- to 13-mers; Fig. 3.5A). A histogram of the distribution of poly(alanine) sequences in *A. yamamai* silk fibroin clearly indicates that 12-mers and 13-mers are the most common lengths of the poly(alanine) sequences (Fig. 3.5B) [36]. The length of poly(alanine) sequence must be

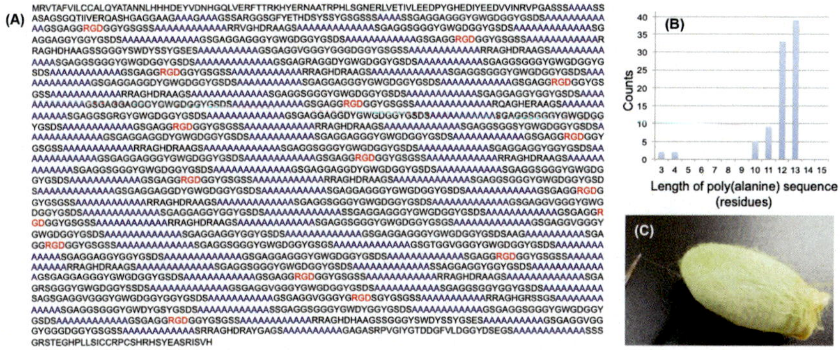

Figure 3.5 (A) Amino acid sequence of *Antheraea yamamai* silk heavy chain. (B) Distribution of the length of poly(alanine) sequences in *A. yamamai* silk. (C) *A. yamamai* silk fibers and cocoon [36].

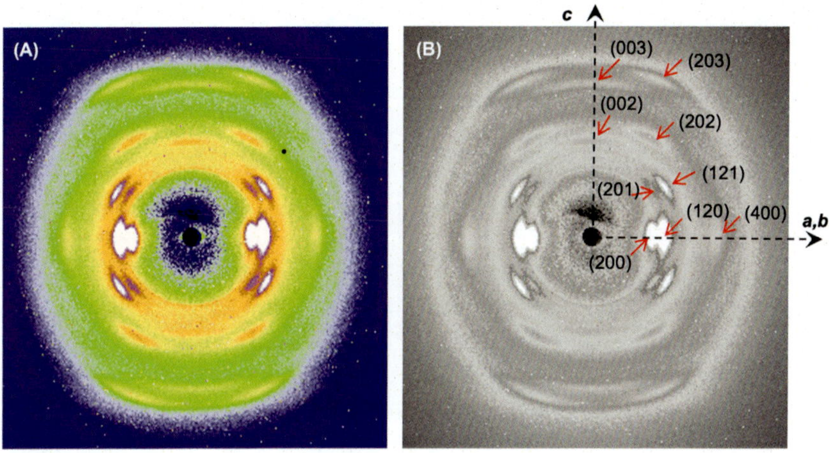

Figure 3.6 Two-dimensional wide-angle X-ray scattering (2D WAXS) patterns of a (A) bundle of *Antheraea yamamai* silk fibers and (B) assignment of the reflections.

optimized in each species by its spinning process and conditions (concentration of ions, shear forces, etc.).

A. yamamai silk fiber bundle was characterized by synchrotron wide-angle X-ray scattering (WAXS). The obtained 2D scattering patterns and reflections are shown in Fig. 3.6. The Miller indices of the reflections were determined according to the crystal structure of *B. mori* silk, which has a unit cell with the space group $P2_1$-C_2^2 (Table 3.1) [60,66]. The Miller indices and *d*-spacing values suggested that an antiparallel beta-pleated sheet with an orthogonal unit cell [dimensions: $a = 10.72$ Å, $b = 9.73$ Å, c (fiber axis) $= 6.80 \pm 0.05$ Å] was reasonably compatible with

Table 3.1 Miller indices and *d*-spacing values obtained from two-dimensional patterns.

Miller index (*hkl*)	Orientation (degrees)	*d*-spacing (Å)
(200)	0	5.36
(120)	0	4.43
(400)	0	2.68
(121)	32	3.71
(201)	38	4.26
(202)	58	2.91
(203)	69	2.07
(002)	90	3.43
(003)	90	2.25

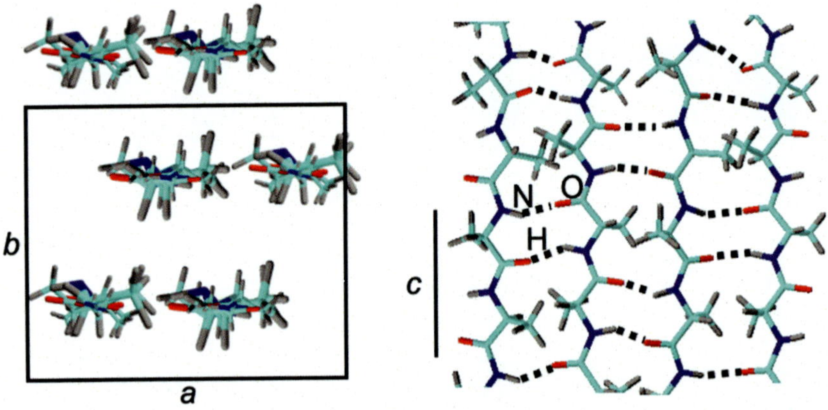

Figure 3.7 Antiparallel beta-pleated sheet structure with an orthogonal unit cell with dimensions $a = 10.72$ Å, $b = 9.73$ Å, c (fiber axis) $= 6.80 \pm 0.05$ Å.

the positions and orientations of the reflections (Fig. 3.7). The reflection along the fiber axis was very weak, indicating the size distribution of the beta-sheet crystals along the fiber axis. The unit cell contains four molecular chains, a pair that symmetrically forms beta-sheet structure via hydrogen bonds, moreover, the up- and down-molecular chains alternate with each other in an antiparallel manner.

Each silk has a different crystal lattice, which can be attributed to differences in the silk amino acid sequences. However, the relationship between the crystal lattices of different silks and the following characteristics of silk fibroin as a biomaterial remains mostly unexplored, even though the fact that crystallinity (i.e., the amount of crystalline region) affects the physical and biological properties of silk-based biomaterials.

3.4 Three-dimensional and self-assembly structures

Natural spider silk is a fascinating material. For thousands of years, unveiling its mystery has been one of the major human endeavors in material science; hence, we envision the scalable production of artificial silk fibers with mechanical performance compared to their natural counterparts. Although multiple theories (liquid crystalline and micelle-like models) have been well established [76–78], the underlying self-assembly mechanism is not yet fully understood. From a physicochemical viewpoint, the complicated natural spinning process can be considered a sophisticated phase separation. The involved solvent is pivotal in the assembly, allowing various biochemical triggers in a specific order to achieve the solidification of fibers from soluble proteins [79]. Importantly, mimicking the natural spinning process is a key factor in producing artificial spider dragline silk with outstanding mechanical properties. Here the current understanding of soluble spider silk proteins' self-assembly to form hierarchical structures during the natural spinning process is reviewed.

In nature, MaSp is stored at high concentrations (30%–50% w/v) inside the spider gland. Despite its high concentration, this protein is highly soluble together with other protein components in the spinning dope. It is transformed into the stable insoluble silk fiber after passing through a pH and salt gradient across the gland as well as experiencing shear forces, dehydration, and changes in the biochemical environment. After spinning, the resulting spider silk fiber forms a beta-sheet structure in the crystalline region and contains a PPII helix population with glycine-rich sequences, which are located in the amorphous region [42,80].

The spinning dope of the major ampullate gland is slightly more basic (pH 7.6) and contains a high concentration of chaotropic ions (Na^+, Cl^-, and K^+), while closer to the spinning duct, the pH becomes more acidic, and the concentrations of chaotropic ions decrease while those of kosmotropic ions (PO_4^{3-}) increase (Fig. 3.8) [81,82]. The conformations of the NTD and *C*-terminal domain (CTD) of spidroin show strong pH dependence, and these domains are essential for controlling the pH dependence of fiber assembly [29,83]. Spidroin terminal domains are critical in the self-assembly of silk fibers. The NTD functions as a pH-sensitive sensor and thus governs silk protein polymerization in vivo [82,84]. As the spidroins travel through the spinning duct, a decline in pH triggers rapid homodimerization via the NTDs, modulated by a subset of conserved charged side chains, producing linked spidroin chains [29,82,85]. The

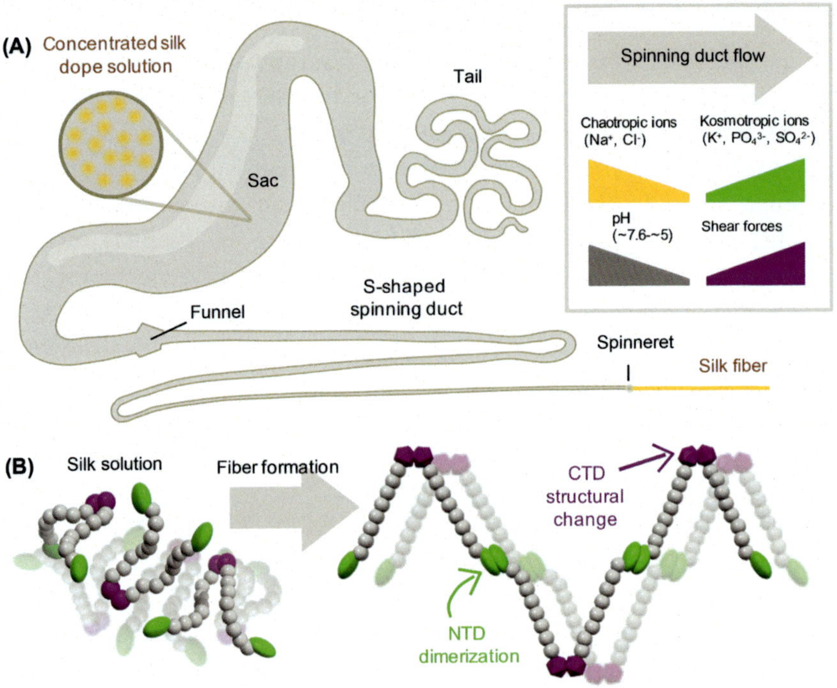

Figure 3.8 Spider gland and ion- and pH-dependent conformations of NTD and CTD. (A) The storage sac contains a highly concentrated silk solution (20%–50% w/v) and other conserved proteins. This content of the sac is slightly basic (pH $\geq$ 7) and has a high concentration of chaotropic ions, while closer to the spinning duct, the pH and concentration of chaotropic ions decrease and the concentration of kosmotropic ions increases. Spidroin will experience shear forces in the spinning duct before finally forming the silk fiber. (B) At a slightly basic pH ($\geq$7) and a high concentration of chaotropic ions, CTD forms a folded dimer and NTD is monomeric, while at an acidic pH and a low concentration of chaotropic ions, CTD becomes unfolded and NTD forms a folded dimer with an antiparallel orientation.

CTD, meanwhile, forms constitutive homodimers and has been implicated in the prearrangement of spidroin chains prior to polymerization [83]; the spidroins are also thought to undergo a structural change upon acidification, possibly beta-sheet nucleation formation in the central repeat regions [82]. Based on sequence conservation, the terminal domains are predicted to play a similar role across the different spidroin classes [86,87]. In the case of the RD, the structural transition depends on the length of the polyAla sequences. For relatively short polyAla sequences, in situ analysis using vibrational circular dichroism (VCD) demonstrated that soluble native spidroin contains random coil and PPII helix populations [88].

Solution-state NMR spectroscopy showed that the recombinant RD of MaSp1 is composed of two major populations: random coil (~66%) and PPII helix (~24%). The PPII helix conformations in the glycine-rich region are proposed as a soluble prefibrillar form, which undergoes intramolecular interactions to form a reverse turn. This study also showed that the conformation of the RD is not influenced by the pH gradient [88]. In contrast to pH, ion concentration plays a significant role in the solubility and fiber formation of spider silk. Chaotropic ions (Na^+, Cl^-, and K^+) improve solubility by preventing intra- and intermolecular interactions among the RDs, while kosmotropic ions (PO_4^{3-}) promote hydrogen bonds among RDs, allowing intra- and intermolecular interactions [32].

The mechanical and physical properties of spider dragline and silkworm silk fibers are known to depend on humidity; according to previous studies [15,89,90], the amorphous and crystalline regions are influenced by water molecules. A skin layer is present around the spidroin-based fiber, which is important for the gas/water vapor barrier. Recently, the skin layer of *Nephila* spider dragline, especially the biological and physical functions of the skin layers, was investigated [91]. The mechanical and physical properties as well as the crystal structures are independent of the skin constituents, suggesting that the protein core region of the spider dragline determines its structural and mechanical properties. Surprisingly, the skin layer does not influence supercontraction; that is, the layer does not have water vapor barrier properties but does protect the fiber from protease degradation. This result implies that the skin layer is more important as a biological barrier than as a physical barrier. Furthermore, under the skin layer, we could detect bindle-like microfibrils (Fig. 3.9), one of the essential hierarchical structures of silk required for its excellent mechanical properties [92].

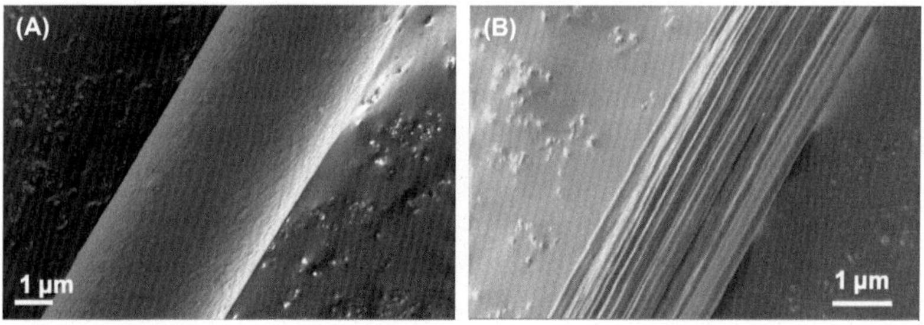

Figure 3.9 Field emission scanning electron microscopy images showing the surface of native *Trichonephila clavata* (formally *Nephila clavata*) spider dragline (A) before and (B) after protease treatment [92].

3.5 Roles of water

3.5.1 Types of water molecules

The role of water molecules in proteins and swollen polymeric materials has been investigated widely. Bound (nonfreezing), bulk (freezing), and intermediate (freezing bound) water molecules exist in proteins and hydrogels [93–95]. Differential scanning calorimetry (DSC) and NMR relaxometry have been used as typical methods to characterize and distinguish the state of water molecules in polymeric hydrogels [96,97]. The difference in the state of water molecules in living organisms has been suggested to influence various aspects of biological structures and functions, including their sizes, in bacterial and vegetative cells [98–100]. The bulk water is not crucial for enzymatic hydration, whereas the bound water plays a vital role in enzymatic catalysis [101,102]. The biological activities of enzymes and other proteins depend on how the water molecules are associated with these bioactive molecules; that is, the activities depend on the bound water content of the enzymes and proteins [103]. The bound water content is considered a significant factor in the control of drug release rate and enzymatic activity in hydrogel-based biomaterials [104]. The relationship between the state of water and the cytotoxicity of protein-based swollen material (e.g., hydrogels) was also confirmed using silkworm silk-based physical hydrogels [105]. These previous studies have indicated that the state of water in hydrogels used as biomaterials directly influences their biocompatibility and biological response against cells.

3.5.2 Biocompatibility and hydration

The effect of water with silk proteins was evaluated in terms of the cell viability. The viability of human mesenchymal stem cells on the silk hydrogels was suggested to influence the ratio of the bulk and bound water. In other words, the improvement of the silk hydrogel's cell viability implies that the cells and the cell-adhesion proteins in the extracellular matrix preferentially expand and adhere to a substrate containing more bound water. This is because the cell-adhesion proteins also may need bound water to exhibit their functions, similar to the other proteins [101,102,104]. The bound water in the silk hydrogel, therefore, accelerates cell-adhesion proteins, such as fibronectins, in the cellular matrix to interact with the surface of the silk hydrogels. In contrast, the bulk water would disturb the cell-adhesion proteins to adhere on the surface of the silk hydrogels, due to the relatively higher mobility of water.

3.5.3 Secondary structure under dry and humid conditions

Secondary structures are affected by the presence of water molecules. The secondary structure of silk at different relative humidities (RHs) and temperatures was evaluated in the crystalline state of silk films [89]. The crystallinity of the silk film prepared at 25°C and RH 58% was 21.3% ± 1.8%. The WAXS result of the silk film incubated at RH 6% demonstrated that the peak with a *d*-spacing value of 0.47 nm originated from the helical structure that appeared at 40°C (Fig. 3.10A). The peaks with *d*-spacing values of 0.45 and 0.37 nm were detected at 220°C, indicating that beta-sheet crystal formation was induced at approximately 220°C [106]. The WAXS profiles of the silk films incubated at RH 58% and RH 75% show that beta-sheet structure was induced at temperatures above 140°C when the silk film was incubated at RH 58%. In contrast, beta-sheet structure was formed without the heating process when silk films were incubated at RH 75% (Fig. 3.10B,C). The silk film at RH 75% contained both beta-sheets and helical structures even at 260°C, suggesting that the thermal treatment induces beta-sheet formation, and also the water molecules induced the formation of helical and beta-sheet structures (Fig. 3.11). According to the previous studies related to silk molecule's structural transition [107–109], bound water disrupts the intermolecular cohesive forces between protein chains, resulting in a reduced steric hindrance that promotes chain movement in the noncrystalline regions and induces beta-sheet crystallization in the silks. Thus this water plasticization promotes beta-sheet formation via helix–helix interactions based on the literature on prion proteins [110]. In the case of silk molecules, the Gly-rich sequence (soft segments) can form helical structures, resulting in helix–helix interactions [31]. The helix–helix

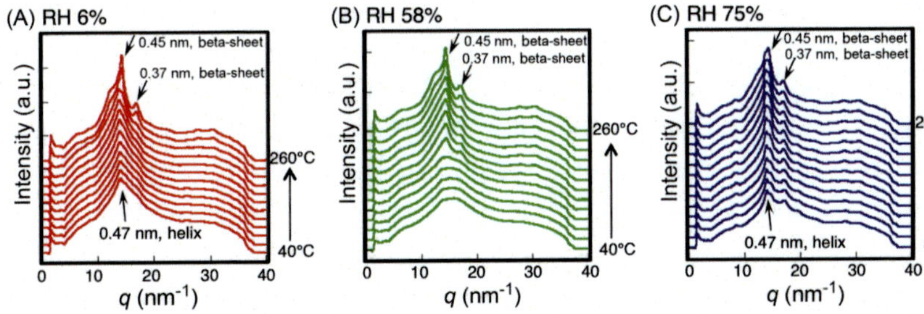

Figure 3.10 WAXS data of *Bombyx mori* silk films incubated at (A) RH 6%, (B) RH 58%, and (C) RH 75% during the heating process. Each WAXS profile was measured at 20°C intervals from 40°C to 260°C [89].

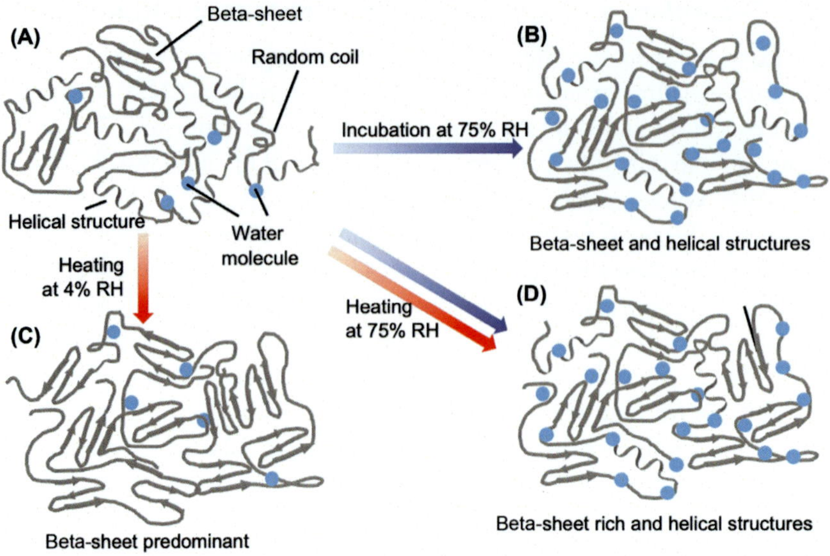

Figure 3.11 Schematic illustration of the water plasticization of silk films. Silk molecules (A) at 40°C and RH 6%; (B) at 40°C and 75% RH; (C) at RH 6% after heating treatment; and (D) at RH 75% after heating [89].

interaction of the soft segments would accelerate the assembly of beta-strands of alanine-rich sequences, yielding the beta-sheet structure. As described here, water molecules play a role as a plasticizer and promote beta-sheet formation. Further, the thermal treatment removed water molecules from the beta-strands and amorphous regions of silk films, inducing the beta-sheet structures.

3.5.4 Plasticization and crystallization

As mentioned in the previous section, plasticization and crystallization are significantly related to the physical properties of protein-based materials. The mechanical properties of silk films are dependent on RHs, that is, the number of water molecules around silk protein molecules [89,111]. The stress–strain curves at different RHs obtained from the tensile tests demonstrate that the silk films at RH 97% obviously differed from those obtained from silk films incubated at RH 84% or less (Fig. 3.12) [89]. Four types of mechanical properties, namely, tensile strength, Young's modulus, elongation at break, and toughness, maintained up to RH 84%. On the other hand, at RH 97%, the mechanical properties of films changed drastically. The elongation at break and toughness of the silk films

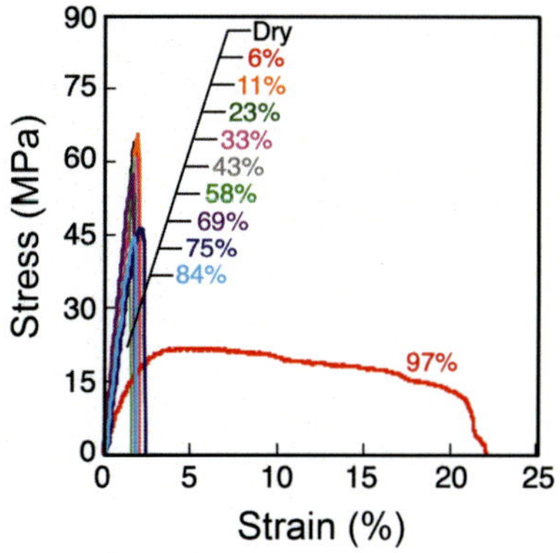

Figure 3.12 Stress–strain curves of silk films incubated at different RHs from RH 0% (dry) to 97% [89].

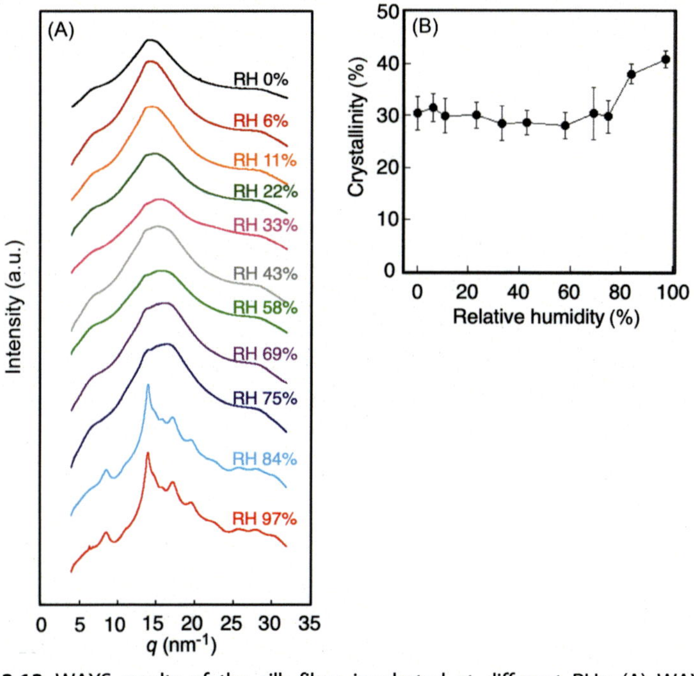

Figure 3.13 WAXS results of the silk films incubated at different RHs: (A) WAXS one-dimensional (1D) profiles; (B) degree of crystallinity calculated from WAXS 1D profiles [89].

incubated at RH 97% increased 15- and 7-fold higher than those incubated at RH 84%. To clear the mechanism behind the significant change in mechanical properties, the crystal structure of the silk films at different RH was characterized by WAXS (Fig. 3.13) [89]. The crystallinity maintained at approximately 30% at up to RH 75%, whereas the crystallinity increased to approximately 40% at over RH 84%. The structural and mechanical characterizations indicated that the silk films prepared at RH 97% showed high crystallinity and elongation at break. There is generally a trade-off relationship between crystallinity and ductility in material science; however, the silk film at RH 97% can simultaneously realize both properties. This surprising phenomenon is mainly due to the plasticizing effect of water molecules on amorphous silk molecules. At RH 97% amorphous regions of silk films were plasticized, resulting in the increase of elasticity of silk films, and also water molecules induced beta-sheet formation via helix–helix interactions.

3.5.5 Molecular network structures

Molecular network and cross-link are a fundamental approach to change the mechanical properties of polymeric materials. Besides, hydration control is important to enhance the mechanical properties like the toughness of protein-based materials. Hydrogel is the best candidate among the various polymeric materials to investigate the network structures [112]. A hydrogel is a polymeric material with a hydrated network structure, whereas cross-linked polymeric materials with less hydration are elastomers, such as natural rubbers and adhesives. The difference between hydrogels and rubbers in terms of material properties makes it evident that the hydration state of the molecular network affects the strength and toughness of polymeric materials. In the case of protein materials, hydration is a key factor in controlling various properties, and hence hydrogel is considered as the best model material to investigate the network structure. The molecular network in hydrogels is a critical structural aspect to maximize their physical properties. To date, fiber-reinforced, interpenetrated network, double-network, and tetra-polyethylene glycol hydrogels are considered to have network-originated physical properties [113–117].

Silk-based materials are expected to demonstrate high toughness and ductility because of the excellent properties of natural spider dragline [30,118–120]. Although silkworm cocoons and spider webs/draglines are used as a tough structural material in nature, silk material has not yet been used as a tough structural material for human life [121]. High-strength silk scaffolds have been developed for bone repair and showed a high compressive strength of approximately 13 MPa [122]. A highly concentrated silk hydrogel exhibited a compressive strength of approximately 6 MPa

[123]. Soft and elastic silk hydrogel and artificial silk fibers have been developed; however, these silk materials are rubber-like elastic materials that lack mechanical strength [124,125]. The specific hierarchical and network structure of natural silk fibers needs to be mimicked to achieve the same toughness and strength as spider dragline silks. One of the important structural features of silk fibers is the network structure. In the network structures of spider dragline silks, the beta-sheet structures are considered to play a role as cross-linking points [126]. In addition, chemical cross-links like covalent bonds have been reported in native silk fibers, which is related to PTM such as dityrosine. Several types of dragline silks have been reported to contain appreciable quantities of dityrosine residues, forming intermolecular networks [20]. Covalent dityrosine bonds have been found in the other organisms as well, such as Tussah silk fibroin, caddisfly larvae, and sandcastle worms [127–129].

In the case of silk hydrogels, two major types of networks have been reported to date, namely, physical and chemical cross-links. In the case of a silk physical gel via fibril formation, the gelation of silk solutions can be induced by pH changes, ultrasonication, vortexing, electric fields, and poor solvent-induced phase separation [130–135]. These gelations are accompanied by the phase separation of the silk solution, which produces silk physical hydrogels through silk microfibril formation. Physical cross-links consist of heterogeneous microfibrillar networks and also show brittle mechanical properties [105]. In contrast, silk hydrogels with chemical cross-links between dityrosine residues in the silk molecules can be prepared by enzymatic oxidation reactions in silk solution [136]. This silk chemical hydrogel is more elastic and transparent than silk physical hydrogel, because phase separation hardly occurs, and the molecular weights between the covalent cross-linking points are higher than those of physical gel.

A silk hydrogel with chemical and physical cross-links, namely, silk chemical–physical hydrogel, was developed by combining the protocols for chemical and physical silk hydrogel [137]. Optimizing the hydration state of the silk chemical–physical hydrogel leads to rubber-like elasticity and silk fiber-like toughness of the silk material (Fig. 3.14) [137]. The water content-controlled silk material was more of a resin rather than a hydrogel. The silk resins with controlled network structures and hydration states successfully achieved the highest toughness and favorable biodegradability as a bulk silk material.

The stress–strain profiles of the silk resins prepared from chemical/physical gel at a high silk concentration (150 g/L) and different RHs were evaluated (Fig. 3.15) [137]. The Young's modulus showed a decreasing

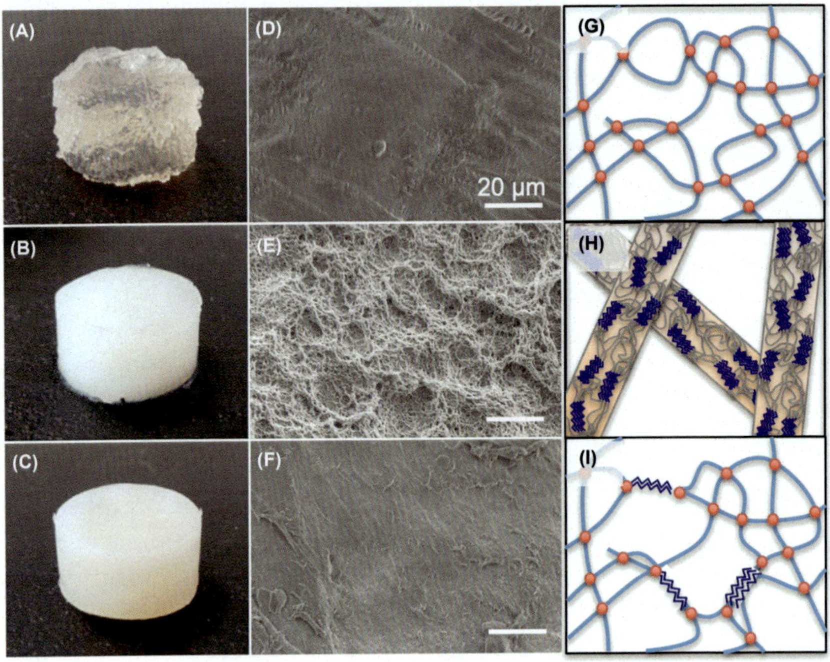

Figure 3.14 (A) Silk chemical gel, (B) physical gel, and (C) chemical–physical gel. The concentration of silk was 100 g/L. (D–F) Scanning electron microscopy images of the inside of the silk hydrogels. Scale bars indicate 20 μm. (G–I) Schematic illustrations of network structures of chemical gel, physical gel, and chemical–physical gel. Red points are the chemical cross-linking points, and blue structures are beta-sheet crystals [137].

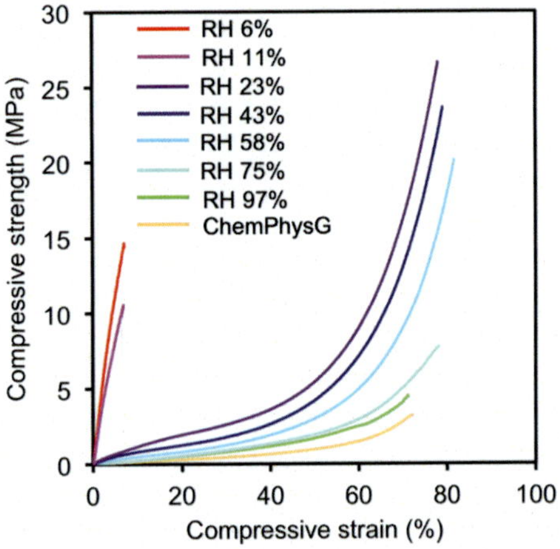

Figure 3.15 Typical stress–strain curves of silk chemical/physical hydrogel and silk resin at different RH [137].

tendency, while the strain at break exhibited an increasing trend with an increase in RH. The mechanical properties of the silk resin incubated at an RH of 23% showed the highest toughness, approximately 5 MJ/m^3. At lower RHs, the silk materials shrank and showed a significantly higher Young's modulus and lower strength, ductility, and toughness. The silk resins incubated at an RH of greater than 75% showed relatively lower compressive strength, resulting in lower toughness. Based on the overall results of silk materials containing chemical–physical cross-links, the network structure and hydration state are essential factors for regulating the mechanical properties of silk materials, including silk hydrogels and resins [137].

References

[1] W.B. Huttner, Sulphation of tyrosine residues—a widespread modification of proteins, Nature 299 (5880) (1982) 273–276.
[2] M.J. Harrington, A. Masic, N. Holten-Andersen, J.H. Waite, P. Fratzl, Iron-clad fibers: a metal-based biological strategy for hard flexible coatings, Science 328 (5975) (2010) 216–220.
[3] H.G. Silverman, F.F. Roberto, Understanding marine mussel adhesion, Mar. Biotechnol. 9 (6) (2007) 661–681.
[4] L.A. Burzio, J.H. Waite, Cross-linking in adhesive quinoproteins: studies with model decapeptides, Biochemistry 39 (36) (2000) 11147–11153.
[5] L.O. Burzio, V.A. Burzio, T. Silva, L.A. Burzio, J. Pardo, Environmental bioadhesion: themes and applications, Curr. Opin. Biotechnol. 8 (3) (1997) 309–312.
[6] M. Yu, T.J. Deming, Synthetic polypeptide mimics of marine adhesives, Macromolecules 31 (15) (1998) 4739–4745.
[7] M.E. Yu, J.Y. Hwang, T.J. Deming, Role of L-3,4-dihydroxyphenylalanine in mussel adhesive proteins, J. Am. Chem. Soc. 121 (24) (1999) 5825–5826.
[8] J.H. Waite, X. Qin, Polyphosphoprotein from the adhesive pads of Mytilus edulis, Biochemistry 40 (9) (2001) 2887–2893.
[9] K. Numata, P.J. Baker, Synthesis of adhesive peptides similar to those found in Blue mussel (Mytilus edulis) using papain and tyrosinase, Biomacromolecules 15 (8) (2014) 3206–3212.
[10] M.P. Ferla, W.M. Patrick, Bacterial methionine biosynthesis, Microbiol-SGM 160 (2014) 1571–1584.
[11] J. Gatesy, C. Hayashi, D. Motriuk, J. Woods, R. Lewis, Extreme diversity, conservation, and convergence of spider silk fibroin sequences, Science 291 (5513) (2001) 2603–2605.
[12] M.B. Hinman, R.V. Lewis, Isolation of a clone encoding a second dragline silk fibroin. Nephila clavipes dragline silk is a two-protein fiber, J. Biol. Chem. 267 (27) (1992) 19320–19324.
[13] C.Y. Hayashi, N.H. Shipley, R.V. Lewis, Hypotheses that correlate the sequence, structure, and mechanical properties of spider silk proteins, Int. J. Biol. Macromol. 24 (2-3) (1999) 271–275.
[14] M. Marhabaie, T.C. Leeper, T.A. Blackledge, Protein composition correlates with the mechanical properties of spider (Argiope trifasciata) dragline silk, Biomacromolecules 15 (1) (2014) 20–29.

[15] A.D. Malay, K. Arakawa, K. Numata, Analysis of repetitive amino acid motifs reveals the essential features of spider dragline silk proteins, PLoS One 12 (8) (2017) e0183397.
[16] D.J. Perry, D. Bittencourt, J. Siltberg-Liberles, E.L. Rech, R.V. Lewis, Piriform spider silk sequences reveal unique repetitive elements, Biomacromolecules 11 (11) (2010) 3000–3006.
[17] P. Geurts, L. Zhao, Y. Hsia, E. Gnesa, S. Tang, F. Jeffery, et al., Synthetic spider silk fibers spun from pyriform spidroin 2, a glue silk protein discovered in orb-weaving spider attachment discs, Biomacromolecules 11 (12) (2010) 3495–3503.
[18] A.A. Van Dijk, L.L. Van Wijk, A. Van Vliet, P. Haris, E. Van Swieten, G.I. Tesser, et al., Structure characterization of the central repetitive domain of high molecular weight gluten proteins. I. Model studies using cyclic and linear peptides, Protein Sci. 6 (3) (1997) 637–648.
[19] P.R. Shewry, N.G. Halford, P.S. Belton, A.S. Tatham, The structure and properties of gluten: an elastic protein from wheat grain, Philos. Trans. Roy. Soc. B 357 (1418) (2002) 133–142.
[20] J.R. dos Santos-Pinto, G. Lamprecht, W.Q. Chen, S. Heo, J.G. Hardy, H. Priewalder, et al., Structure and post-translational modifications of the web silk protein spidroin-1 from Nephila spiders, J. Proteom. 105 (2014) 174–185.
[21] J. dos Santos-Pinto, H. Arcuri, G. Lubec, M. Palma, Structural characterization of the major ampullate silk spidroin-2 protein produced by the spider Nephila clavipes, Biochim. Biophys. Acta 1864 (10) (2016) 1444–1454.
[22] J. dos Santos-Pinto, H. Arcuri, H. Priewalder, H. Salles, M. Palma, G. Lubec, Structural model for the spider silk protein spidroin-1, J. Proteome Res. 14 (9) (2015) 3859–3870.
[23] R. Jensen, D. Morse, The bioadhesive of Phragmatopoma californica tubes—a silk-like cement containing L-DOPA, J. Comp. Physiol. B 158 (3) (1988) 317–324.
[24] R. Stewart, C. Wang, H. Shao, Complex coacervates as a foundation for synthetic underwater adhesives, Adv. Colloid Interface Sci. 167 (1-2) (2011) 85–93.
[25] B. Partlow, M. Bagheri, J. Harden, D. Kaplan, Tyrosine templating in the self-assembly and crystallization of silk fibroin, Biomacromolecules 17 (11) (2016) 3570–3579.
[26] H. Craig, S. Blamires, M. Sani, M. Kasumovic, A. Rawal, J. Hook, DNP NMR spectroscopy reveals new structures, residues and interactions in wild spider silks, Chem. Commun. 55 (32) (2019) 4687–4690.
[27] J. dos Santos-Pinto, H. Arcuri, F. Esteves, M. Palma, G. Lubec, Spider silk proteome provides insight into the structural characterization of Nephila clavipes flagelliform spidroin, Sci. Rep. 8 (2018) 14674.
[28] M. Landreh, G. Askarieh, K. Nordling, M. Hedhammar, A. Rising, C. Casals, et al., A pH-dependent dimer lock in spider silk protein, J. Mol. Biol. 404 (2) (2010) 328–336.
[29] G. Askarieh, M. Hedhammar, K. Nordling, A. Saenz, C. Casals, A. Rising, et al., Self-assembly of spider silk proteins is controlled by a pH-sensitive relay, Nature 465 (7295) (2010) 236–238.
[30] K. Numata, P. Cebe, D.L. Kaplan, Mechanism of enzymatic degradation of beta-sheet crystals, Biomaterials 31 (10) (2010) 2926–2933.
[31] N.A. Oktaviani, A. Matsugami, A.D. Malay, F. Hayashi, D.L. Kaplan, K. Numata, Conformation and dynamics of soluble repetitive domain elucidates the initial beta-sheet formation of spider silk, Nat. Commun. 9 (1) (2018) 2121.
[32] N.A. Oktaviani, A. Matsugami, F. Hayashi, K. Numata, Ion effects on the conformation and dynamics of repetitive domains of a spider silk protein: implications for solubility and beta-sheet formation, Chem. Commun. (Camb) 55 (66) (2019) 9761–9764.

[33] M. Andersson, J. Johansson, A. Rising, Silk spinning in silkworms and spiders, Int. J. Mol. Sci. 17 (8) (2016) 1290.
[34] D. Ebrahimi, O. Tokareva, N.G. Rim, J.Y. Wong, D.L. Kaplan, M.J. Buehler, Silk—its mysteries, how it is made, and how it is used, ACS Biomater. Sci. Eng. 1 (10) (2015) 864—876.
[35] C. Fu, Z. Shao, V. Fritz, Animal silks: their structures, properties and artificial production, Chem. Commun. 43 (2009) 6515—6529.
[36] K. Numata, R. Sato, K. Yazawa, T. Hikima, H. Masunaga, Crystal structure and physical properties of Antheraea yamamai silk fibers: long poly(alanine) sequences are partially in the crystalline region, Polymer 77 (2015) 87—94.
[37] T. Asakura, M. Okonogi, K. Horiguchi, A. Aoki, H. Saito, D.P. Knight, et al., Two different packing arrangements of antiparallel polyalanine, Angew. Chem. Int. Ed. Engl. 51 (5) (2012) 1212—1215.
[38] F.J. Chen, D. Porter, F. Vollrath, Silk cocoon (Bombyx mori): multi-layer structure and mechanical properties, Acta Biomater. 8 (7) (2012) 2620—2627.
[39] Y. Liu, Z.Z. Shao, F. Vollrath, Relationships between supercontraction and mechanical properties of spider silk, Nat. Mater. 4 (12) (2005) 901—905.
[40] S. Keten, Z. Xu, B. Ihle, M.J. Buehler, Nanoconfinement controls stiffness, strength and mechanical toughness of beta-sheet crystals in silk, Nat. Mater. 9 (4) (2010) 359—367.
[41] T. Lefevre, M.E. Rousseau, M. Pezolet, Protein secondary structure and orientation in silk as revealed by Raman spectromicroscopy, Biophys. J. 92 (8) (2007) 2885—2895.
[42] J.D. van Beek, S. Hess, F. Vollrath, B.H. Meier, The molecular structure of spider dragline silk: folding and orientation of the protein backbone, Proc. Natl. Acad. Sci. U S A 99 (16) (2002) 10266—10271.
[43] O. Liivak, A. Blye, N. Shah, L.W. Jelinski, A microfabricated wet-spinning apparatus to spin fibers of silk proteins. Structure-property correlations, Macromolecules 31 (9) (1998) 2947—2951.
[44] C. Riekel, C. Branden, C. Craig, C. Ferrero, F. Heidelbach, M. Muller, Aspects of X-ray diffraction on single spider fibers, Int. J. Biol. Macromol. 24 (2-3) (1999) 179—186.
[45] F. Vollrath, D. Porter, Silks as ancient models for modern polymers, Polymer 50 (24) (2009) 5623—5632.
[46] N. Becker, E. Oroudjev, S. Mutz, J.P. Cleveland, P.K. Hansma, C.Y. Hayashi, et al., Molecular nanosprings in spider capture-silk threads, Nat. Mater. 2 (4) (2003) 278—283.
[47] C.Y. Hayashi, R.V. Lewis, Molecular architecture and evolution of a modular spider silk protein gene, Science 287 (5457) (2000) 1477—1479.
[48] T. Lefevre, M. Pezolet, Unexpected beta-sheets and molecular orientation in flagelliform spider silk as revealed by Raman spectromicroscopy, Soft Matter 8 (23) (2012) 6350—6357.
[49] T. Asakura, A. Kuzuhara, R. Tabeta, H. Saito, Conformation characterization of Bombyx-Mori silk fibroin in the solid-state by high-frequency C-13 cross polarization magic angle spinning NMR, X-ray-diffraction, and infrared-spectroscopy, Macromolecules 18 (10) (1985) 1841—1845.
[50] T. Asakura, J. Yao, T. Yamane, K. Umemura, A.S. Ulrich, Heterogeneous structure of silk fibers from Bombyx mori resolved by 13C solid-state NMR spectroscopy, J. Am. Chem. Soc. 124 (30) (2002) 8794—8795.
[51] J.E. Jenkins, S. Sampath, E. Butler, J. Kim, R.W. Henning, G.P. Holland, et al., Characterizing the secondary protein structure of black widow dragline silk using solid-state NMR and X-ray diffraction, Biomacromolecules 14 (10) (2013) 3472—3483.

[52] H. Yoshimizu, T. Asakura, The structure of Bombyx-Mori silk fibroin membrane swollen by water studied with ESR, C-13-NMR, and FT-IR spectroscopies, J. Appl. Polym. Sci. 40 (9-10) (1990) 1745–1756.

[53] J.E. Trancik, J.T. Czernuszka, F.I. Bell, C. Viney, Nanostructural features of a spider dragline silk as revealed by electron and X-ray diffraction studies, Polymer 47 (15) (2006) 5633–5642.

[54] S. Ling, Z. Qi, D.P. Knight, Z. Shao, X. Chen, Synchrotron FTIR microspectroscopy of single natural silk fibers, Biomacromolecules 12 (9) (2011) 3344–3349.

[55] M.E. Rousseau, T. Lefevre, M. Pezolet, Conformation and orientation of proteins in various types of silk fibers produced by Nephila clavipes spiders, Biomacromolecules 10 (10) (2009) 2945–2953.

[56] J. Sirichaisit, V.L. Brookes, R.J. Young, F. Vollrath, Analyis of structure/property relationships in silkworm (Bombyx mori) and spider dragline (Nephila edulis) silks using Raman spectroscopy, Biomacromolecules 4 (2) (2003) 387–394.

[57] C.J. Benmore, T. Izdebski, J.L. Yarger, Total X-ray scattering of spider dragline silk, Phys. Rev. Lett. 108 (17) (2012).

[58] A. Bram, C.I. Branden, C. Craig, I. Snigireva, C. Riekel, X-ray diffraction from single fibres of spider silk, J. Appl. Crystallogr. 30 (1997) 390–392.

[59] D.T. Grubb, L.W. Jelinski, Fiber morphology of spider silk: the effects of tensile deformation, Macromolecules 30 (10) (1997) 2860–2867.

[60] R.E. Marsh, R.B. Corey, L. Pauling, An investigation of the structure of silk fibroin, Biochim. Biophys. Acta 16 (1) (1955) 1–34.

[61] A. Martel, M. Burghammer, R.J. Davies, E. Di Cola, C. Vendrely, C. Riekel, Silk fiber assembly studied by synchrotron radiation SAXS/WAXS and Raman spectroscopy, J. Am. Chem. Soc. 130 (50) (2008) 17070–17074.

[62] A. Martel, M. Burghammer, R.J. Davies, C. Riekel, Thermal behavior of Bombyx mori silk: evolution of crystalline parameters, molecular structure, and mechanical properties, Biomacromolecules 8 (11) (2007) 3548–3556.

[63] C. Riekel, B. Madsen, D. Knight, F. Vollrath, X-ray diffraction on spider silk during controlled extrusion under a synchrotron radiation X-ray beam, Biomacromolecules 1 (4) (2000) 622–626.

[64] C. Riekel, M. Muller, F. Vollrath, In situ X-ray diffraction during forced silking of spider silk, Macromolecules 32 (13) (1999) 4464–4466.

[65] C. Riekel, F. Vollrath, Spider silk fibre extrusion: combined wide- and small-angle X-ray microdiffraction experiments, Int. J. Biol. Macromol. 29 (3) (2001) 203–210.

[66] Y. Takahashi, M. Gehoh, K. Yuzuriha, Structure refinement and diffuse streak scattering of silk (Bombyx mori), Int. J. Biol. Macromol. 24 (2-3) (1999) 127–138.

[67] Z. Yang, D.T. Grubb, L.W. Jelinski, Small-angle X-ray scattering of spider dragline silk, Macromolecules 30 (26) (1997) 8254–8261.

[68] T. Lefevre, F. Paquet-Mercier, S. Lesage, M.E. Rousseau, S. Bedard, M. Pezolet, Study by Raman spectromicroscopy of the effect of tensile deformation on the molecular structure of Bombyx mori silk, Vib. Spectrosc. 51 (1) (2009) 136–141.

[69] T. Seydel, K. Kolln, I. Krasnov, I. Diddens, N. Hauptmann, G. Helms, et al., Silkworm silk under tensile strain investigated by synchrotron X-ray diffraction and neutron spectroscopy, Macromolecules 40 (4) (2007) 1035–1042.

[70] I. Krasnov, I. Diddens, N. Hauptmann, G. Helms, M. Ogurreck, T. Seydel, et al., Mechanical properties of silk: interplay of deformation on macroscopic and molecular length scales, Phys. Rev. Lett. 100 (4) (2008) 048104.

[71] J. Sirichaisit, R.J. Young, F. Vollrath, Molecular deformation in spider dragline silk subjected to stress, Polymer 41 (3) (2000) 1223–1227.

[72] V.L. Brookes, R.J. Young, F. Vollrath, Deformation micromechanics of spider silk, J. Mater. Sci. 43 (10) (2008) 3728–3732.

[73] T. Lefevre, F. Paquet-Mercier, J.F. Rioux-Dube, M. Pezolet, Review structure of silk by raman spectromicroscopy: from the spinning glands to the fibers, Biopolymers 97 (6) (2012) 322–336.
[74] C.Y. Hayashi, R.V. Lewis, Evidence from flagelliform silk cDNA for the structural basis of elasticity and modular nature of spider silks, J. Mol. Biol. 275 (5) (1998) 773–784.
[75] A.D. Malay, R. Sato, K. Yazawa, H. Watanabe, N. Ifuku, H. Masunaga, et al., Relationships between physical properties and sequence in silkworm silks, Sci. Rep. 6 (2016) 27573.
[76] M. Heim, D. Keerl, T. Scheibel, Spider silk: from soluble protein to extraordinary fiber, Angew. Chem. Int. Ed. 48 (20) (2009) 3584–3596.
[77] T.Y. Lin, H. Masunaga, R. Sato, A.D. Malay, K. Toyooka, T. Hikima, et al., Liquid crystalline granules align in a hierarchical structure to produce spider dragline microfibrils, Biomacromolecules 18 (4) (2017) 1350–1355.
[78] H.J. Jin, D.L. Kaplan, Mechanism of silk processing in insects and spiders, Nature 424 (6952) (2003) 1057–1061.
[79] S. Rammensee, U. Slotta, T. Scheibel, A.R. Bausch, Assembly mechanism of recombinant spider silk proteins, Proc. Natl. Acad. Sci. 105 (18) (2008) 6590–6595.
[80] A.H. Simmons, C.A. Michal, L.W. Jelinski, Molecular orientation and two-component nature of the crystalline fraction of spider dragline silk, Science 271 (5245) (1996) 84–87.
[81] D.P. Knight, F. Vollrath, Changes in element composition along the spinning duct in a Nephila spider, Die Naturwissenschaften 88 (4) (2001) 179–182.
[82] M. Andersson, G. Chen, M. Otikovs, M. Landreh, K. Nordling, N. Kronqvist, et al., Carbonic anhydrase generates CO_2 and H^+ that drive spider silk formation via opposite effects on the terminal domains, PLoS Biol. 12 (8) (2014) e1001921.
[83] F. Hagn, L. Eisoldt, J.G. Hardy, C. Vendrely, M. Coles, T. Scheibel, et al., A conserved spider silk domain acts as a molecular switch that controls fibre assembly, Nature 465 (7295) (2010) 239–242.
[84] F. Vollrath, D.P. Knight, X.W. Hu, Silk production in a spider involves acid bath treatment, Proc. R. Soc. London. Ser. B Biol. Sci. 265 (1398) (1998) 817–820.
[85] N. Kronqvist, M. Otikovs, V. Chmyrov, G. Chen, M. Andersson, K. Nordling, et al., Sequential pH-driven dimerization and stabilization of the N-terminal domain enables rapid spider silk formation, Nat. Commun. 5 (2014) 3254.
[86] A. Rising, G. Hjälm, W. Engström, J. Johansson, N-terminal nonrepetitive domain common to dragline, flagelliform, and cylindriform spider silk proteins, Biomacromolecules 7 (11) (2006) 3120–3124.
[87] M.A. Collin, T.H. Clarke, N.A. Ayoub, C.Y. Hayashi, Genomic perspectives of spider silk genes through target capture sequencing: conservation of stabilization mechanisms and homology-based structural models of spidroin terminal regions, Int. J. Biol. Macromol. 113 (2018) 829–840.
[88] N.A. Oktaviani, A. Matsugami, A.D. Malay, F. Hayashi, D.L. Kaplan, K. Numata, Conformation and dynamics of soluble repetitive domain elucidates the initial β-sheet formation of spider silk, Nat. Commun. 9 (1) (2018) 1–11.
[89] K. Yazawa, K. Ishida, H. Masunaga, T. Hikima, K. Numata, Influence of water content on the beta-sheet formation, thermal stability, water removal, and mechanical properties of silk materials, Biomacromolecules 17 (3) (2016) 1057–1066.
[90] C.J. Fu, D. Porter, Z.Z. Shao, Moisture effects on Antheraea pernyi silk's mechanical property, Macromolecules 42 (20) (2009) 7877–7880.
[91] K. Yazawa, A.D. Malay, H. Masunaga, K. Numata, Role of skin layers on mechanical properties and supercontraction of spider dragline silk fiber, Macromol. Biosci. 19 (3) (2019) e1800220.

[92] H. Sogawa, K. Nakano, A. Tateishi, K. Tajima, K. Numata, Surface analysis of native spider draglines by FE-SEM and XPS, Front. Bioeng. Biotechnol. 8 (2020) 231.
[93] C.A. Fyfe, A.I. Blazek, Investigation of hydrogel formation from hydroxypropylmethylcellulose (HPMC) by NMR spectroscopy and NMR imaging techniques, Macromolecules 30 (20) (1997) 6230–6237.
[94] H. Hatakeyama, T. Hatakeyama, Interaction between water and hydrophilic polymers, Thermochim. Acta 308 (1-2) (1998) 3–22.
[95] X. Qu, A. Wirsen, A.C. Albertsson, Novel pH-sensitive chitosan hydrogels: swelling behavior and states of water, Polymer 41 (12) (2000) 4589–4598.
[96] M. Carenza, G. Cojazzi, B. Bracci, L. Lendinara, L. Vitali, M. Zincani, et al., The state of water in thermoresponsive poly(acryloyl-L-proline methyl ester) hydrogels observed by DSC and H-1-NMR relaxometry, Radiat. Phys. Chem. 55 (2) (1999) 209–218.
[97] V.J. McBrierty, S.J. Martin, F.E. Karasz, Understanding hydrated polymers: the perspective of NMR, J. Mol. Liq. 80 (2-3) (1999) 179–205.
[98] Y. Ishihara, H. Saito, J. Takano, Differences in the surface membranes and water content between the vegetative cells and spores of Bacillus subtilis, Cell Biochem. Funct. 17 (1) (1999) 9–13.
[99] Y. Ishihara, J. Takano, S. Mashimo, M. Yamamura, Determination of the water necessary for survival of Bacillus-Subtilis vegetative cells and spores, Thermochim. Acta 235 (2) (1994) 153–160.
[100] A.J. Westphal, P.B. Price, T.J. Leighton, K.E. Wheeler, Kinetics of size changes of individual Bacillus thuringiensis spores in response to changes in relative humidity, Proc. Natl. Acad. Sci. U S A 100 (6) (2003) 3461–3466.
[101] G. Careri, A. Giansanti, J.A. Rupley, Proton percolation on hydrated lysozyme powders, Proc. Natl. Acad. Sci. U S A 83 (18) (1986) 6810–6814.
[102] G. Careri, E. Gratton, P.H. Yang, J.A. Rupley, Correlation of IR-spectroscopic, heat-capacity, diamagnetic susceptibility and enzymatic measurements on lysozyme powder, Nature 284 (5756) (1980) 572–573.
[103] H.K.T.F. Pessen, Measurements of protein hydration by various techniques, in: C. H.W. Hirs, S.N. Timasheff (Eds.), Methods in Enzymology, Academic Press, New York, 1985, p. 219.
[104] A.S. Hoffman, W.R. Gombotz, S. Uenoyama, L.C. Dong, G. Schmer, Immobilization of enzymes and antibodies to radiation grafted polymers for therapeutic and diagnostic applications, Radiat. Phys. Chem. 27 (4) (1986) 265–273.
[105] K. Numata, T. Katashima, T. Sakai, State of water, molecular structure, and cytotoxicity of silk hydrogels, Biomacromolecules 12 (6) (2011) 2137–2144.
[106] L.F. Drummy, D.M. Phillips, M.O. Stone, B. Farmer, R.R. Naik, Thermally induced α-helix to β-sheet transition in regenerated silk fibers and films, Biomacromolecules 6 (6) (2005) 3328–3333.
[107] J. Guan, D. Porter, F. Vollrath, Thermally induced changes in dynamic mechanical properties of native silks, Biomacromolecules 14 (3) (2013) 930–937.
[108] C. Mo, P. Wu, X. Chen, Z. Shao, The effect of water on the conformation transition of Bombyx mori silk fibroin, Vib. Spectrosc. 51 (1) (2009) 105–109.
[109] G.R. Plaza, G.V. Guinea, J. Pérez-Rigueiro, M. Elices, Thermo-hygro-mechanical behavior of spider dragline silk: glassy and rubbery states, J. Polym. Sci. Part B Polym. Phys. 44 (6) (2006) 994–999.
[110] M. Morillas, D.L. Vanik, W.K. Surewicz, On the mechanism of alpha-helix to beta-sheet transition in the recombinant prion protein, Biochemistry 40 (23) (2001) 6982–6987.

[111] K. Yazawa, A.D. Malay, H. Masunaga, Y. Norma-Rashid, K. Numata, Simultaneous effect of strain rate and humidity on the structure and mechanical behavior of spider silk, Commun. Mater. 1 (1) (2020).
[112] H. Kamata, X. Li, U.I. Chung, T. Sakai, Design of hydrogels for biomedical applications, Adv. Healthc. Mater. 4 (16) (2015) 2360–2374.
[113] L. Ambrosio, R. De Santis, L. Nicolais, Composite hydrogels for implants, Proc. Inst. Mech. Eng. H. 212 (2) (1998) 93–99.
[114] J.P. Gong, Y. Katsuyama, T. Kurokawa, Y. Osada, Double-network hydrogels with extremely high mechanical strength, Adv. Mater. 15 (14) (2003) 1155.
[115] M. Santin, S.J. Huang, S. Iannace, L. Ambrosio, L. Nicolais, G. Peluso, Synthesis and characterization of a new interpenetrated poly(2-hydroxyethylmethacrylate)-gelatin composite polymer, Biomaterials 17 (15) (1996) 1459–1467.
[116] L. Ambrosio, R. De Santis, S. Iannace, P.A. Netti, L. Nicolais, Viscoelastic behavior of composite ligament prostheses, J. Biomed. Mater. Res. 42 (1) (1998) 6–12.
[117] T. Sakai, T. Matsunaga, Y. Yamamoto, C. Ito, R. Yoshida, S. Suzuki, et al., Design and fabrication of a high-strength hydrogel with ideally homogeneous network structure from tetrahedron-like macromonomers, Macromolecules 41 (14) (2008) 5379–5384.
[118] D.L. Kaplan, C.M. Mello, S. Arcidiacono, S. Fossey, K. Senecal, W. Muller, Silk, Protein-Based Materials, Springer, 1997, pp. 103–131.
[119] E.M. Pritchard, D.L. Kaplan, Silk fibroin biomaterials for controlled release drug delivery, Expert. Opin. Drug. Deliv. 8 (6) (2011) 797–811.
[120] F. Vollrath, D. Porter, Spider silk as a model biomaterial, Appl. Phys. A 82 (2) (2006) 205–212.
[121] K. Numata, Poly(amino acid)s/polypeptides as potential functional and structural materials, Polym. J. 47 (8) (2015) 537–545.
[122] B.B. Mandal, A. Grinberg, E.S. Gil, B. Panilaitis, D.L. Kaplan, High-strength silk protein scaffolds for bone repair, Proc. Natl. Acad. Sci. U S A 109 (20) (2012) 7699–7704.
[123] K. Numata, S. Yamazaki, T. Katashima, J.-A. Chuah, N. Naga, T. Sakai, Silk-pectin hydrogel with superior mechanical properties, biodegradability, and biocompatibility, Macromol. Biosci. 14 (6) (2014) 799–806.
[124] A. Heidebrecht, L. Eisoldt, J. Diehl, A. Schmidt, M. Geffers, G. Lang, et al., Biomimetic fibers made of recombinant spidroins with the same toughness as natural spider silk, Adv. Mater. 27 (13) (2015) 2189.
[125] A.M. Hopkins, L. De Laporte, F. Tortelli, E. Spedden, C. Staii, T.J. Atherton, et al., Silk hydrogels as soft substrates for neural tissue engineering, Adv. Funct. Mater. 23 (41) (2013) 5140–5149.
[126] Y. Termonia, Molecular modeling of spider silk elasticity, Macromolecules 27 (25) (1994) 7378–7381.
[127] D. Balasubramanian, R. Kanwar, Molecular pathology of dityrosine cross-links in proteins: structural and functional analysis of four proteins, Mol. Cell Biochem. 234-235 (1-2) (2002) 27–38.
[128] D.J. Raven, C. Earland, M. Little, Occurrence of dityrosine in Tussah silk fibroin and keratin, Biochim. Biophys. Acta 251 (1) (1971) 96–99.
[129] C.S. Wang, N.N. Ashton, R.B. Weiss, R.J. Stewart, Peroxinectin catalyzed dityrosine crosslinking in the adhesive underwater silk of a casemaker caddisfly larvae, Hysperophylax occidentalis, Insect Biochem. Mol. Biol. 54 (2014) 69–79.
[130] U.J. Kim, J. Park, C. Li, H.J. Jin, R. Valluzzi, D.L. Kaplan, Structure and properties of silk hydrogels, Biomacromolecules 5 (3) (2004) 786–792.
[131] N. Kojic, M.J. Panzer, G.G. Leisk, W.K. Raja, M. Kojic, D.L. Kaplan, Ion electrodiffusion governs silk electrogelation, Soft Matter 8 (26) (2012) 2897–2905.

[132] Y. Lin, X. Xia, K. Shang, R. Elia, W. Huang, P. Cebe, et al., Tuning chemical and physical cross-links in silk electrogels for morphological analysis and mechanical reinforcement, Biomacromolecules 14 (8) (2013) 2629–2635.
[133] A. Matsumoto, J. Chen, A.L. Collette, U.J. Kim, G.H. Altman, P. Cebe, et al., Mechanisms of silk fibroin sol-gel transitions, J. Phys. Chem. B 110 (43) (2006) 21630–21638.
[134] X.Q. Wang, J.A. Kluge, G.G. Leisk, D.L. Kaplan, Sonication-induced gelation of silk fibroin for cell encapsulation, Biomaterials 29 (8) (2008) 1054–1064.
[135] T. Yucel, P. Cebe, D.L. Kaplan, Vortex-induced injectable silk fibroin hydrogels, Biophys. J. 97 (7) (2009) 2044–2050.
[136] B.P. Partlow, C.W. Hanna, J. Rnjak-Kovacina, J.E. Moreau, M.B. Applegate, K.A. Burke, et al., Highly tunable elastomeric silk biomaterials, Adv. Funct. Mater. 24 (29) (2014) 4615–4624.
[137] K. Numata, N. Ifuku, H. Masunaga, T. Hikima, T. Sakai, Silk resin with hydrated dual chemical-physical cross-links achieves high strength and toughness, Biomacromolecules 18 (6) (2017) 1937–1946.

Questions for this chapter

1. Draw the chemical structures of the polar uncharged amino acids.
2. What is the secondary structure of a protein?
3. Why are hierarchical structures of proteins important?
4. How can we characterize the crystal structures of protein? Which methodology provides what kinds of structural information?
5. How can we evaluate the amorphous state of protein/peptide? Please explain the methodology and the information we can obtain by each method.

CHAPTER 4

Physical properties

The relationship between the structure and property of structural proteins has been discussed and been one of the long-standing scientific subjects for polymer scientists, material researchers, chemists, physicists, and biologists. In particular, spider dragline silks have been studied widely due to their outstanding mechanical properties. In the case of silk material, there have been various discussions regarding the extent to which the amino acid sequences affect the physical properties of different silk types. On the one hand, there is clearly a relationship between certain sequence motifs and the ability of the fibroin polypeptides to adopt the typical structure consisting of stacked beta-sheets and amorphous regions characteristic of the semicrystalline silk polymer [1,2]. Both GAGAGS and AAAAAA motifs can adopt stacked beta-sheet conformations to constitute the main crystalline component of the fibers, which lead to predictable changes in material properties, such as differences in thermal stability and the packing of the beta-sheets [3,4]. Certain combinations of amino acid sequences are considered to express specific functions under biological conditions [5]. Furthermore, the genes encoding fibroins are clearly under selective pressure to preserve specific sequence features, as seen in the fact that certain combinations of amino acid motifs appear tandemly in the sequences of silk proteins from divergent taxa [6]. On the other hand, ample empirical evidence has shown that extrinsic parameters can also significantly impact the material properties of silk fibers [7,8], even though further investigations are needed to resolve this issue. In this chapter, the current understanding of the physical properties of structural proteins is summarized and introduced.

4.1 Mechanical property

4.1.1 Stress–strain curve

Mechanical properties are the most important properties for structural proteins because structural proteins are expected to contribute to the mechanical properties of tissues, organs, or material. Here, strength, modulus, elasticity, toughness, and resilience are introduced as representative

Biopolymer Science for Proteins and Peptides
DOI: https://doi.org/10.1016/B978-0-12-820555-6.00006-9

mechanical properties. Those mechanical properties are evaluated by tensile, bending, or compression tests. Fig. 4.1 shows the typical stress–strain curves of a polymeric material and a silk fiber. Based on the stress–strain curve, much mechanical information can be calculated and obtained. The deformation process of structural proteins, as well as polymeric materials, includes reversible and irreversible deformations. This plastic deformation occurs when the material is stretched. When the yield point is reached, the polymer chains are more mobile and viscoelastic. Further stretching causes strain softening and subsequently strain hardening, which is often

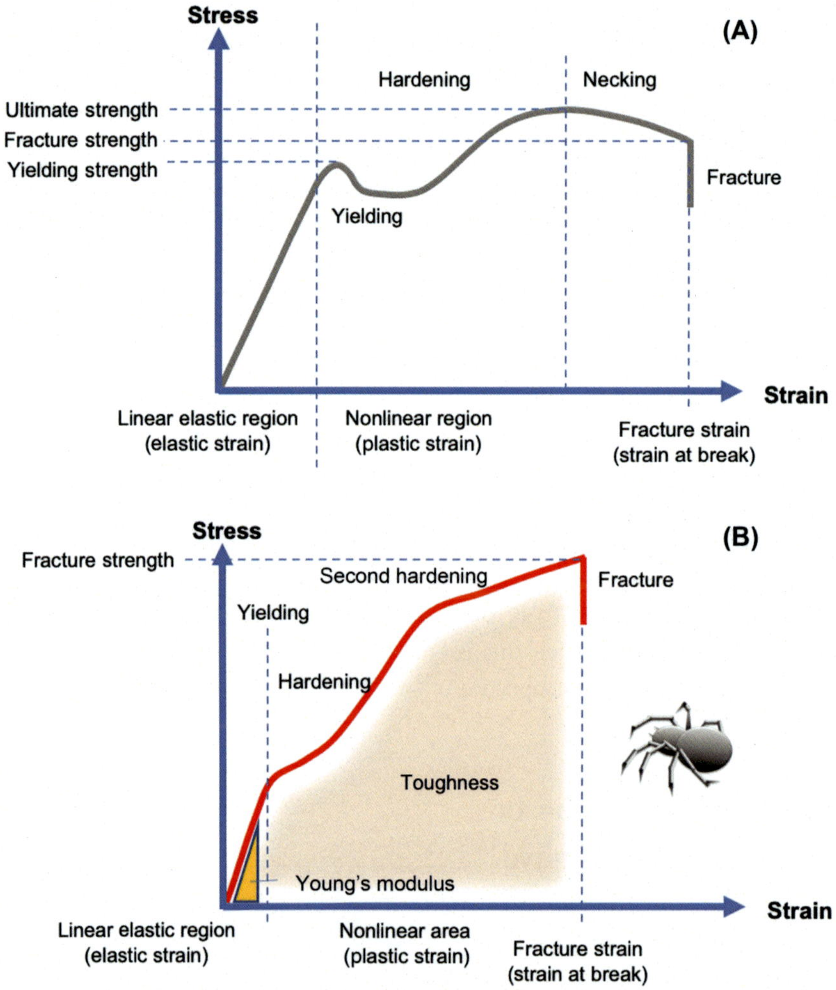

Figure 4.1 Typical stress–strain curves of (A) polymeric material and (B) a silk fiber.

the case with synthetic polymers such as plastics. Necking is a phenomenon in which deformation-mediated molecular diffusion starts after the yielding point. The deformed sample becomes smaller in the transverse direction, which is a direction perpendicular to the stretching direction, resulting in a neck formation. The molecules flow and the outer shape changes, resulting in shoulders being formed and being visible. Necking may occur after the yield point, as the reduced area gives the neck the largest local stress compared with other areas. The neck eventually becomes a fracture after excess strain is applied. According to the necking molecular mechanism of polymeric materials, necking is associated with the plastic deformation yielding. However, necking has not been studied and discussed in structural proteins, which might be due to their excess intermolecular hydrogen bonds. In the case of natural rubber, strain-induced crystallization enhances its strength during stretching deformation. Similar to natural rubber, spider silk shows the increase in strength together with the elongation, namely, stretching deformation.

In general, the ductility of polymeric materials depends strongly on how well the material is deformed largely before it breaks. Tough polymeric materials improve toughness by dissipating a significant amount of energy during the yield process to failure. Brittle polymeric materials, on the other hand, are broken without the molecules exhibiting fluidity. The fracture energy defined by the area between the curve and the axis of elongation (the *x*-axis) in the stress–strain curve is the most suitable physical property value representing the ductility of the material. It is important to note that to understand the mechanical properties of polymer materials, these physical property values should not be treated individually but should be considered comprehensively based on the stress–strain curve.

The tensile deformation tests of various structural proteins such as spider dragline, viscid silk, silkworm silk, collagen, resilin, and elastin are shown below with conventional polymer materials (Fig. 4.2A). Kevlar shows the highest strength with a low fracture strain, whereas spider dragline shows relatively higher fracture strength and strain, resulting in its excellent toughness. Viscid silk, which is a capture silk in orb webs, can stretch and absorb energy so that it can be used to make strong and optically characteristic spider webs [10,11]. Viscid capture silk is predominantly amorphous and is not converted to a crystalline state even when it is stretched by 500%.

Fig. 4.2B shows the stress–strain curves of silkworm silk fibers [9]. The results of tensile deformation tests showing the ultimate tensile strength, extensibility (elongation to break), Young's modulus, and toughness (energy

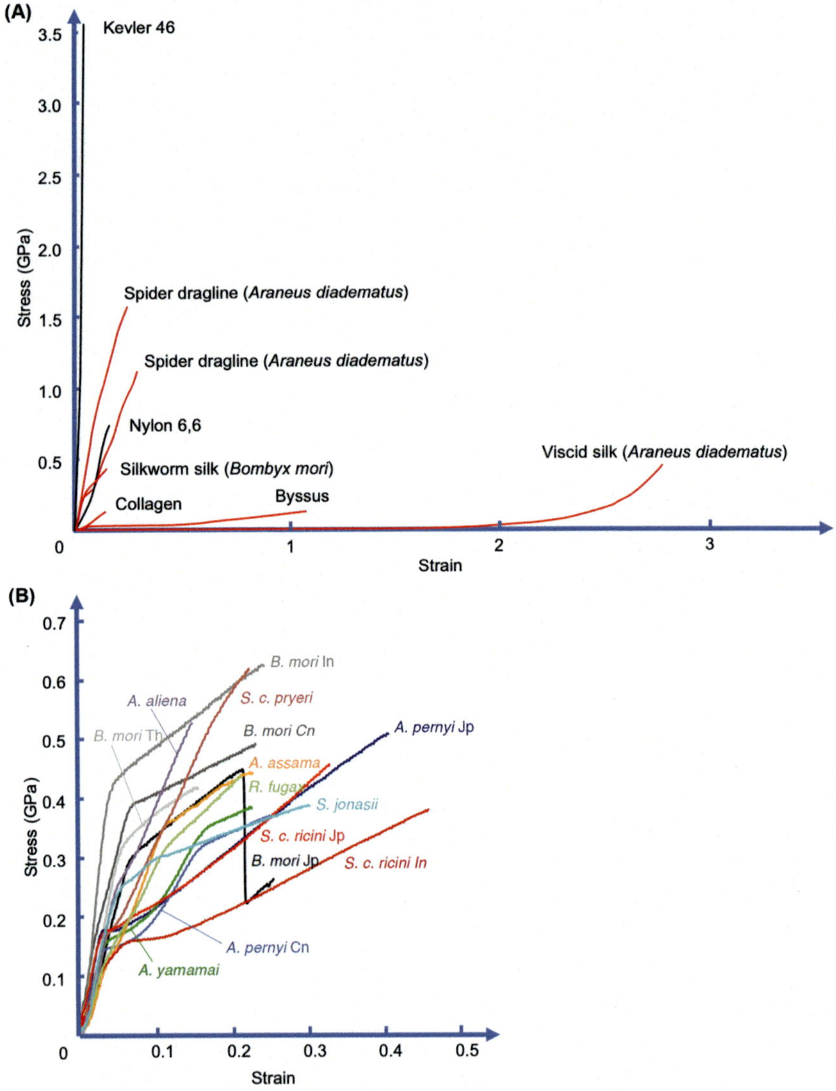

Figure 4.2 Examples of stress–strain curves of (A) various polymer materials and (B) different species of silkworm silk fibers [9].

to break) of individual silk fibers are summarized in Table 4.1. Even the same silkworm silks demonstrate different mechanical properties because all silks are biological samples and show some natural variations. The overall results fall within the range of expected values, although with an extensive data spread and a generally low level of reproducibility between fibers from

Table 4.1 Summary of mechanical properties of various structural proteins and controls.

Polymer material	Origin	Modulus (GPa)	Strength (GPa)	Extensibility	Toughness ($MJ \cdot m^{-3}$)	Resilience (%)
Elastin	Bovine ligament [12]	0.0011	0.002	1.5	1.6	90
Resilin	Dragonfly tendon [13]	0.002	0.004	1.9	4	92
Collagen	Mammalian tendon [14,15]	1.2	0.12	0.13	6	90
Mussel byssus	*Mytilus californianus* [16]	0.87	0.075	1.09	45	28
Dragline silk	*Araneus diadematus* [17]	10	1.1	0.3	160	35
Viscid silk	*A. diadematus* [17]	0.003	0.5	2.7	150	35
Silkworm silk	*Bombyx mori* (Japan)	5.13	0.40	0.268	0.071	–
	B. mori (China)	7.23	0.46	0.241	0.081	–
	B. mori (Thailand)	6.66	0.55	0.232	0.091	–
	B. mori (India)	8.61	0.57	0.245	0.103	–
	Antheraea pernyi	4.72	0.34	0.267	0.060	–
	Antheraea yamamai	4.58	0.39	0.356	0.092	–
	Antheraea assama	4.31	0.36	0.292	0.068	–
	S. c. pryeri	4.71	0.53	0.241	0.078	–
	S. c. ricini	4.61	0.38	0.339	0.078	–
	Rhodinia fugax	4.09	0.43	0.223	0.058	–
	Actias aliena	4.69	0.39	0.149	0.030	–
	Saturnia jonasii	7.10	0.37	0.196	0.050	–
Kevlar	Nonprotein	130	3.6	0.027	50	–
Carbon fiber	Nonprotein	300	4	0.013	25	–

Sources: (Overall) J.M. Gosline, P.A. Guerette, C.S. Ortlepp, K.N. Savage, The mechanical design of spider silks: from fibroin sequence to mechanical function, J. Exp. Biol. 202 (Pt 23) (1999) 3295–3303 [18]; (silkworm silk) A.D. Malay, R. Sato, K. Yazawa, H. Watanabe, N. Ifuku, H. Masunaga, et al., Relationships between physical properties and sequence in silkworm silks, Sci. Rep. 6 (2016) 27573 [9].

the same cocoon, as previously noted [19–23]. Specific trends were recognized, for example, the higher initial elastic modulus observed for Indian *Bombyx mori* silk (with a mean value of 8.6 GPa) than the other *B. mori* samples, which may reflect differences in the local rearing environments. However, taken as a whole, no obvious correlations could be made between phylogeny and either the tensile strength or elastic modulus of the different samples, with mean values falling between 0.34–0.57 GPa and 4–8.6 GPa, respectively.

On the other hand, in terms of extensibility, the samples could be classified into two groups. Those exhibiting higher strain to break included *Bombyx mori*, *Antheraea*, and *Samia* silks, with an overall mean range of 23%–36%, whereas *Actias aliena*, *Rhodinia fugax*, and *Saturnia jonasii*, displayed low extensibility values, ranging on average from 14% to 22%. As a consequence of the differences in extensibility, the toughness were also divergent between the two groups ranging from 0.06 to 0.1 $GJ \cdot m^{-3}$ for the high extensibility silks and 0.03–0.06 $GJ \cdot m^{-3}$ for the low extensibility silks.

During extension deformation, the different crystallization behaviors of different structural proteins are closely related to their physical properties and natural roles [24]. Spider dragline silk undergoes significant structural changes during stretching deformation, particularly during the initial phase, which resulted in it exhibiting the highest strength among the silk fibers tested. The effects of the reeling and stretching speeds on the structure and mechanical properties of spider dragline silk have been studied [25,26]. The mechanism underlying the significant increase noticed in the toughness of spider dragline silk at high extension rates has been investigated by many groups; however, the molecular mechanism of toughness has remained a challenging issue. Spider capture silk was confirmed to be entirely amorphous before and during the deformation [24]; this is consistent with the requirement that spider webs must maintain their optical properties and elasticity when under physical attack and following environmental deformation by the elements.

4.1.2 Effects of humidity and water

Water molecules function as plasticizers and/or nuclear agents for protein/polypeptide-based materials. Here is the effect of RH on the mechanical properties of silk films. The stress–strain curves can be obtained from the tensile tests (Fig. 4.3A) [27]. According to the

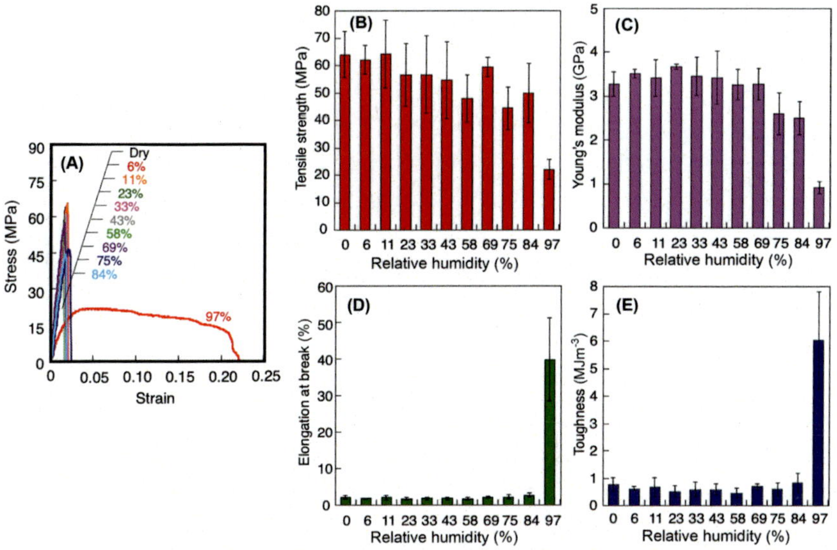

Figure 4.3 Stress–strain curves of silk films incubated at different RHs from RH 0% (dry) to RH 97% (A) (same as Fig. 3.12). Effects of humidity on mechanical properties: (B) tensile strength, (C) Young's modulus, (D) elongation at break, and (E) toughness of silk films incubated at different RHs from RH 0% (dry) to RH 97% [27].

stress–strain curves, four types of mechanical properties, namely, tensile strength, Young's modulus, elongation at break, and toughness, were determined (Fig. 4.3B–E) [27]. The mechanical properties of the silk films did not change significantly up to RH 84%, whereas the mechanical properties changed drastically in the case of films incubated at RH 97%. The elongation at break and toughness of the silk films incubated at RH 97% increased 15- and 7-fold higher than those incubated at RH 84%. This drastic change in mechanical properties is because larger crystallites were formed in very humid conditions. Wide-angle X ray scattering (WAXS) characterization clarified that the silk films prepared at RH 97% demonstrated high crystallinity and elongation at break [27]. Even though there is generally a trade-off relationship between crystallinity and ductility in material science, the silk film at RH 97% can realize both the properties simultaneously. These simultaneous changes are mainly because of the plasticizing effect of water molecules on amorphous and crystalline sequences of silk molecules; namely, amorphous regions of silk films were plasticized, increasing the elasticity of silk films. At the same time, the beta-sheet formation was also induced by the presence of water molecules at RH 97%.

The relationship between different strain rates and RHs in terms of mechanical properties was also investigated [26]. The stress–strain curves were obtained with the strain rates ranged from $3.3 \times 10^{-5}\ s^{-1}$ to $3.3\ s^{-1}$ at RHs of 0%, 75%, and 97% (Fig. 4.4). The tensile strength and elongation at break tended to increase with an increase in strain rate. The tensile strength, elongation at break, and Young's modulus increased linearly as the strain rate increased, independent of the experimental RH conditions. The tensile strength increased about 1.4-fold, 2.3-fold, and 1.5-fold as the strain rate increased from $3.3 \times 10^{-5}\ s^{-1}$ to $3.3\ s^{-1}$ at RH 0%, RH 43%, and RH 97%, respectively. The RH effects on the mechanical properties were also recognized; namely, the tensile strength decreased as the RH increased [26]. The elongation at break increased as the RH increased. Regardless of the applied strain rates, the elongation at break at RH 97% was approximately 2-fold higher than that at RH 0%. The Young's modulus linearly decreased as the RH increased. As described in the previous paragraph, the decrease in Young's modulus originated from the plasticization effect of water molecules at high RH conditions [28–32]. The toughness of the dragline silk was not significantly changed by the RH conditions at strain rates ranging from $3.3 \times 10^{-5}\ s^{-1}$ to $3.3 \times 10^{-1}\ s^{-1}$. At a strain rate of $3.3\ s^{-1}$, the toughness was highest at RH 43%, followed by the toughness at RH 75%. Overall, in dry condition RH 0%, the toughness at high deformation rates originated from the increase in yield stress. In contrast, the dragline at RH 75% and RH 97% exhibited rubber-like behavior because of the increase in rubbery components induced by water plasticization.

In addition to the physical properties, the state of the water molecules in a silk hydrogel affects cell viability; human cell lines and cell-adhesion

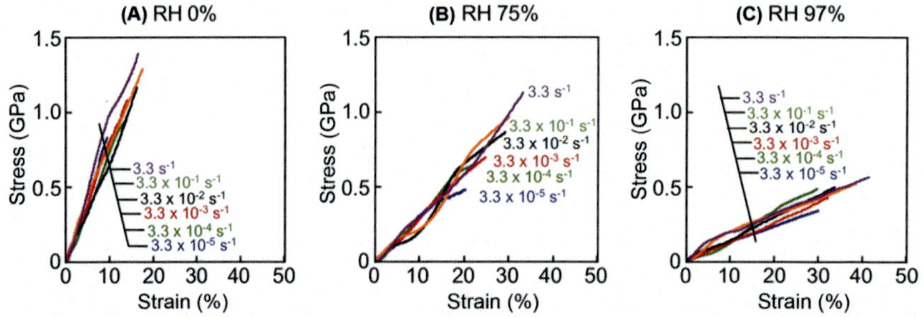

Figure 4.4 Stress–strain curves of spider dragline silk fibers at (A) RH 0%, (B) RH 75%, and (C) RH 97% under different strain rates ranging from $3.3 \times 10^{-5}\ s^{-1}$ to $3.3\ s^{-1}$ [26].

proteins in the extracellular matrix preferentially expand and adhere to a substrate containing more bound water [33]. Cebe et al. have found that hot-water vapor annealing induces the crystallization of silk films and that water processing can be replaced by conventional methods using organic solvents [34].

4.1.3 Poisson's ratio

Poisson's ratio is an important indicator for discussing the mechanical properties of a material and is a value used to characterize the deformation of a material based on its expansion or contraction in the direction perpendicular to the direction of the load. The value of Poisson's ratio is the negative of the ratio of lateral strain to axial strain. Poisson's ratio values for an isotropic material range from 0.0 to 0.5. Poisson's ratio of incompressible materials such as natural rubber is considered to be 0.5.

The dragline silk from *Trichonephila clavipes* (formerly known as *Nephila clavipes*) was reported to have a Poisson's ratio of 1.52, which is remarkably high and indicates that the silk is an anisotropic and elastic material [35]. Besides, the dragline silk of *Trichonephila edulis* was reported to have a moderate Poisson's ratio of approximately 0.37 (Table 4.2) [37]. Thus the Poisson's ratios of spider dragline are an interesting subject, but have not been studied in detail. According to a previous report [36], the average Poisson's ratio of *Trichonephila clavata* dragline was determined to be 0.26 based on the scanning electron microscopy (SEM) observation of the dragline fibers at 20% strain. In contrast, the average Poisson's ratio of the microfibrils

Table 4.2 Poisson's ratios of silks and other materials.

Material	Poisson's ratio
Spider dragline (*Trichonephila clavata*)	0.2–0.3 (average: 0.26) [36]
Nanofibrils in spider dragline (*T. clavata*)	1.9–3.7 (average: 2.9) [36]
Spider dragline (*Trichonephila edulis*)	0.37–0.38 [37]
Spider major ampullate silk (*Trichonephila clavipes*)	0–1.52 [35]
Spider minor ampullate silk (*T. clavipes*)	0–1.50 [35]
Collagen fibril	2.1 ± 0.7 [38]
Cartilage: matrix of primarily collagen (type II) and proteoglycans	0.1–0.4 [39]
Natural rubber	0.4999 [40]
Metamaterials	Approximately 4.5 (the maximum value) [41]

(components of dragline) exceeded 0.5, which is not a typical value for an isotropic material. Some reports have indicated that spring-like zigzag structures in metamaterials and helical structures in collagen can produce atypical Poisson's ratios [38,41]. The relationship between the microfibrils and dragline silk in Poisson's ratio is not clear on the basis of the current understandings.

4.1.4 Deformation rate effect of spider dragline silk

Spider dragline silk is a unique biological and structural material noted for its excellent toughness, ductility, and strength [10,42]. The dragline silk fiber is used as a major component in web construction, and thus critical for prey capture, and also functions as a lifeline for the spider [18]. The effect of strain rate on the mechanical properties of dragline silk fibers has been investigated by several groups [17,18,43,44]. Denny characterized the dragline silk derived from *Araneus sericatus* and found that the toughness of the silk increased 3-fold when the strain rate varied from $0.0005\ s^{-1}$ to $0.024\ s^{-1}$, with concomitant increase in tensile strength and Young's modulus, under RH conditions of 48%–52% [17]. In contrast, Cunniff et al. analyzed the dragline silk derived from *T. clavipes* and found that increasing the strain rate from $0.1\ s^{-1}$ to $3000\ s^{-1}$ did not influence the toughness, Young's modulus, or elongation at break, at constant RH 50% [43]. In addition, Gosline et al. found that the toughness of the dragline silk from *Araneus diadematus* increased 10-fold upon varying the strain rate from $0.0005\ s^{-1}$ to $30\ s^{-1}$ (the experimental RH condition was not reported) [18]. More recently, Hudspeth et al. studied the dragline silk of *T. clavipes* and reported that the toughness increased 3-fold by varying the strain rate from $0.001\ s^{-1}$ to $1700\ s^{-1}$ at RH 34% [44].

The effects of strain rate on mechanical properties of spider dragline silk have been studied as described in the previous paragraph. Here, the effect of humidity on the strain rate-dependent mechanical parameters is discussed due to the sensitivity of spider dragline in its physical properties [28–30,32,45–50]. To clear the relationships between deformation rate and humidity on the mechanical and nanoscale structural properties of spider dragline, simultaneous WAXS and tensile tests under different strain rates and RH conditions were performed [26]. Tensile tests of dragline silk fibers were conducted at RH 43% under different strain rates from $3.3 \times 10^{-5}\ s^{-1}$ to $3.3\ s^{-1}$. The resultant stress–strain curves are shown in Fig. 4.5 [26]. The tensile strength and elongation at break were dependent

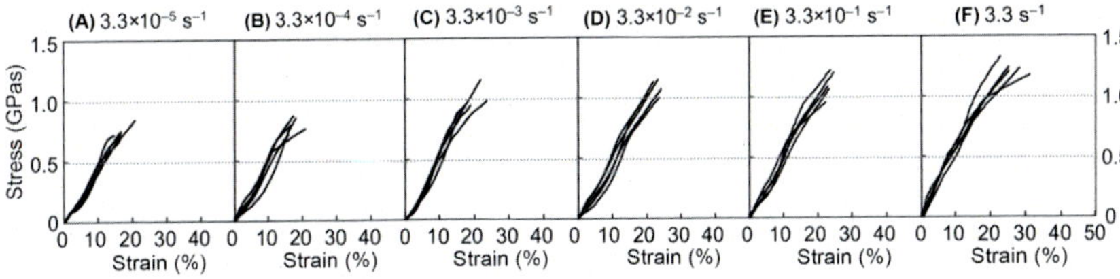

Figure 4.5 Stress–strain curves of spider dragline silk fibers at RH 43% under different strain rates: (A) 3.3×10^{-5} s^{-1}, (B) 3.3×10^{-4} s^{-1}, (C) 3.3×10^{-3} s^{-1}, (D) 3.3×10^{-2} s^{-1}, (E) 3.3×10^{-1} s^{-1}, and (F) 3.3 s^{-1} [26].

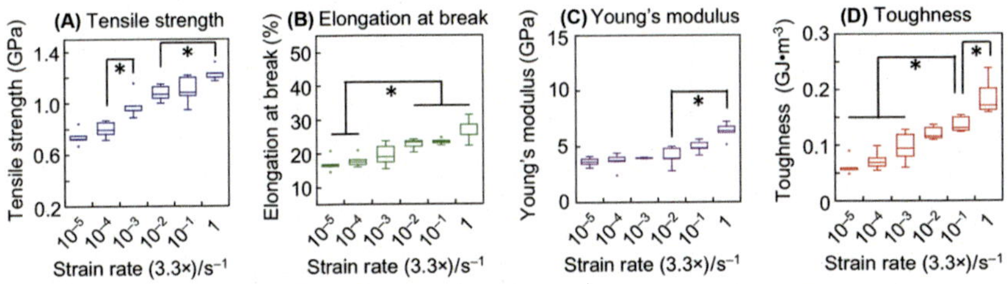

Figure 4.6 Mechanical properties of dragline silk fibers at RH 43% under strain rates ranging from 3.3×10^{-5} s^{-1} to 3.3 s^{-1}: (A) tensile strength, (B) elongation at break, (C) Young's modulus, and (D) toughness. *Significant differences between groups at $P < 0.05$ [26].

on the strain rates, and also increased linearly with increasing strain rate. The mechanical properties of the dragline silk fiber at RH 43% under different strain rates are displayed in Fig. 4.6 [26]. The tensile strength and elongation at break exhibited a 1.7-fold and 1.6-fold increase as the strain rate increased from 3.3×10^{-5} s^{-1} to 3.3 s^{-1}, respectively. The increase in tensile strength could be attributed to the more crystalline beta-sheet orientation at the higher strain rate, contributing to higher Young's modulus [44]. The increase in elongation at break at the higher strain rate indicates that the tensile force could be applied more uniformly to both crystalline beta-sheet and amorphous region of dragline silk fibers [18,44]. As a result, the toughness exhibited a 3-fold increase as the strain rate increased from 3.3×10^{-5} s^{-1} to 3.3 s^{-1}. The increase in toughness was in agreement with the previous study that found a 3-fold increase in toughness when varying the strain rate from 1×10^{-3} s^{-1} to 1.7×10^{3} s^{-1} [44].

The effect of deformation rate on the fracture behavior of silk fibers has been studied partially. The fracture surfaces of the dragline silk fibers after tensile tests at RH 43% were observed via SEM [26]. The fracture surfaces of the dragline silk fibers were relatively smooth in the case of strain rates ranging from $3.3 \times 10^{-5}\ s^{-1}$ to $3.3 \times 10^{-3}\ s^{-1}$, whereas a relatively rough fracture surface was observed at over $3.3 \times 10^{-2}\ s^{-1}$ of the strain rates. The dragline silk consists of a hierarchically arranged protein core surrounded by outer skin layers [51]. The protein core is composed of fibrils formed by a bundle-like assembly of microfibrils [36,52]. The microfibrils (diameter less than 100 nm) form one hierarchical level and could break when the dragline silk fiber is fractured by stretching deformation. Accordingly, a reasonable explanation of the difference in the fracture surface is that the dragline silk fibers could break at more macroscopic structural defects rather than microfibrils at the slower strain rates, while breaking at microfibrils at the faster strain rates.

Spider dragline is used as a spider's lifeline to prevent unexpected falling and to capture prey, such as flying insects. Strain rates greater than $1\ s^{-1}$ are considered to be applied to the dragline silk fibers in the natural state [26]. This natural range of strain rates is identical to the range of strain rates shown in Fig. 4.5, wherein the toughness of the dragline silk at RH 43% and RH 75% was higher than that at RH 0% and RH 97%. The effect of RH on toughness at a strain rate of $3.3\ s^{-1}$ implies that the spider dragline might evolve to exhibit toughness at a moderate RH, that is, a naturally humid environment and the relatively high deformation rates that occur in natural spider behaviors.

4.2 Supercontraction

Water molecules influence all structural proteins. In the natural state, the dragline fibers need to function under a wide range of conditions; for instance, RH varies depending on the environmental conditions, which can have a considerable effect on performance [49,53]. Spider dragline silk shows particular water sensitivity, due to its unique character, supercontraction (Fig. 4.7) [55,56]. The spider dragline silk absorbs water at high humidity, altering its material properties and shrinking with an increase in overall volume. This process generates substantial stress in spider dragline silk when it is restrained and the potential to perform work. Supercontraction is considered to provide a mechanism that tensions a spider web as it become loaded with dew or rain, according to previous literature [57,58]. However, the molecular mechanics of supercontraction itself has been studied but not fully understood yet [31,59,60]. It seems that diglutamine and proline play key roles for

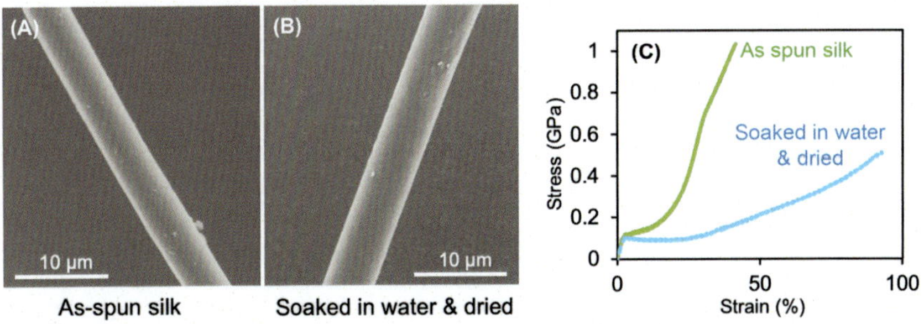

Figure 4.7 Supercontraction of *Trichonephila clavata* spider dragline. SEM images of the spider draglines before (A) and after the immersion in water (B). (C) Stress–strain curves of the spider draglines before and after the immersion in water [54].

supercontraction, and also the noncrystalline region is mainly plasticized by water molecules [9,27].

4.3 Thermal property

Thermal properties of proteins are studied in terms of glass transition (T_g), melting (T_m), and degradation temperatures. In addition, proteins are easily associated with water molecules, resulting in water-related multiple thermal transitions. To remove the effects of water, the author prefers to use perfectly dried samples. One of the common methods to characterize the thermal properties of polymeric materials, including protein polymers, is differential scanning calorimetry (DSC) (Fig. 4.8).

4.3.1 Glass transition

Water molecules are known to influence the thermal properties of proteins, including silk [33,34,46,61–71]. As described in Section 3.5.1, water molecules in protein-based materials including silk materials can primarily be categorized into two types: free water and bound water. Free water is unbound water that behaves like bulk water, whereas bound water strongly interacts with protein molecules and demonstrates different characters from bulk water [33,72–74].

To date, the effects of bound water on the thermal, mechanical, and biological properties of silks have been widely studied [33,34,46,61–71]. Asakura et al. have found that the hydration of *B. mori* silk fibroin results in the stabilization of silk I form by solid-state nuclear magnetic resonance [69–71], and the crystalline fraction of *B. mori* silk fibers does not change

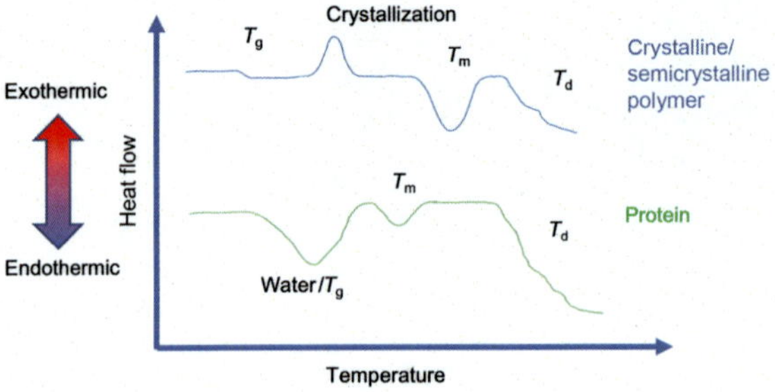

Figure 4.8 Typical DSC profiles of a crystalline polymer and a structural protein polymer like silk.

after hydration [75]. Cebe et al. and Shao et al. have reported that bound water affects the T_g of *B. mori* silk films, and T_g decreases as the water content of the silk films increases [63,66,76]. The storage modulus and loss tangent of *B. mori* silk and *T. edulis* dragline [62,68] and the elastic modulus of *Antheraea pernyi* silk [46] and *Argiope trifasciata* spider silk [67] are also influenced by bound water, since bound water plays a role in disrupting the hydrogen bonds between amorphous silk molecules, which enhances the mobility of the silk molecules and contributes to glass transitions [61–63,66,76]. In addition to the T_g associated with the transition from the solid-state to the liquid-state in silks, a T_g derived from the removal of water molecules has been detected in the thermal analysis of films of *B. mori* silk fibroin [63,76] and spider silk-like proteins, which primarily consist of genetic variants of MaSp1 from the dragline of the spider *T. clavipes* [64,65].

4.3.2 Transition and degradation

In addition to glass transition, there are multiple thermal transitions in protein-based materials. Here is an example of the thermal properties of silkworm silks [9]. The cocoon materials were subjected to thermogravimetric analyses (TGAs), where changes in sample mass were measured as a function of a temperature gradient up to 500°C (Fig. 4.9). The saturniid silks show higher thermal stability than *B. mori*, as seen in the values for the thermal decomposition peaks (T_d), in agreement with previous reports [77–79]. An initial weight loss is detected below 100°C, which corresponds to the evaporation of adsorbed water. The exception is *R. fugax*, which produced a water evaporation peak at around 105°C, presumably due to the dense packing of fibers in the cocoon, impeding the loss of

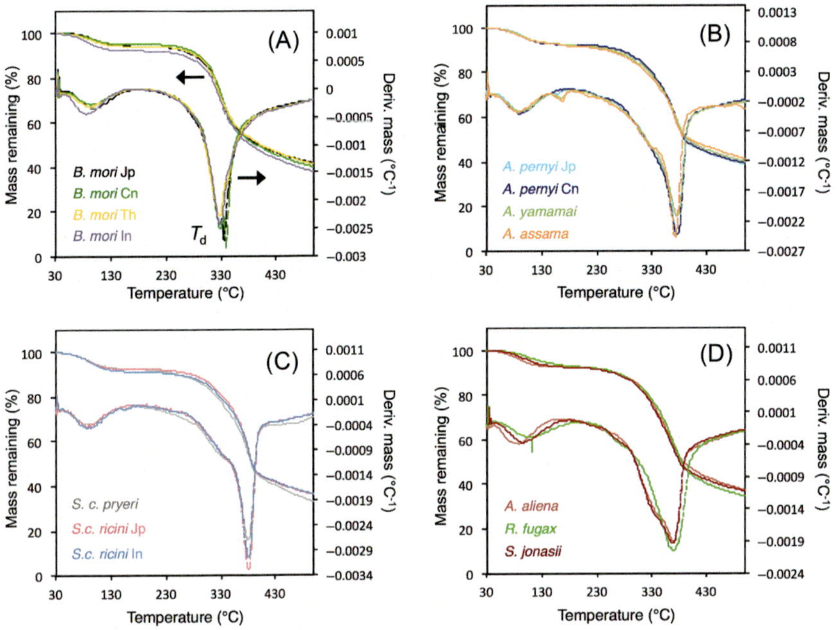

Figure 4.9 Thermogravimetric analyses of the different cocoon samples. Each panel combines the percent mass loss data (TGA: top curve, left axis) and the first derivative plots of the percent mass remaining (DTG: bottom curve, right axis) versus temperature. Temperatures corresponding to peaks or transitions are indicated. (A) *Bombyx mori*; (B) *Antheraea*; (C) *Samia*; (D) *A. aliena*, *R. fugax*, and *S. jonasii* [9]. *Cn*, China; *In*, India; *Jp*, Japan; *Th*, Thailand.

water molecules. In the *B. mori samples*, no significant changes are observed after the water peak until above 200°C. An abrupt decrease occurred beyond 280°C, producing a T_d peak at around 335°C in the differential TGA plots (DTG; Fig. 4.9). Approximately 40% of the initial weight remains at the end of the run at 500°C. For the saturniid silks, *Antheraea* and *Samia* samples showed qualitatively similar TGA profiles, even though the latter produced higher T_d values and had somewhat sharper overall features. In contrast to *B. mori*, the thermal degradation followed a two-step regime, producing a shoulder at approximately 330°C–340°C in the DTG plots, followed by a more drastic decrease in mass. A similar multistep profile has been detected for regenerated film derived from degummed *A. pernyi* fibers [77]. A small but distinct peak was sometimes observed at approximately 160°C–170°C in the derivative plots, presumably corresponding to the degradation of calcium oxalate crystals [80,81]. The wild silk samples from *A. aliena*, *R. fugax*, and *S. jonasii* show broader transitions and generally less distinct features than the other saturniid silks.

Thermal properties of silkworm silks were reported based on their DSC profiles (Fig. 4.10) [9]. Consistent with the TGA data (Fig. 4.9), the saturniid silks showed more highly defined profiles with a greater number of transitions than *B. mori*. In *B. mori* samples, aside from the T_w and T_d peaks, an indistinct endothermic peak was seen at approximately 230°C (denoted as T_{en1}). The saturniid silks, in contrast, typically produced two small, distinct endotherms (T_{en1} and T_{en2}), which have been attributed to molecular motion within the amorphous or laterally ordered regions of fibroin [82]. Another small endotherm sometimes appeared between 160°C and 165°C, attributed to the decomposition of calcium oxalate crystals (T_{co}) [80]. *A. aliena*, *R. fugax*, and *S. jonasii* displayed less defined peaks than the other saturniid samples, possibly reflecting a reduction in concerted molecular motions during heating.

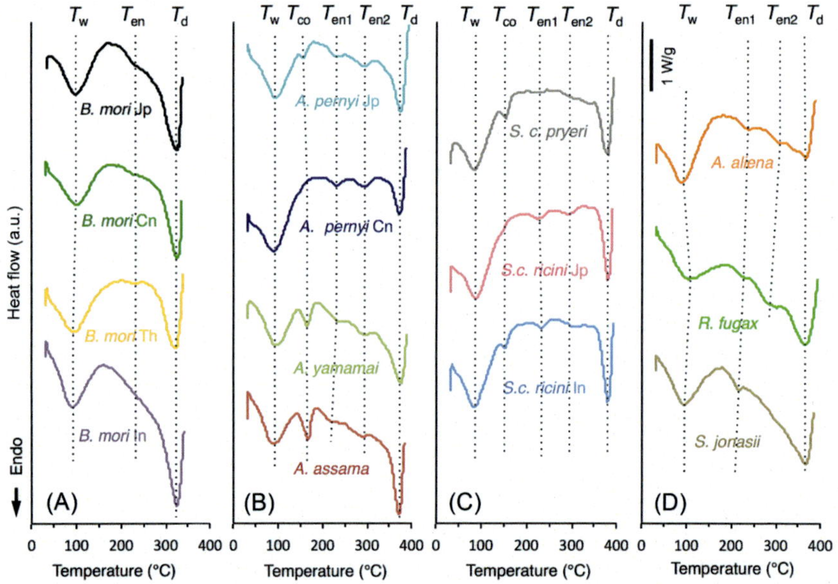

Figure 4.10 DSC measurements of the different cocoon samples. Temperatures corresponding to the peaks or transitions are indicated: T_{co}, peak attributed to the degradation of calcium oxalate crystals; T_d, thermal degradation peak; $T_{en(1,2)}$, endothermic peaks; T_w, water evaporation peak. (A) *Bombyx mori*; (B) *Antheraea*; (C) *Samia*; (D) *A. aliena*, *R. fugax*, and *S. jonasii* [9]. *Cn*, China; *In*, India; *Jp*, Japan; *Th*, Thailand.

4.3.3 Effects of water molecules

Water molecules affect the thermal properties of protein materials, due to the amide bonds of their chemical structures. The effects of RH and temperature on the retardation of silk films were studied previously [27]. Thermal treatment induces the beta-sheet formation of silk proteins, whereas the water molecules induced the formation of the helical structure as well as beta-sheet structure. According to the literature related to the relationship between water and silk molecules [62,66,67], bound water disrupts the intermolecular cohesive forces between protein chains; this results in a reduced steric hindrance that promotes chain movement in the noncrystalline regions and induces beta-sheet crystallization in the silks. As explained in Section 4.1.2, a reasonable explanation for the results of this study is that water plasticization promotes beta-sheet formation via helix–helix interactions, as suggested by the literature on prion proteins [83]. In the case of silk molecules, either the longer poly(Ala) sequence or the Gly-rich sequence forms helical structures, resulting in helix–helix interactions to initiate the intermolecular interactions. The helix–helix interaction would accelerate the assembly of poly(Ala) sequence, yielding the beta-sheet structure. Thus water molecules acted as a plasticizer and promoted the beta-sheet formations via helix–helix interactions. Conversely, the thermal treatment removed water molecules from the beta-strands and amorphous regions of silk films and then induced the formation of the beta-sheet structures [27].

TGA of the silk films incubated at different RHs showed a two-step weight loss (Fig. 4.11A); namely, the first weight decrease was due to the water removal (evaporation) and was dependent on the RH. The water content of the silk samples depended on the weight decrease in the first step (Fig. 4.11B). The water content of the films increased with an incrase in humidity. The highest water content was 12.6 ± 1.6 wt.% and occurred when the film was incubated under an RH of 84% [27]. In Fig. 4.1B, the dried silk films still have a negligible amount of water, meaning that in the silk films dried in a vacuum oven at 40°C for 24 h, water molecules bound to silks as reported previously by Agarwal et al. [61] The second decrease in weight, which was attributed to thermal degradation, was detected above 250°C, when the silk film turned into black char [27]. The derivative plots in Fig. 4.11C demonstrate the temperature where the water removal and degradation occurred at the fastest rates. The degradation temperature over 200°C was independent of the water content,

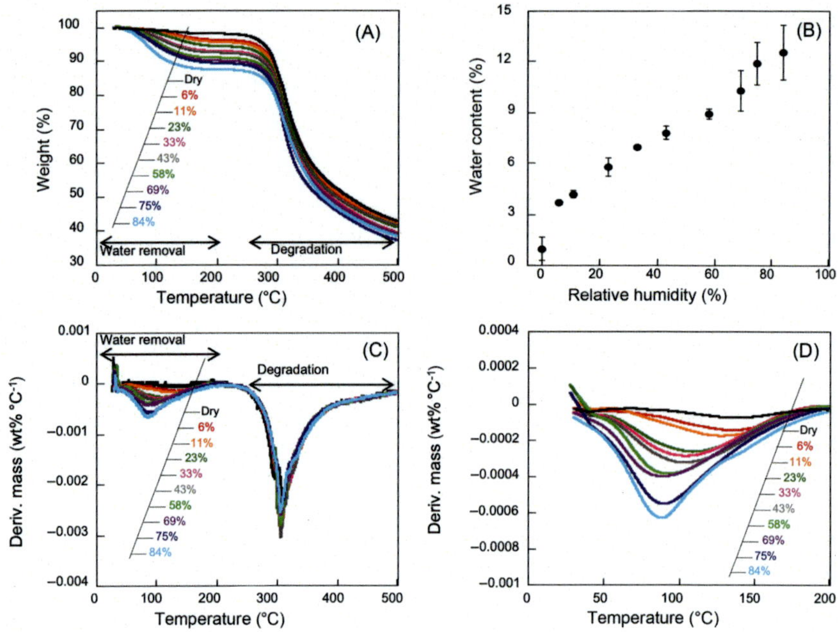

Figure 4.11 TGA of the *Bombyx mori* silk film incubated at different RHs and heated from 30°C–500°C at 20°C/min. (A) Weight change in the film. (B) Water content of the film. (C) Derivative plot of (A). (D) Magnified plot of (C) at temperatures up to 200°C. Error bars represent the standard deviation of the samples ($n = 3$) [27].

suggesting that the water molecules were completely removed at 200°C, and the breakage of the main chain in the silk molecules proceeded without water. The thermal degradation of silk films is therefore independent of their water content.

On the other hand, water removal was reported to depend on the water content of the silk films [27]. The temperature of the water removal decreased with an increase in water content (Fig. 4.11D). The decrease in the water removal temperature can be explained by the following two reasons. First, a larger amount of free water in silks at a higher RH resulted in water removal at a lower temperature. Secondly, the higher water content induced more water plasticization, particularly in the amorphous phase, resulting in higher mobility of the silk molecules and a lower T_g. Thus higher chain mobility can facilitate the more water removal process.

The surface morphology and retardation of the silk film during the heating process were characterized with the silk films incubated at RH 6% and 75% after heating to 40°C, 160°C, 180°C, and 240°C (Fig. 4.12).

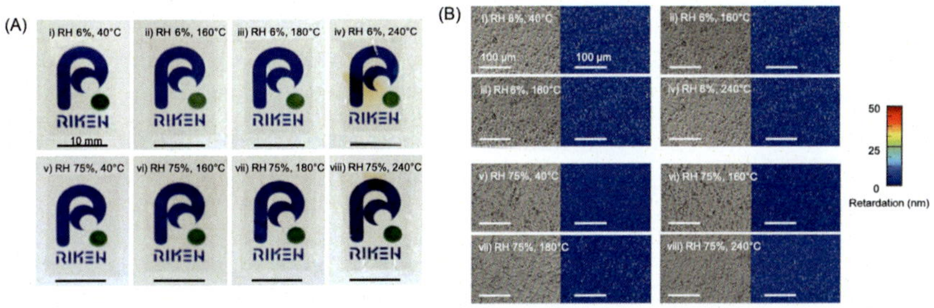

Figure 4.12 (A) Pictures and (B) optical micrographs and retardation color mapping of *Bombyx mori* silk films incubated at RH 6% (i–iv) and RH 75% (v–viii) after heating to 40°C (i, v), 160°C (ii, vi), 180°C (iii, vii), and 240°C (iv, viii). Each scale bar in (A) and (B) denotes 10 mm and 100 μm, respectively. Retardation values ranged from 0 to 50 nm, as shown in the color scale [27].

The surface conditions of the silk films at RH 6% were maintained up to 240°C, and then the colorless and transparent silk film turned brown in color at approximately 240°C due to air oxidation (Fig. 4.12Ai–iv). The color of the silk film at RH 75% did not change significantly (Fig. 4.12Av–viii), indicating that the excess water molecules in the silk film could inhibit its air oxidation. The retardation of the silk films was also maintained during heating up to 240°C at RH 75% (Fig. 4.12B), with the result that the humidity and thermal treatments did not affect the alignment of crystalline components on the microscale.

The viscosities of raw cocoon fibers and degummed fibers were studied to monitor the molecular weight changes, since the molecular weight of crystalline silk fibroins cannot be determined by other methods. The molecular weight of silk cocoon raw fibers and degummed fibers was maintained after heat treatment at 240°C, suggesting that the silk molecule is thermally stable and tunable by the material shape and the crystalline/amorphous states. Unlike the thermal degradation behaviors of silk materials, the water removal process is dependent on the water content as well as environmental humidity. The derivative plots of the water removal for silk films, raw cocoon fibers, and degummed fibers were obtained from the TGA profiles. Water molecules in the silk films evaporated at a higher temperature than raw cocoon fibers and degummed fibers, indicating the presence of more bound water in the silk films [27]. The amorphous region of a polymer material contains more bound water than the crystalline region, which was reported for the bound water in other biological materials [75,84,85]. The silk films contained more amorphous regions than both fiber samples; thus more water molecules could exist in the bound state in silk films, resulting in water removal at a higher temperature than that for the fibers. Thus the water removal behavior of silk materials is influenced by the ratio of bound water to free water, strongly related to the crystallinity of silk materials.

4.3.4 Thermal structural changes

Thermal treatment removes the water molecules from proteins, which results in beta-sheet formation and/or denaturation (melting) of proteins. Fibrillar proteins often form more beta-sheet structures by dehydration treatments. Here is an example of silk proteins in different material forms. The silk cocoon raw fibers (Fig. 4.13A–C) and degummed fibers (Fig. 4.13D–F), prepared at RH 6%, 58%, and 75%, were characterized by WAXS [27]. The WAXS results showed that the beta-sheet structure was predominant in all fiber samples before heat treatment and was independent of the water content. The thermal treatment induced additional beta-sheet structures; however, a clear transition from helical to beta-sheet

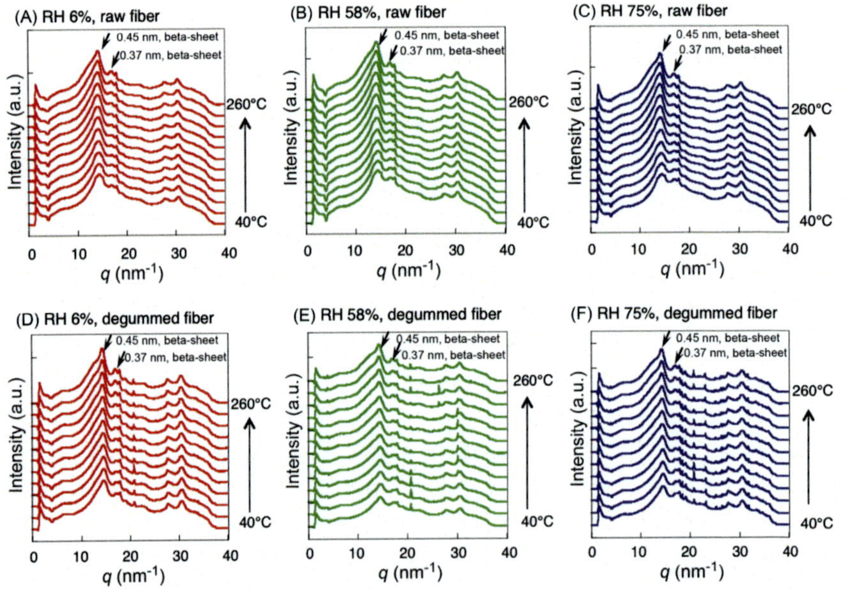

Figure 4.13 WAXS profiles of the *Bombyx mori* cocoon (A–C) raw fibers and (D–F) degummed fibers measured in 20°C intervals from 40°C to 260°C and incubated at RH 6% (A, D), RH 58% (B, E), and RH 75% (C, F) [27].

structures was not detected, which was different from the observations made with the silk films. Thus the silk fibers show more resistance to the water plasticization effect than do the silk films [27].

4.4 Rheological property

Rheology provides structural and physical information of the flow of matter, mainly, soft matters including polymer liquid and solid materials. Based on the rheological characterization with oscillatory measurements of polymer materials, we can obtain information on network structure, chain entanglements, relaxation rates, and loss/storage modulus even at the transition phase between liquid and solid states. The liquid state protein, such as native silk dope, was studied [86].

4.4.1 Native silk dope

Native silk dope shows shear-thinning; namely, its viscosity decreases with an increase in shear rate, together with flow-induced molecular orientation [87,88]. Similar to the concentrated polymer solution, silk dope demonstrates viscoelasticity, with the elastic (storage) modulus (G') exceeding the viscous (loss) modulus (G'') at high frequency and a crossover to viscous behavior ($G' < G''$) at low frequency. The first report on the rheological properties of the native silk dope from spiders and silkworms at the native concentrations was done by Holland et al. [86]. The conversion of the liquid dope to the solid fiber by spider and silkworm occurs via a liquid–liquid phase separation and also strain-induced phase separation [89]. With rheological analysis, the flow and deformation of silk proteins in the spinning forces applied to the dope can be studied to reproduce artificial silk fibers. Oscillating the dope samples over a range of frequencies allows characterization of the dope state over various conditions and periods [86]. Based on the applied stress and the resulting strain, elastic G' (storage) and viscous G'' (loss) moduli can be calculated. Fig. 4.14 shows the angular frequency (ω) dependence of both G' and G'' for silkworm dope [90]. The silk dopes responded like a liquid over a long timescale, and hence the moduli depended on the frequency with G'' being higher than G'. The moduli were relatively independent of frequency above the crossover point. In addition, G' was greater than G'', which suggests that the dopes, over short timescales, behave more like a solid [86]. These rheological results would suggest both silks behave like

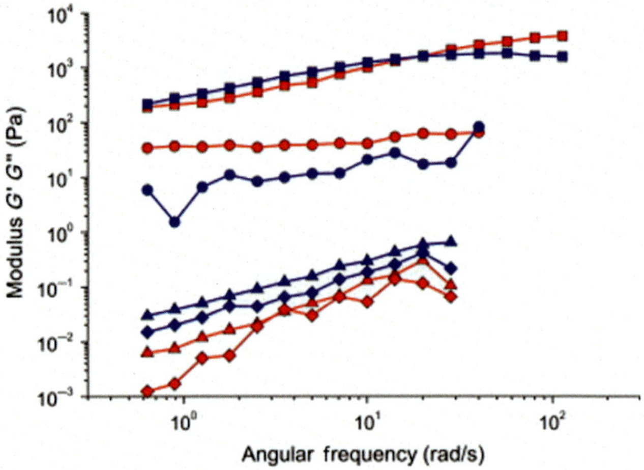

Figure 4.14 Representative oscillatory measurements of natural and reconstituted silkworm dope. Dry weight concentrations of natural dope are 18.6% (*squares*) and 4.6% (*circles*) and of reconstituted dope they are 18.5% (*triangles*) and 4.5% (*diamonds*). Elastic modulus G' (*red*) and viscous modulus G' (*blue*). *Source: Adapted from C. Holland, A.E. Terry, D. Porter, F. Vollrath, Natural and unnatural silks, Polymer 48 (12) (2007) 3388–3392 [90].*

concentrated polymers of high molecular weight, which is similar to weak hydrogels.

4.4.2 Gelation

Gelation is the transition from a polymer solution to a gel, which is a swollen polymer network material. The silk dope is similar to silk hydrogel but is not identical in terms of several properties, including transition behaviors. Here introduces the silk physical hydrogel induced by phase separations via hydrophobic interactions, namely, the silk hydrogel prepared from mixing silk solution and ethanol [33]. The silk protein aqueous solution with a 63 g/L concentration was mixed with ethanol at various ratios, resulting in the silk solutions with different concentrations. The time evolution of storage modulus G' and loss modulus G'' of the samples were measured at 37°C to determine the gelation point (t_{gel}). The storage modulus G' and the loss modulus G'' of a mixture of silk solution and ethanol (7/3 ratio) with a final silk concentration of 44 g/L are shown in Fig. 4.15 [33]. The G' and G'' increased with time and crossover occurred at around 1150 s. Because the gelation threshold is defined as the intermediate point between sol and gel, the t_{gel} is estimated as this point. The samples prepared at silk

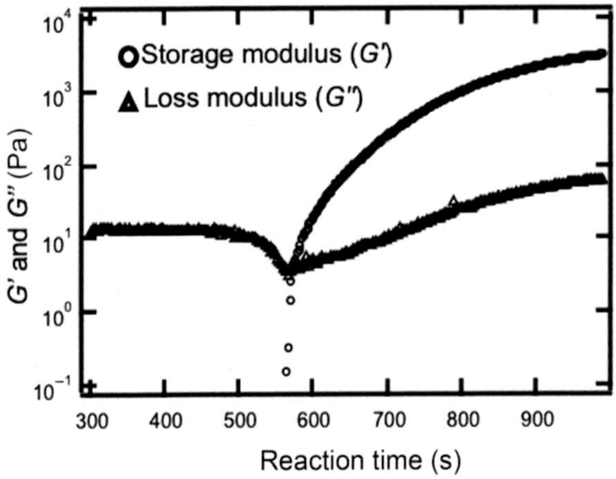

Figure 4.15 The storage modulus (G') and the loss modulus (G'') during gelation of the silk solution induced with ethanol were measured at 37°C [33].

concentrations of 32 and 25 g/L showed relatively shorter t_{gel}, while the samples with relatively higher (57 g/L) and lower concentration (13 and 7 g/L) of silk proteins demonstrated longer t_{gel}.

The gelation of the silk solution can be described as a ternary phase-separation system of silk polymer, water (solvent), and ethanol (nonsolvent), as schematically shown in Fig. 4.16 [33,91,92]. By increasing the solution/ethanol ratio, the phase is expected to change from solution phase (I) to aggregation of silk polymer phase (III) by way of gel phase (II). The retardation of t_{gel} at the higher (57 g/L, silk solution/ethanol = 9/1) and lower concentration (13 g/L, silk solution/ethanol = 2/8) are corresponding to the boundary between phase (I) and (II), and phase (II) and (III), respectively. The sample prepared at a silk concentration of 7 g/L (silk solution/ethanol = 1/9), which showed no gelation, is in the phase (III). The slight decrease in loss modulus G'' just before the gelation, which is unusual and different from that of synthetic polymers [93], also indicates the aggregation of silk polymers with the formation of a beta-sheet structure and subsequent gelation from aggregated silk polymers. Based on the structural and morphological characterizations of the silk hydrogels by attenuated total reflection-Fourier transform infrared, wide-angle X-ray diffraction (WAXD), and differential interference contrast microscopy, the silk molecules were assembled to form fibrillar and heterogeneous

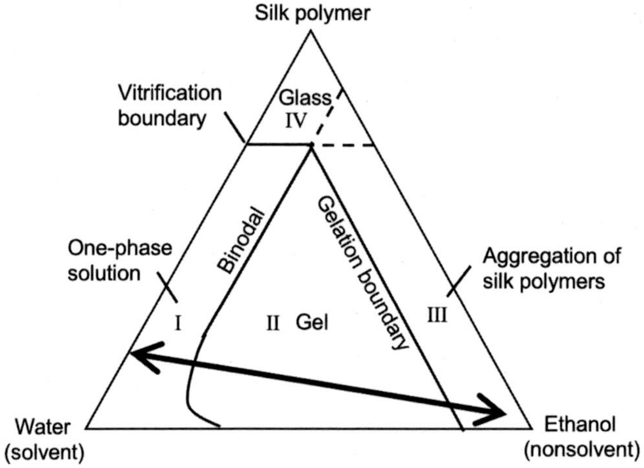

Figure 4.16 Schematic ternary phase diagram of the system: silk polymer—water (solvent)—ethanol (nonsolvent). Solutions in region I are a homogeneous liquid state. Silk polymer solutions in region II separate into two phases and their gelation occurs. In region III, silk polymer aggregates. Silk solutions in region IV are in a glass state. A bold arrow denotes the state described in the text [33].

networks with beta-sheet structures, which are the prerequisite for the present gelation.

4.5 Optical property

The optical properties of structural proteins have not been studied very well. Here are two examples, namely, the optical properties of silk and reflectin.

4.5.1 Optical property of silk and spider web-related compound

Viscid capture silk and its glues are noncrystalline and can only be crystallized partially under the unnatural conditions [94]. Recently, the WAXD structural analysis of spider capture silk during stretching deformation revealed that its structure is predominantly amorphous and is not converted to a crystalline state even when the silk is stretched by 500% [24]. The amino acid sequence and composition of the capture silk of *T. clavata* differ from those of *T. clavata* dragline silk and *B. mori* silk. The amino acid sequences for the β-sheet formations found in *T. clavata* dragline silk and *B. mori* silk are AAAAAA and GAGAGS, respectively; however, these are not found in *T. clavata* capture silk, which mostly contains glycine- and proline-rich β-spiral structures [95,96]. The capture spiral silk does not exhibit any crystalline components and nor does it undergo crystallization before and during the stretching deformation. This allows it to retain its elastic and optical properties during deformation. Spider webs made of capture silk are known to display multiple colors during the reflection, refraction, and transmission of sunlight [97,98]. According to the WAXD structural analysis of viscid capture silk, it is entirely amorphous during stretching deformation [24]. This is because the crystallization does not affect the optical properties of spider webs of this silk. These attributes of capture silk, which is one of the main components of spider webs, ensure that it is better suited than other types of silks for catching prey.

Thread color is a striking example of the effect of low-molecular-weight compounds in draglines. Golden orb-weaving spiders (*Trichonephila*), a genus of araneomorph spiders, produce a bright yellow thread that absorbs light with wavelengths less than 500 nm (Fig. 4.17) [99]. Based on evidence that pigments in *Trichonephila clavipes* dragline silk

Figure 4.17 An orb web produced by a *Nephila* spider at night.

are malleable and depend on light environments. The yellow color functions as a lure [100,101]; the color is mainly hypothesized to serve the purpose of attracting prey or camouflaging the orb web [102]. The yellow pigment of *T. clavipes* silk is contained in the low-molecular-weight fraction in dissolved silks. Arakawa's group at Keio University reported that the major pigment of golden dragline derived from *T. clavata* was identified as xanthurenic acid, and also xanthurenic acid shows a slight antibacterial effect. However, the function of the golden color for the spider dragline has not been clarified until now. It is speculated that golden orb-weaving spiders use the golden pigment for other purposes, such as to attract their prey in the sunlight.

4.5.2 Reflectin

Reflectivity in biological tissues is achieved by the alternative layer structure of flat platelets of high refractive index and low refractive index region, resulting in reflection functions and structural colors. In most cases of aquatic animals, reflector platelets consist of purine crystals, particularly guanine and hypoxanthine [103,104]. In contrast, cephalopod reflector platelets are composed of a unique protein with high-refractive index, which is reflectin [105]. The self-assembly structure of reflectin differs among different species [106], according to a previous literature about in vitro assays [107]. Recently, a new role of reflectin in color dynamics, namely, the "cephalopod-blue" can be from wavelength-dependent light scattering rather than reflection [108]. Thus the structure–function relationship of reflectin has not been perfectly clarified, but structural protein-based optics has a huge potential to create the next-generation bioinspired optics.

References

[1] C.L. Craig, C. Riekel, Comparative architecture of silks, fibrous proteins and their encoding genes in insects and spiders, Comp. Biochem. Physiol. Part B Biochem. Mol. Biol. 133 (4) (2002) 493–507.

[2] F. Sehnal, M. Zurovec, Construction of silk fiber core in Lepidoptera, Biomacromolecules 5 (3) (2004) 666–674.

[3] J.M. Ageitos, K. Yazawa, A. Tateishi, K. Tsuchiya, K. Numata, The benzyl ester group of amino acid monomers enhances substrate affinity and broadens the substrate specificity of the enzyme catalyst in chemoenzymatic copolymerization, Biomacromolecules 17 (1) (2016) 314–323.

[4] J.O. Warwicker, Comparative studies of fibroins. II. Crystal structures of various fibroins, J. Mol. Biol. 2 (6) (1960) 350–362.

[5] J. Gatesy, C. Hayashi, D. Motriuk, J. Woods, R. Lewis, Extreme diversity, conservation, and convergence of spider silk fibroin sequences, Science 291 (5513) (2001) 2603–2605.
[6] E.S. Lintz, T.R. Scheibel, Dragline, egg stalk and byssus: a comparison of outstanding protein fibers and their potential for developing new materials, Adv. Funct. Mater. 23 (36) (2013) 4467–4482.
[7] P. Calvert, Materials science: silk and sequence, Nature 393 (6683) (1998) 309–311.
[8] Z.Z. Shao, F. Vollrath, Materials: surprising strength of silkworm silk, Nature 418 (6899) (2002) 741.
[9] A.D. Malay, R. Sato, K. Yazawa, H. Watanabe, N. Ifuku, H. Masunaga, et al., Relationships between physical properties and sequence in silkworm silks, Sci. Rep. 6 (2016) 27573.
[10] S.W. Cranford, A. Tarakanova, N.M. Pugno, M.J. Buehler, Nonlinear material behaviour of spider silk yields robust webs, Nature 482 (7383) (2012) 72–76.
[11] J. Gosline, M. Lillie, E. Carrington, P. Guerette, C. Ortlepp, K. Savage, Elastic proteins: biological roles and mechanical properties, Philos. Trans. R. Soc. London. Ser. B Biol. Sci. 357 (1418) (2002) 121–132.
[12] B.B. Aaron, J.M. Gosline, Elastin as a random-network elastomer—a mechanical and optical analysis of single elastin fibers, Biopolymers 20 (6) (1981) 1247–1260.
[13] K. Bailey, T. Weis-Fogh, Amino acid composition of a new rubber-like protein, reslin, Biochim. Biophys. Acta 48 (1961) 452–459.
[14] C.M. Pollock, R.E. Shadwick, Allometry of muscle, tendon, and elastic energy storage capacity in mammals, Am. J. Physiol. 266 (3 Pt 2) (1994) R1022–R1031.
[15] C.M. Pollock, R.E. Shadwick, Relationship between body mass and biomechanical properties of limb tendons in adult mammals, Am. J. Physiol. 266 (3 Pt 2) (1994) R1016–R1021.
[16] E.C. Bell, J.M. Gosline, Mechanical design of mussel byssus: material yield enhances attachment strength, J. Exp. Biol. 199 (4) (1996) 1005–1017.
[17] M. Denny, The physical properties of spider's silk and their role in the design of orb-webs, J. Exp. Biol. 65 (2) (1976) 483–506.
[18] J.M. Gosline, P.A. Guerette, C.S. Ortlepp, K.N. Savage, The mechanical design of spider silks: from fibroin sequence to mechanical function, J. Exp. Biol. 202 (Pt 23) (1999) 3295–3303.
[19] F. Chen, D. Porter, F. Vollrath, Structure and physical properties of silkworm cocoons, J. R. Soc. Interface 9 (74) (2012) 2299–2308.
[20] J. Perez-Rigueiro, M. Elices, J. Llorca, C. Viney, Tensile properties of silkworm silk obtained by forced silking, J. Appl. Polym. Sci. 82 (8) (2001) 1928–1935.
[21] R. Rajkhowa, J. Kaur, X.G. Wang, W. Batchelor, Intrinsic tensile properties of cocoon silk fibres can be estimated by removing flaws through repeated tensile tests, J. R. Soc. Interface 12 (107) (2015).
[22] K. Sen, M.K. Babu, Studies on Indian silk. II. Structure-property correlations, J. Appl. Polym. Sci. 92 (2) (2004) 1098–1115.
[23] Y. Wang, J. Guan, N. Hawkins, D. Porter, Z. Shao, Understanding the variability of properties in Antheraea pernyi silk fibres, Soft Matter 10 (33) (2014) 6321–6331.
[24] K. Numata, H. Masunaga, T. Hikima, S. Sasaki, K. Sekiyama, M. Takata, Use of extension-deformation-based crystallisation of silk fibres to differentiate their functions in nature, Soft Matter 11 (31) (2015) 6335–6342.
[25] B.L. Thiel, K.B. Guess, C. Viney, Non-periodic lattice crystals in the hierarchical microstructure of spider (major ampullate) silk, Biopolymers 41 (7) (1997) 703–719.
[26] K. Yazawa, A.D. Malay, H. Masunaga, Y. Norma-Rashid, K. Numata, Simultaneous effect of strain rate and humidity on the structure and mechanical behavior of spider silk, Commun. Mater. 1 (1) (2020) 10.

[27] K. Yazawa, K. Ishida, H. Masunaga, T. Hikima, K. Numata, Influence of water content on the beta-sheet formation, thermal stability, water removal, and mechanical properties of silk materials, Biomacromolecules 17 (3) (2016) 1057−1066.
[28] T.A. Blackledge, C. Boutry, S.-C. Wong, A. Baji, A. Dhinojwala, V. Sahni, et al., How super is supercontraction? Persistent versus cyclic responses to humidity in spider dragline silk, J. Exp. Biol. 212 (13) (2009) 1981−1989.
[29] C. Boutry, T.A. Blackledge, Wet webs work better: humidity, supercontraction and the performance of spider orb webs, J. Exp. Biol. 216 (19) (2013) 3606−3610.
[30] M. Elices, G.R. Plaza, J. Perez-Rigueiro, G.V. Guinea, The hidden link between supercontraction and mechanical behavior of spider silks, J. Mech. Behav. Biomed. Mater. 4 (5) (2011) 658−669.
[31] G.V. Guinea, M. Elices, J. Perez-Rigueiro, G.R. Plaza, Stretching of supercontracted fibers: a link between spinning and the variability of spider silk, J. Exp. Biol. 208 (Pt 1) (2005) 25−30.
[32] Y. Liu, Z. Shao, F. Vollrath, Relationships between supercontraction and mechanical properties of spider silk, Nat. Mater. 4 (2005) 901.
[33] K. Numata, T. Katashima, T. Sakai, State of water, molecular structure, and cytotoxicity of silk hydrogels, Biomacromolecules 12 (6) (2011) 2137−2144.
[34] X. Hu, K. Shmelev, L. Sun, E.S. Gil, S.H. Park, P. Cebe, et al., Regulation of silk material structure by temperature-controlled water vapor annealing, Biomacromolecules 12 (5) (2011) 1686−1696.
[35] K.J. Koski, P. Akhenblit, K. McKiernan, J.L. Yarger, Non-invasive determination of the complete elastic moduli of spider silks, Nat. Mater. 12 (3) (2013) 262−267.
[36] T.Y. Lin, H. Masunaga, R. Sato, A.D. Malay, K. Toyooka, T. Hikima, et al., Liquid crystalline granules align in a hierarchical structure to produce spider dragline microfibrils, Biomacromolecules 18 (4) (2017) 1350−1355.
[37] F. Vollrath, B. Madsen, Z. Shao, The effect of spinning conditions on the mechanics of a spider's dragline silk, Proc. Biol. Sci. 268 (1483) (2001) 2339−2346.
[38] H.C. Wells, K.H. Sizeland, H.R. Kayed, N. Kirby, A. Hawley, S.T. Mudie, et al., Poisson's ratio of collagen fibrils measured by small angle X-ray scattering of strained bovine pericardium, J. Appl. Phys. 117 (4) (2015) 044701.
[39] P. Kiviranta, J. Rieppo, R.K. Korhonen, P. Julkunen, J. Toyras, J.S. Jurvelin, Collagen network primarily controls Poisson's ratio of bovine articular cartilage in compression, J. Orthop. Res. 24 (4) (2006) 690−699.
[40] M.L. Anderson, P.H. Mott, C.M. Roland, The compression of bonded rubber disks, Rubber Chem. Technol. 77 (2) (2004) 293−302.
[41] M. Eidini, G.H. Paulino, Unraveling metamaterial properties in zigzag-base folded sheets, Sci. Adv. 1 (8) (2015) e1500224.
[42] C. Holland, K. Numata, J. Rnjak-Kovacina, F.P. Seib, The biomedical use of silk: past, present, future, Adv. Health. Mater. 8 (1) (2019) 1800465.
[43] P.M. Cunniff, S.A. Fossey, M.A. Auerbach, J.W. Song, D.L. Kaplan, W.W. Adams, et al., Mechanical and thermal properties of dragline silk from the spider Nephila clavipes, Polym. Adv. Technol. 5 (8) (1994) 401−410.
[44] M. Hudspeth, X. Nie, W. Chen, R. Lewis, Effect of loading rate on mechanical properties and fracture morphology of spider silk, Biomacromolecules 13 (8) (2012) 2240−2246.
[45] C.P. Brown, J. MacLeod, H. Amenitsch, F. Cacho-Nerin, H.S. Gill, A.J. Price, et al., The critical role of water in spider silk and its consequence for protein mechanics, Nanoscale 3 (9) (2011) 3805−3811.
[46] C. Fu, D. Porter, Z. Shao, Moisture effects on Antheraea pernyi silk's mechanical property, Macromolecules 42 (20) (2009) 7877−7880.
[47] T. Giesa, R. Schuetz, P. Fratzl, M.J. Buehler, A. Masic, Unraveling the molecular requirements for macroscopic silk supercontraction, ACS Nano 11 (10) (2017) 9750−9758.

[48] B.D. Opell, K.E. Buccella, M.K. Godwin, M.X. Rivas, M.L. Hendricks, Humidity-mediated changes in an orb spider's glycoprotein adhesive impact prey retention time, J. Exp. Biol. 220 (Pt 7) (2017) 1313–1321.
[49] T. Vehoff, A. Glisovic, H. Schollmeyer, A. Zippelius, T. Salditt, Mechanical properties of spider dragline silk: humidity, hysteresis, and relaxation, Biophys. J. 93 (12) (2007) 4425–4432.
[50] F. Vollrath, D.T. Edmonds, Modulation of the mechanical properties of spider silk by coating with water, Nature 340 (1989) 305.
[51] K. Yazawa, A.D. Malay, H. Masunaga, K. Numata, Role of skin layers on mechanical properties and supercontraction of spider dragline silk fiber, Macromol. Biosci. 19 (3) (2019) e1800220.
[52] N. Du, X.Y. Liu, J. Narayanan, L. Li, M.L.M. Lim, D. Li, Design of superior spider silk: from nanostructure to mechanical properties, Biophys. J. 91 (12) (2006) 4528–4535.
[53] S.J. Blamires, W.I. Sellers, Modelling temperature and humidity effects on web performance: implications for predicting orb-web spider (Argiope spp.) foraging under Australian climate change scenarios, Conserv. Physiol. 7 (1) (2019) coz083.
[54] G.V. Guinea, M. Elices, G.R. Plaza, G.B. Perea, R. Daza, C. Riekel, et al., Minor ampullate silks from Nephila and Argiope spiders: tensile properties and microstructural characterization, Biomacromolecules 13 (7) (2012) 2087–2098.
[55] T.A. Blackledge, C. Boutry, S.C. Wong, A. Baji, A. Dhinojwala, V. Sahni, et al., How super is supercontraction? Persistent versus cyclic responses to humidity in spider dragline silk, J. Exp. Biol. 212 (13) (2009) 1980–1988.
[56] R.W. Work, A comparative-study of the super-contraction of major ampullate silk fibers of orb-web-building spiders (Araneae), J. Arachnol. 9 (3) (1981) 299–308.
[57] M. Elices, J. Perez-Rigueiro, G. Plaza, G.V. Guinea, Recovery in spider silk fibers, J. Appl. Polym. Sci. 92 (6) (2004) 3537–3541.
[58] G.V. Guinea, M. Elices, J. Perez-Rigueiro, G. Plaza, Self-tightening of spider silk fibers induced by moisture, Polymer 44 (19) (2003) 5785–5788.
[59] J. Perez-Rigueiro, M. Elices, G.V. Guinea, Controlled supercontraction tailors the tensile behaviour of spider silk, Polymer 44 (13) (2003) 3733–3736.
[60] J. Perez-Rigueiro, M. Elices, G. Plaza, J.I. Real, G.V. Guinea, The effect of spinning forces on spider silk properties, J. Exp. Biol. 208 (Pt 14) (2005) 2633–2639.
[61] N. Agarwal, D.A. Hoagland, R.J. Farris, Effect of moisture absorption on the thermal properties of Bombyx mori silk fibroin films, J. Appl. Polym. Sci. 63 (3) (1997) 401–410.
[62] J. Guan, D. Porter, F. Vollrath, Thermally induced changes in dynamic mechanical properties of native silks, Biomacromolecules 14 (3) (2013) 930–937.
[63] X. Hu, D. Kaplan, P. Cebe, Effect of water on the thermal properties of silk fibroin, Thermochim. Acta 461 (1) (2007) 137–144.
[64] W. Huang, S. Krishnaji, O.R. Tokareva, D. Kaplan, P. Cebe, Influence of water on protein transitions: morphology and secondary structure, Macromolecules 47 (22) (2014) 8107–8114.
[65] W. Huang, S. Krishnaji, O.R. Tokareva, D. Kaplan, P. Cebe, Influence of water on protein transitions: thermal analysis, Macromolecules 47 (22) (2014) 8098–8106.
[66] C. Mo, P. Wu, X. Chen, Z. Shao, The effect of water on the conformation transition of Bombyx mori silk fibroin, Vib. Spectrosc. 51 (1) (2009) 105–109.
[67] G.R. Plaza, G.V. Guinea, J. Pérez-Rigueiro, M. Elices, Thermo-hygro-mechanical behavior of spider dragline silk: glassy and rubbery states, J. Polym. Sci. Part B Polym. Phys. 44 (6) (2006) 994–999.
[68] Q. Yuan, J. Yao, L. Huang, X. Chen, Z. Shao, Correlation between structural and dynamic mechanical transitions of regenerated silk fibroin, Polymer 51 (26) (2010) 6278–6283.

[69] T. Asakura, M. Demura, Y. Watanabe, K. Sato, H-1 pulsed NMR-study of Bombyx-Mori silk fibroin—dynamics of fibroin and of absorbed water, J. Polym. Sci. Pol. Phys. 30 (7) (1992) 693–699.
[70] M. Ishida, T. Asakura, M. Yokoi, H. Saito, Solvent-induced and mechanical-treatment-induced conformational transition of silk fibroins studied by high-resolution solid-state C-13 NMR-spectroscopy, Macromolecules 23 (1) (1990) 88–94.
[71] H. Yoshimizu, T. Asakura, The structure of *Bombyx Mori* silk fibroin membrane swollen by water studied with ESR, C-13-NMR, and FT-IR spectroscopies, J. Appl. Polym. Sci. 40 (9-10) (1990) 1745–1756.
[72] Y.S. Kim, L. Dong, M.A. Hickner, T.E. Glass, V. Webb, J.E. McGrath, State of water in disulfonated poly(arylene ether sulfone) copolymers and a perfluorosulfonic acid copolymer (Nafion) and its effect on physical and electrochemical properties, Macromolecules 36 (17) (2003) 6281–6285.
[73] I.D. Kuntz, Hydration of macromolecules. III. Hydration of polypeptides, J. Am. Chem. Soc. 93 (2) (1971) 514–516.
[74] K.Y. Lee, W.S. Ha, DSC studies on bound water in silk fibroin/S-carboxymethyl kerateine blend films, Polymer 40 (14) (1999) 4131–4134.
[75] T. Asakura, K. Isobe, A. Aoki, S. Kametani, Conformation of crystalline and non-crystalline domains of [3-13C]Ala-, [3-13C]Ser-, and [3-13C]Tyr-Bombyx mori silk fibroin in a hydrated state studied with 13C DD/MAS NMR, Macromolecules 48 (22) (2015) 8062–8069.
[76] X. Hu, D. Kaplan, P. Cebe, Dynamic protein – water relationships during β-sheet formation, Macromolecules 41 (11) (2008) 3939–3948.
[77] H.Y. Kweon, I.C. Um, Y.H. Park, Thermal behavior of regenerated Antheraea pernyi silk fibroin film treated with aqueous methanol, Polymer 41 (20) (2000) 7361–7367.
[78] S. Mazzi, E. Zulker, J. Buchicchio, B. Anderson, X. Hu, Comparative thermal analysis of Eri, Mori, Muga, and Tussar silk cocoons and fibroin fibers, J. Therm. Anal. Calorim. 116 (3) (2014) 1337–1343.
[79] M. Tsukada, M. Obo, H. Kato, G. Freddi, F. Zanetti, Structure and dyeability of Bombyx mori silk fibers with different filament sizes, J. Appl. Polym. Sci. 60 (10) (1996) 1619–1627.
[80] G. Freddi, Y. Gotoh, T. Mori, I. Tsutsui, M. Tsukada, Chemical structure and physical properties of Antheraea assama silk, J. Appl. Polym. Sci. 52 (6) (1994) 775–781.
[81] R.L. Frost, M.L. Weier, Thermal treatment of whewellite—a thermal analysis and Raman spectroscopic study, Thermochim. Acta 409 (1) (2004) 79–85.
[82] M. Tsukada, Y. Goto, G. Freddi, M. Matsumura, H. Shiozaki, H. Ishikawa, Structure and physical properties of epoxide-treated tussah silk fibers, J. Appl. Polym. Sci. 44 (12) (1992) 2203–2211.
[83] M. Morillas, D.L. Vanik, W.K. Surewicz, On the mechanism of alpha-helix to beta-sheet transition in the recombinant prion protein, Biochemistry 40 (23) (2001) 6982–6987.
[84] K. Nakamura, T. Hatakeyama, H. Hatakeyama, Effect of bound water on tensile properties of native cellulose, Text. Res. J. 53 (11) (1983) 682–688.
[85] S. Vyas, S. Pradhan, N. Pavaskar, A. Lachke, Differential thermal and thermogravimetric analyses of bound water content in cellulosic substrates and its significance during cellulose hydrolysis by alkaline active fungal cellulases, Appl. Biochem. Biotechnol. 118 (1-3) (2004) 177–188.
[86] C. Holland, A.E. Terry, D. Porter, F. Vollrath, Comparing the rheology of native spider and silkworm spinning dope, Nat. Mater. 5 (11) (2006) 870–874.
[87] P.R. Laity, C. Holland, Thermo-rheological behaviour of native silk feedstocks, Eur. Polym. J. 87 (2017) 519–534.

[88] P.R. Laity, S.E. Gilks, C. Holland, Rheological behaviour of native silk feedstocks, Polymer 67 (2015) 28–39.
[89] A.D. Malay, T. Suzuki, T. Katashima, N. Kono, K. Arakawa, K. Numata, Spider silk self-assembly via modular liquid-liquid phase separation and nanofibrillation, Sci. Adv. 6 (45) (2020).
[90] C. Holland, A.E. Terry, D. Porter, F. Vollrath, Natural and unnatural silks, Polymer 48 (12) (2007) 3388–3392.
[91] J. Arnauts, H. Berghmans, Amorphous thermoreversible gels of atactic polystyrene, Polym. Commun. 28 (3) (1987) 66–68.
[92] P.J. Flory, Principles of Polymer Chemistry, Cornell University Press, Ithaca, NY, 1953.
[93] M. Kurakazu, T. Katashima, M. Chijiishi, K. Nishi, Y. Akagi, T. Matsunaga, et al., Evaluation of gelation kinetics of tetra-PEG gel, Macromolecules 43 (8) (2010) 3935–3940.
[94] C.L. Craig, Spiderwebs and silk: tracing evolution from molecules to genes to phenotypes, Oxford Univesity Press, Oxford England; New York, 2003.
[95] N. Becker, E. Oroudjev, S. Mutz, J.P. Cleveland, P.K. Hansma, C.Y. Hayashi, et al., Molecular nanosprings in spider capture-silk threads, Nat. Mater. 2 (4) (2003) 278–283.
[96] C.Y. Hayashi, R.V. Lewis, Molecular architecture and evolution of a modular spider silk protein gene, Science 287 (5457) (2000) 1477–1479.
[97] G.R.S. Deb, M. Kane, N. Naidoo, D.J. Little, M.E. Herberstein, Optics of spider "sticky" orb webs, Proc. SPIE 7975 (2011).
[98] C.L. Craig, Alternative foraging modes of orb web weaving spiders, Biotropica 21 (3) (1989) 257–264.
[99] C.L. Craig, R.S. Weber, G.D. Bernard, Evolution of predator-prey systems: spider foraging plasticity in response to the visual ecology of prey, Am. Nat. 147 (2) (1996) 205–229.
[100] Y. Henaut, S. Machkour-M'Rabet, P. Winterton, S. Calme, Insect attraction by webs of Nephila clavipes (Araneae: Nephilidae), J. Arachnol. 38 (1) (2010) 135–138.
[101] T.E. White, R.L. Dalrymple, M.E. Herberstein, D.J. Kemp, The perceptual similarity of orb-spider prey lures and flower colours, Evol. Ecol. 31 (1) (2017) 1–20.
[102] J.E. Carrel, Spiderwebs and silk—tracing evolution from molecules to genes to phenotypes, Science 303 (5655) (2004) 175.
[103] K.M. Cooper, R.T. Hanlon, B.U. Budelmann, Physiological color-change in squid iridophores. II. Ultrastructural mechanisms in *Lolliguncula brevis*, Cell Tissue Res. 259 (1) (1990) 15–24.
[104] R.T. Hanlon, K.M. Cooper, B.U. Budelmann, T.C. Pappas, Physiological color-change in squid iridophores.1. Behavior, morphology and pharmacology in *Lolliguncula brevis*, Cell Tissue Res. 259 (1) (1990) 3–14.
[105] W.J. Crookes, L.L. Ding, Q.L. Huang, J.R. Kimbell, J. Horwitz, M.J. McFall-Ngai, Reflectins: the unusual proteins of squid reflective tissues, Science 303 (5655) (2004) 235–238.
[106] R.A. Cloney, S.L. Brocco, Chromatophore organs, reflector cells, iridocytes and leucophores in cephalopods, Am. Zool. 23 (3) (1983) 581–592.
[107] Z. Guan, T.T. Cai, Z.M. Liu, Y.F. Dou, X.S. Hu, P. Zhang, et al., Origin of the reflectin gene and hierarchical assembly of its protein, Curr. Biol. 27 (18) (2017) 2833.
[108] T.T. Cai, K. Han, P.L. Yang, Z. Zhu, M.C. Jiang, Y.Y. Huang, et al., Reconstruction of dynamic and reversible color change using reflectin protein, Sci. Rep 9 (2019).

Questions for this chapter

1. Describe how to determine and calculate toughness.
2. Why does spider dragline show higher toughness than carbon fiber?
3. Explain the difference in terms of melting point between protein and plastic polymers.
4. Why is the Poisson's ratio of natural rubber approximately 0.5?
5. Explain "supercontraction" of a spider dragline.

CHAPTER 5

Biological properties with cells

Cells are highly structured by lipids, carbohydrates (polysaccharides), and proteins. The role(s) of structural proteins in cells are not only to form the cytoskeleton but also to contribute to mechanical properties. Another important and essential property of structural proteins is a biological one.

5.1 Cell adhesion and proliferation

An excellent biomaterial must meet many requirements that include, but are not limited to, biocompatibility, biodegradability, mechanical robustness and durability, and amenability to the processing under ambient aqueous conditions, like a physiological condition [1,2]. Collagen is one of the most abundant proteins in mammals and is a major structural component of extracellular matrices [3]. Due to the origin of collagen and elastin, they demonstrate excellent biocompatibility and cell adhesion and can promote cell proliferation and differentiation.

5.1.1 Collagen

Collagen is one of the most abundant proteins in mammals and is a major structural component of extracellular matrices [3,4]. Thus collagen has excellent biocompatibility and cell adhesion and can promote cell proliferation and differentiation. As explained in Chapter 3, Structure, collagen usually forms fibrils with a triple helical structure to express its biological functions. The primary function of collagen is the connector between cells in tissues, even though there are many types of collagens. Collagen is classified into roughly two groups, namely, fibrillar (types I, II, III, V, and XI) and nonfibrillar (other types) collagens. Type I collagen, the most abundant fibrillar collagen in human body, contributes to tendon, skin, vasculature, and soft parts of bones. Type II collagen is cross-linking to proteoglycans and the major fibrillar network component in cartilage, contributing to the strength, ductility, and toughness for large-scale deformation. Type IV collagen is nonfibrillar and fibril-associated collagen to connect the fibrils of other types of collagen. In most mammalian connective tissues, type VI

Biopolymer Science for Proteins and Peptides
DOI: https://doi.org/10.1016/B978-0-12-820555-6.00003-3

collagen is bound to the sides of type I fibrils and may bind them together to form thicker collagen fibers. The collagen-containing network structure is a fibrillar protein-based hydrogel. Therefore their elasticity, modulus, and strength are originated from the network sizes and structures. The overall properties of collagen are summarized in Section 7.6.

5.1.2 Silk

In contrast to collagen, silk fibroin is not a cellular matrix component and so it is not expected to be biocompatible for mammalian cells. Silk has been used as a suture for a long time, supporting the practical use of silk as load-bearing biomaterial, for example, degummed *Bombyx mori* silk fibers processed into a knitted surgical mesh (SERI Surgical Scaffold manufactured by Sofregen Inc., Medford, MA, United States), silk sutures (coated with waxes, Ethicon Inc. and several other manufacturers) [2,5,6]. Recently, chemically synthetic materials have been widely used suture material, but silk sutures are still in demand for specialized applications where precise handling is of paramount importance (e.g., eye surgery). Silk sutures are strong, easy to handle, lie flat on the tissue surface, and allow for secure knots [2].

In the case of *B. mori* silkworm silk, the silk filaments are still coated with sericin and additional waxes. There is an ongoing debate about the potential role of sericin in these adverse functions; however, sericin shows a low allergenic and immunogenic profile in mice; in fact, this profile is similar to that seen for silk fibroin or alginate, according to a previous literature [7]. These reports are supported by in vitro data with macrophages: extracted sericin from *B. mori* silk cocoons showed no significant release of the inflammatory marker tumor necrosis factor alpha (TNF-α); similar observations were made with silk fibroin, although extracted sericin combined with bacterial lipopolysaccharide-induced TNF-α release. Besides, recoating of silk fibroin with sericin showed no macrophage response, while virgin silk induced a high level of TNF-α release, suggesting that other leachable compound(s), or these compounds combined with sericin, may be responsible for the adverse clinical reactions reported for silk [8]. Based on these reports on sericin and silk, complete and reproducible sericin removal from *B. mori* silk, which is generally called as "degaming," is an essential step in silkworm silk utilization to reduce the inflammation risks. For example, SERI Surgical Scaffold is described by the manufacturer as highly purified silk with $\geq 95\%$ purity [2].

Dedicated biocompatibility assessment is critical when generating novel silk formats, such as nano- and microparticles, hydrogels, scaffolds, films, and coatings [2,9]. Direct in vivo comparison of silk with collagen and synthetic biopolyesters such as polycaprolactone, polylactic acid, and poly[lactide-*co*-glycolic acid], indicates that *B. mori* silkworm silk fibroin is typically at least as good as these synthetic materials and often superior to other natural biopolymers [9]. As new applications for silk emerge, appropriate biocompatibility studies must be performed to support these developments. These new applications and their approval situations are summarized in the recent review [2,10] and also in Chapter 8, Biopolymer Material and Composite.

5.2 Cytotoxicity/neurotoxicity of degradation products

5.2.1 Cytotoxicity for protein aggregation

Antiparallel beta-pleated sheet is a key structure found in not only disease proteins but also most common proteins, functioning as stabilizing physical cross-links via hydrogen bonding and van der Waals interactions [11]. Beta-sheets have a widespread presence and importance in various proteins and are fully degradable. This is important in the fundamental context of how proteins are remodeled in vivo, as well as in the context of the design of protein-based biodegradable materials. Besides, such insight appears to be fundamental to help explain the profound differences between native antiparallel beta-sheets and disease-associated beta-sheets such as amyloid-beta (Aβ) fibrils.

Insight into the mechanism of degradation of antiparallel beta-pleated sheets will clarify the effects of beta-sheet crystals on biological remodeling and cytotoxicity [11]. These insights would also have direct implications in protein-based biomaterials designs with excellent biocompatibility and biodegradability. The modulation of beta-sheet content has become a useful approach to control degradation in vitro and in vivo in the context of regenerative medicine [12].

Abnormal aggregations or deposition of misfolded proteins, such as amyloid fibrils with beta-sheet structures, have been recognized as the molecular pathogenesis of Alzheimer's disease, Parkinson's disease, and Huntington disease [13]. Amyloids are generally described as stacked beta-sheet structures aligned perpendicular to the fibril axis known as cross-beta-sheets, and are often based on glutamine-rich or hydrophobic interactions [14]. In contrast to the silks containing antiparallel beta-pleated sheets, the amyloids do not degrade at reasonable rates in vivo and lead to significant physiological complications due to toxicity [15,16]. Aβ peptides in Alzheimer's disease have been studied to clarify the mechanism of neurotoxicity by Aβ fibrils. Monomers, intermediates (oligomers), and fibrils of Aβ peptides with different molecular weights have been investigated for neurotoxicity, and recently a spherical amyloid intermediate of 15–35 nm diameter, which had beta-sheet structures predominantly, demonstrated higher toxicity than Aβ monomers and fibrils [17].

5.2.2 Beta-sheet crystal

B. mori silkworm silk has been widely studied to clarify its crystalline structures, especially, a beta-sheet structure composed of the highly repetitive Gly-Ala-Gly-Ala-Gly-Ser domain, because of the remarkable mechanical properties that originate from a combination of crystalline and less crystalline domains. The antiparallel beta-pleated sheets structure of the silkworm silk was reported in 1955 by Marsh et al. [18], and the crystal structure was suggested to be four molecular chains in the rectangular unit cell with parameters, $a = 9.38$ Å, $b = 9.49$ Å, and c (fiber axis) = 6.98 Å [19]. These crystals in antiparallel beta-sheet structures are composed of alanine–glycine or alanine repeats. An increase in the fraction of beta-sheets also reduces enzymatic degradability of silk fibroin [20]. Several proteolytic enzymes (proteases) have been studied for the digestion of silk fibroin. Protease XIV (a mixture of multiple enzymes) is considered to show high activity toward beta-sheet structures in silk-based fibers, films, and scaffolds [21–23]. In contrast, alpha-chymotrypsin prefers to digest amorphous silk, the less crystalline regions of the assembled silk structures, but does not degrade the crystalline region containing beta-sheet structures. Beta-sheet crystals of silk fibroins are one of the most characterized beta-sheet crystals of proteins.

To evaluate the cytotoxicity of the beta-sheet crystals as well as to reveal the degradation pattern of *B. mori* silkworm silk heavy chain by the proteolytic enzymes, the degradation products (DPs) of crystalline silk films were characterized using various proteases (Fig. 5.1) [11]. Protease XIV and alpha-chymotrypsin are serine proteases that hydrolyze preferentially peptide bonds *C*-terminal to aromatic amino acids. The digestion patterns of the silk heavy chain molecule with a random-coil structure by either protease XIV or alpha-chymotrypsin were estimated according to the substrate specificity of the proteases. The resultant digestion patterns

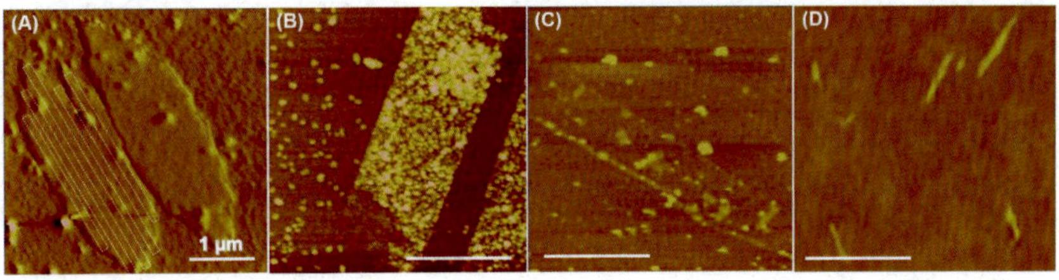

Figure 5.1 Atomic force microscopy (AFM) images of beta-sheet crystalline films of *Bombyx mori* silkworm silk before (A) and after the enzymatic treatments of alpha-chymotrypsin for 24 h (B), proteinase XIV for 12 h (C) and 24 h (D) [11].

of a random-coil silk molecule by protease XIV showed DPs with molecular weights of 146.2–7317.2 Da. In the case of alpha-chymotrypsin, the DPs were almost identical to those from protease XIV in terms of the distribution of molecular weights of DPs.

The DPs from protease XIV contained several low-molecular-weight (less than 50 kDa) fragments, and also approximately 30 high-molecular-weight (more than 240 kDa) components, implying that protease XIV digested several sites at the middle of the primary sequence. The DPs from alpha-chymotrypsin exhibited fewer high-molecular-weight components. Considering the estimation of the enzymatic digestion patterns described in the previous paragraph, enzymatic digestion patterns of random-coil silk molecules by protease XIV and alpha-chymotrypsin were nearly identical, while the digestion patterns of the crystalline silk molecules by the two enzymes exhibited significant differences. This difference in the digestion patterns of silk molecules suggested that the degradation mechanism and the digestion pattern are significantly dependent on the secondary, tertiary, and final assembled structures of silk molecules. In particular, beta-sheet structures are the main component to influence the enzymatic degradation behaviors.

To investigate the secondary structure effects of the enzymatic degradation behaviors, CD analyses were performed with the supernatant of DPs after the enzymatic degradation of silk crystals by protease XIV or alpha-chymotrypsin for 24 h at 37°C (Fig. 5.2). The CD spectrum of the soluble DPs by alpha-chymotrypsin indicated random-coil (unordered) structures with a negative peak at 203 nm. The DPs by protease XIV demonstrated beta-sheet structure with a negative at 216 nm shoulder (Fig. 5.2). Estimating the secondary structure contents of the soluble fragments yielded 37% beta-strand and 50% unordered structures for the soluble DPs by protease XIV. Their secondary structures were estimated to be

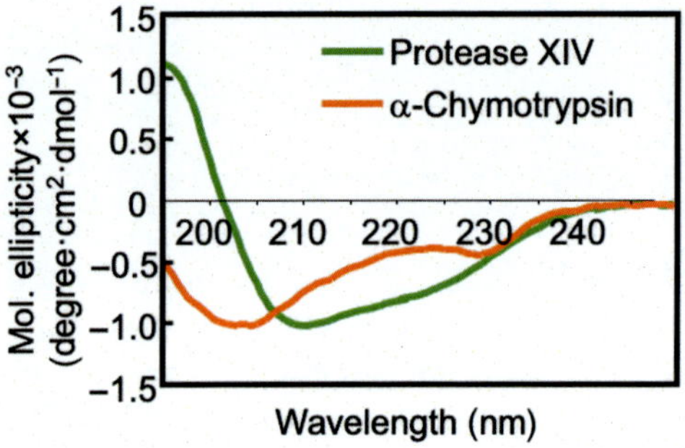

Figure 5.2 CD analyses of soluble degradation products after enzymatic degradations of silk crystals for 24 h. Green (black in print version) and orange (gray in print version) lines show the soluble degradation products by protease XIV and alpha-chymotrypsin, respectively [11].

5% beta-strand and 57% unordered in the case of the soluble fragments from alpha-chymotrypsin. A comparison of these secondary structures suggests that only protease XIV produces soluble beta-sheet fragments from the enzymatic degradation of silk crystals. According to the digestion pattern of silk molecules by protease XIV described above, the soluble DPs contain several hydrophobic domains, so these hydrophobic sequences, $(GAGAGS)_n$, retained beta-sheet structure in a buffer solution.

As explained above, degrading enzymes including proteases preferentially digest biomacromolecules in the amorphous region rather than the crystalline phase. A model of enzymatic degradation of silk crystalline regions due to protease activity was proposed (Fig. 5.3) [11]: tight-chain-packing regions and looser chain-packing regions exist randomly in the silk crystalline region (Fig. 5.3A). The tight-chain-packing regions are composed of ordered beta-sheet crystals. At the same time, the looser chain-packing region consists of less ordered assemblies of beta-sheets and the other structures such as turn and random-coil structures. After the degradation of the loose-chain-packed regions by protease XIV, the tight-chain-packed regions left result in nanofibrils around 5 nm thick and 80−100 nm wide (Fig. 5.3B). With further degradation, silk nanofilaments composed of beta-sheets are liberated and observed by atomic force microscopy (AFM) (Fig. 5.3C). According to the crystal structure and the molecular weight of the heavy chain of silk fibroin, the dimensions of one beta-sheet layer is around 1 nm thick and 56−212 nm long [24]. In Fig. 5.3, the minimum units of nanofilament length and thickness were 160 nm and 2 nm. Hence, a nanofilament consists of a few of the beta-

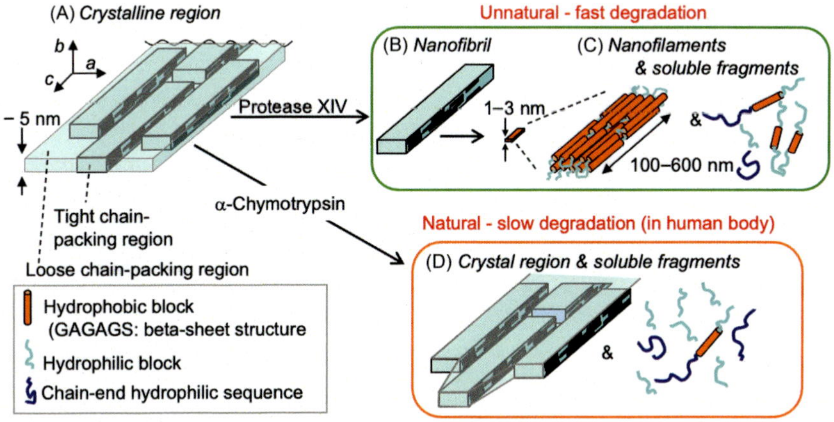

Figure 5.3 Model of enzymatic degradation of crystalline regions of silk fibroin. (A) Crystalline region of silk fibroin. Tight- and loose-chain-packing regions coexist in the crystalline region. (B) Crystalline nanofibril composed of the tight-chain-packing regions after degradation of the loose-chain-packing region by protease XIV. (C) Nanofilament composed of beta-sheet structure and soluble fragments containing beta-sheet structure after the enzymatic degradation by protease XIV. (D) Crystalline region and soluble fragments without beta-sheet structure after degradation of edges and ends of loose-chain-packing region by alpha-chymotrypsin [11].

sheet layers judging from the dimensions observed by AFM. In the case of a nanofilament more than 160 nm long, several hydrophobic blocks along the fiber axis must be involved in the formation. In addition to the nanofilaments, soluble fragments of silk molecules with a beta-sheet structure and a middle-molecular-weight (240 kDa or less) are generated (Fig. 5.5C). Finally, the nanofilaments will collapse into single silk fibroin molecules, followed by complete proteolytic digestion as the process continues. This fast degradation of beta-sheet crystal by protease XIV is not present in the human body. In contrast, alpha-chymotrypsin is not capable of digesting crystalline silk with a beta-sheet structure. After the treatment of silk crystalline region by alpha-chymotrypsin, which is similar to reactions in the human body, only loose-chain-packed regions around tight-chain-packed regions degrade, resulting in a large silk crystal left and soluble fragments generated with random-coil and few beta-sheet structures (Fig. 5.5D). The DPs from alpha-chymotrypsin show no toxicity because beta-sheet fractions in the soluble products are generated much more slowly in comparison with the degradation by protease XIV.

The degradation model described provides new views concerning both the fundamental mechanism of beta-sheet structures as well as for silk-based biomaterials [11]. As mentioned above, the nanofilaments assemble and form nanofibrils and play a role as nucleators of the crystalline regions, an important feature of the system, which can be exploited to design silk-based materials with predictable biodegradability and mechanical properties. This relationship is fundamental in materials science, where the sizes of crystals are important elements for the mechanical properties and degradation rates of bulk polymers.

5.2.3 Cytotoxicity of degradation products

The DPs of silk crystal region (beta-sheet structures) by protease XIV or alpha-chymotrypsin were evaluated for cytotoxicity to differentiated rat pheochromocytoma (PC12) cells using the MTS [3-(4,5-dimethylthiazol-2-yl)-5-(3-carboxymethoxyphenyl)-2-(4-sulfophenyl)-2H-tetrazolium] assay. Silk crystals before enzymatic degradation, DPs at 24 h from protease XIV and alpha-chymotrypsin, bovine serum albumin (BSA), and Aβ peptide 1−40 ($A\beta_{1-40}$) assembly were characterized at 55, 110, and 220 μg/mL (Fig. 5.4) [11]. The positive control, DPs from protease XIV and $A\beta_{1-40}$, showed significant cytotoxicity to neuronal cells. Also, the DPs from protease XIV showed lower cell viability in comparison to the

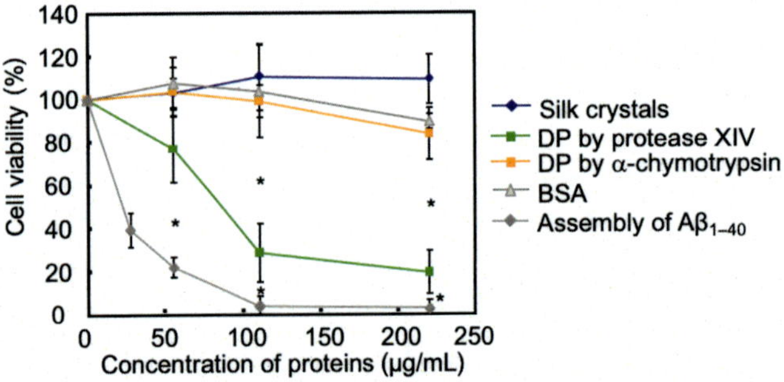

Figure 5.4 Cell viability on the enzymatic degradation products (DPs) of silk crystals on differentiated PC12 cells. Dependence of cell viability (percentage of active cells as compared to controls) on the silk crystals measured by MTS assay. Silk (*blue diamonds*, black in print version) shows cell viability percentages of silk crystals before the enzymatic degradation. DPs by protease XIV (*green squares*, dark gray in print version) and alpha-chymotrypsin (*orange squares*, white in print version) show the percentages of each DPs with different concentrations of silk. BSA (*gray triangles*, light gray in print version) is a negative control. The assembly of $A\beta_{1-40}$ (*gray diamonds*, gray in print version), which was obtained by the incubation of $A\beta_{1-40}$ solution at 4°C for 52 h, is a positive control as reported previously [16]. Data are represented as mean ± standard deviation ($n = 8$). *Significant difference between two groups at $P < .05$ [11].

DPs from alpha-chymotrypsin. This is likely because the DPs from alpha-chymotrypsin, insoluble silk crystal, and soluble hydrophilic domains, show no cytotoxicity. In contrast, soluble fragments with beta-sheet structures from protease XIV digestion have some cytotoxicity on PC12 cells. These beta-sheets in the sequence of silk are considered nontoxic structures in biology, in contrast, for example, to Aβ structures formed in disease states. The soluble beta-sheet fragments from protease XIV in this study demonstrated a more significant impact on cell viability when compared to the beta-sheets in the sequence of natural silk. These results are instructive, as protease XIV, which is not present in the human body, digested the silk crystals much faster than chymotrypsin, which is found in the human body. This difference in rate is important, as this results in a wider molecular weight distribution of DPs. Therefore the products (protease XIV digest) with unnaturally high molecular weight likely contain hydrophobic interactions with higher-molecular-weight beta-sheet structures. On the other hand, compared with the toxicity of $A\beta_{1-40}$, the nanofilaments and soluble fragments of silk, which were the DPs at 24 h from protease XIV, exhibited significantly lower toxicity. This difference is likely due to the differences in sequences chemistry between the $(GAGAGS)_n$ in silks and the glutamine-rich or hydrophobic sequences in disease proteins and peptides. Even though beta-sheet structure-rich DPs of silk slightly showed cytotoxicity, in nature and biological conditions, it is almost impossible to detect the cytotoxicity from the protease-mediated DPs [11].

Aside from direct implications in the design and implementation of silk-based biomaterials for regenerative medicine, insight into relationships between amyloid structures and their toxicity are also suggested [11,25]. Wide-angle X-ray structural analysis of recombinant spider silk fibrils and amyloid-like fibrils of prion protein reflected significant overlaps and minor differences [26]. Prion protein fibrils showed mostly beta-sheet and beta-turn structures, while the silk fibrils contained an additional 3_1-helical structure and/or random-coil conformation [27]. Amyloid fibrils with beta-pleated structure play a role as seeds for fibril formation by the interaction between fibrils used as seeds and serum amyloid protein, similar to prion protein assembly [28,29]. It will be instructive to further compare and contrast enzymatic responses of beta-sheet structures containing protein structures as an experimental route to further elucidate the mechanism for the differences in biological impact between the antiparallel beta-sheet silk fibroins versus the cross-beta-amyloid structures that are inherently more resistant to degradation and also toxic in vivo. These differences may provide new views on modes to alter the biological stability of Aβ structures.

5.2.4 Comparison with amyloid-beta peptides

The antiparallel beta-pleated sheet is a fundamental secondary structure in proteins and a major component in silk fibers generated by several insects including spiders, silkworm, and bagworm, with a key role in stabilizing these proteins' structures via beta-sheet structure-mediated physical cross-links. Silk fibrils were reported to have molecular-level similarity to amyloid fibrils, and also to enhance amyloidosis of amyloid protein due to cross-seeding effects as a disease mechanism. As mentioned in the previous section 5.2.3, however, silk nanofibrils and nanofilaments composed of beta-sheets, which influence various properties of silk fibers, show no significant cytotoxicity in vitro to neuronal cells [11]. Importantly, these beta-sheets are fully degradable and nontoxic structures in biology, in contrast, for example, to β-amyloid structures formed in Aβ fibrils found in brain tissue of persons with Alzheimer's disease. Amyloids are generally described as stacked beta-sheet structures aligned perpendicular to the fibril axis, known as cross-beta-sheets (Fig. 5.5) [30]. Monomers, intermediates (oligomers), and microfibrils of Aβ peptides with different nanoassembly structures have been investigated for neurotoxicity, and recently a spherical amyloid intermediate of 15–35 nm diameter, which had predominantly beta-sheet structures, demonstrated higher toxicity than Aβ monomers and fibrils [17].

The relationships between amino acid sequences, nanoassembled structures, and cytotoxicity against neurons using beta-sheet peptides originating from *Araneus ventricosus* spider silk, Aβ(12–28) (β1), Aβ(28–42) (β2), and full-length Aβ(1–42) were also studied [30]. Morphology of the peptide

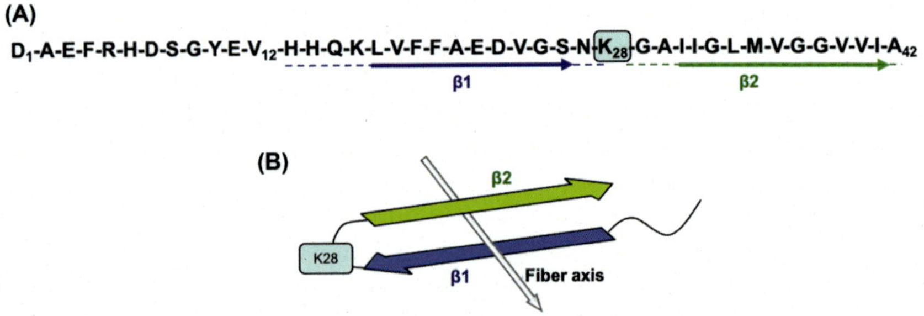

Figure 5.5 Schematic models of Aβ peptide. (A) Amino acid sequence of Aβ(1–42). (B) Aβ(1–42) peptide with cross-beta structure [30].

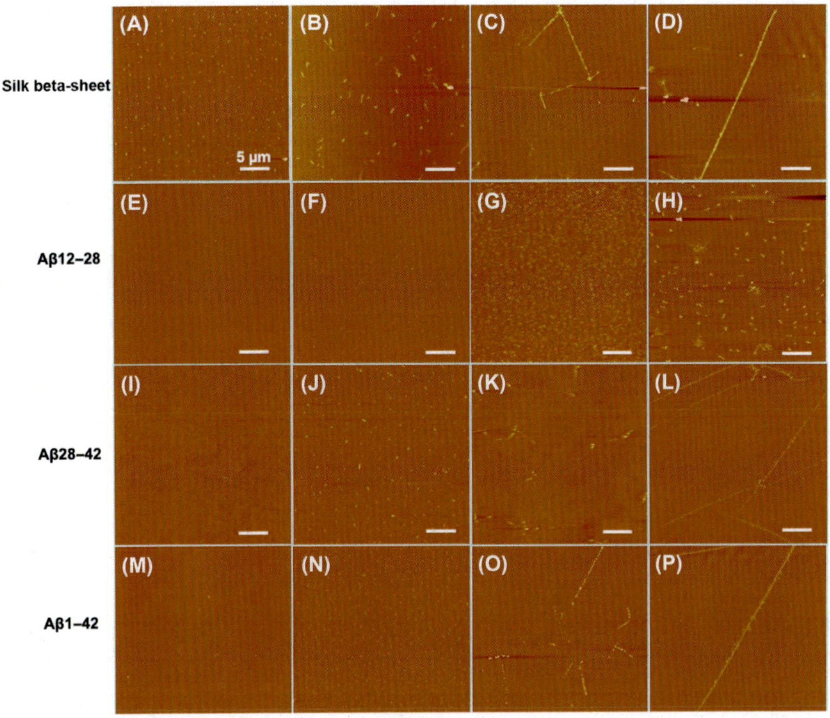

Figure 5.6 AFM height images of nanoassemblies and microfibrils of spider silk beta-sheet (A–D), Aβ(12–28) (E–H), Aβ(28–42) (I–L), and Aβ(1–42) (M–P) before (A, E, I, M) and after incubation for 24 (B, F, J, N), 48 (C, G, K, O), and 72 h (D, H, L, P). Each scale denotes 5 μm. The color scale represents 100 nm height [30].

nanoassembly and microfibrils was characterized by AFM (Fig. 5.6) [30]. Aβ (12–28) is composed of not only nonpolar hydrophobic amino acids (Leu, Val, Phe, Ala) but also basic (His, Lys), polar (Gln, Ser), and acidic amino acids (Glu, Asp), and hence was not capable of forming microfibrils via hydrophobic interactions between the side chains within a relatively short time (72 h). The most notable result from AFM observations is that one partial beta-sheet sequence of Aβ(1–42), namely Aβ(28–42) composed nearly wholly of hydrophobic amino acids, was capable of forming microfibrils without the other Aβ sequences.

To determine the secondary structure of the peptides in solution, circular dichroism (CD) analysis of each peptide solution was performed before and after the incubation for 24, 48, and 72 h [30]. The CD spectrum of all samples showed beta-sheet structure with a negative dip at a 216 nm shoulder, and the beta-sheet content of all samples was constant

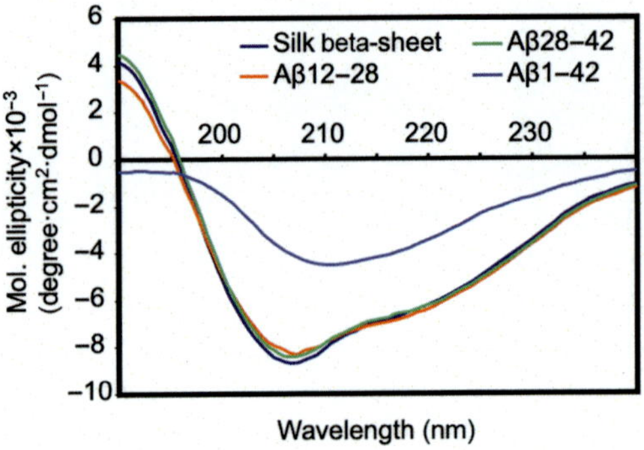

Figure 5.7 CD spectra of the peptides after the incubation for 72 h [30].

for 72 h (Fig. 5.9). Silk beta-sheet, Aβ(12−28), and Aβ(28−42) showed the nearly same beta-sheet content (approximately 26%), while Aβ(1−42) showed around 40% beta-sheet content. Thus the peptide solutions prepared for the present study consisted of a high-content of beta-sheet structure. Furthermore, the CD spectra of the samples after 72 h incubation were identical among the silk beta-sheet, Aβ(12−28), and Aβ(28−42) (Fig. 5.7) [30].

The nanoassemblies and microfibrils of the peptides were evaluated for cytotoxicity against differentiated PC12 cells using the MTS assay [30]. The peptide solutions before and after incubation were tested at different concentrations (55, 110, and 220 μg/mL) (Fig. 5.8). The silk beta-sheet peptide solution before the incubation (0 h) at the highest concentration showed little cytotoxicity (83% ± 3%). The silk beta-sheet peptide solution incubated for 24, 48, and 72 h, containing nanoassemblies and microfibrils, demonstrated no significant cytotoxicity up to 220 μg/mL. In contrast, the cytotoxicity of Aβ(12−28), Aβ(28−42), and Aβ(1−42) peptide solutions increased with an increase in incubation time for the formation of nanoassemblies and microfibrils. Besides, Aβ(12−28) and Aβ(28−42) at 220 μg/mL and 72 h demonstrated significantly higher cytotoxicity (65% and 61% cell viability) in comparison to the full-length Aβ(1−42) (81%). The sequence of β1 of Aβ, Aβ(12−28), is composed of not only hydrophobic amino acids but also basic, polar, and acidic amino acids. Therefore it took a longer time to form nanoassembly-like oligomers due to the less hydrophobic interactions, and to show cytotoxicity to cells in comparison to β2 of Aβ, Aβ(28−42), which contained more hydrophobic amino acids.

These data about nanoassembled structures, beta-sheet content, and cytotoxicity of the peptides suggest that peptide cytotoxicity is dependent

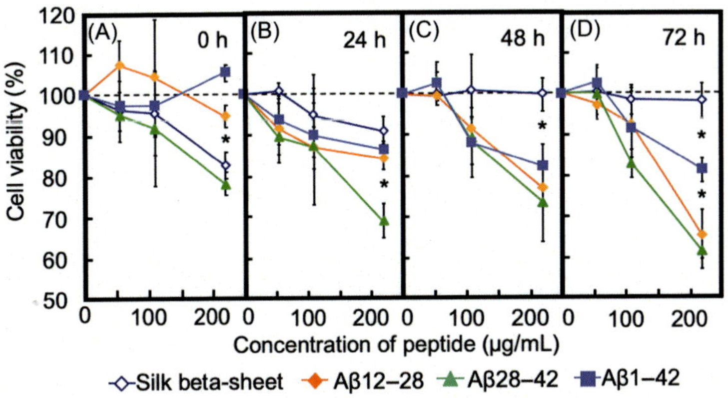

Figure 5.8 Cell viability (percentage of active cells compared to controls) with the peptides on differentiated PC12 cells for 24 h measured by MTS assay. The added peptide solutions were before (A) and after incubation for 24 (B), 48 (C), and 72 h (D). Data are represented as mean ± standard deviation ($n = 8$). *Significant difference between two groups at $P < .05$ [30].

on a combination of the nanoassembly structure as well as the amino acid sequences of beta-sheet forming peptides. Although silk beta-sheet peptides form nanoassemblies and microfibrils with high beta-sheet content, they show no significant cytotoxicity to neurons. These data concerning silk beta-sheet demonstrate that peptides composed of Gly and Ala do not express significant cytotoxicity, most likely because of the lack of charged groups. On the other hand, nanoassemblies of Aβ peptides with the same beta-sheet content and assembly structures as silk beta-sheets, demonstrate cytotoxicity to cells, indicating that the amino acid sequence of amyloid-beta-sheet has a specific effect, which is not possessed by the silk beta-sheets, on cells. As mentioned above, there are significant structural differences between silk beta-sheet and Aβ peptides, including the distribution of charged amino acids in the sequences, namely the composition of charged and hydrophilic amino acids. The Aβ fragments with Lys28 at the *N*-terminus have been reported to be more toxic than the other Aβ fragments, because of the charged Lys. Lys28 was also suggested to be a feasible key to the collapse of the cross-beta structure and playing a significant role in determining the neurotoxicity of Aβ peptides. Aβ solution without any assembly usually shows no significant cytotoxicity to neurons [17], likely because water-soluble charged peptides may not preferentially bind to the surface of cells, in contrast to Aβ nanoassembly. The peptide cytotoxicity is therefore strongly caused by electrically charged nanoassemblies in addition to beta-sheet formation via hydrophobic interactions [30]. Also, the comparison of Aβ(12–28) with silk beta-sheet peptides demonstrates that the property of peptides to form fibrils does not alone implicate cytotoxicity to neurons. These findings provide options to address structural mechanisms of amyloids for various disease proteins related to cytotoxicity, as well as noncytotoxic protein-based biomaterial designs.

5.3 Hydration state for cell viability

To improve the cell viability on polymeric materials, including structural proteins, hydration is one of the most important conditions. As a typical hydrated polymeric material, hydrogel is an attractive biomaterial for regenerative medicine and tissue engineering because of its excellent biocompatibility, which is attributed to its high water content of over 90%. The role of water molecules in hydrogels has been investigated by many researchers, with the result that bound (nonfreezing), bulk (freezing), and intermediate (freezing bound) water have been shown to exist in hydrogels [31–33]. Differential scanning calorimetry (DSC) and nuclear magnetic resonance relaxometry have been used as typical methods to characterize and distinguish the state of water molecules in polymeric hydrogels [34,35]. The difference in the state of water in organisms has been suggested to affect various biological structures, including their size, in bacterial and vegetative cells [36–38]. The bulk water is not crucial for enzymatic hydration, whereas the bound water plays an important role in enzymatic catalysis [39,40]. The biological activities of enzymes and proteins have been reported to depend on how the water molecules associate with these bioactive molecules, that is, the activities depend on the bound water content of the enzymes and proteins [41]. The bound water content has been considered a significant factor in the control of drug release rate as well as enzymatic activity in hydrogel-based biomaterials [42].

The state of water molecules is vital for structural protein-based materials as well. As described above, silk fibroins have been successfully used in the biomedical field as sutures for several decades, and have also been explored as biomaterials for cell culture, tissue engineering, and drug delivery systems, earning Food and Drug Administration approval for such expanded utility because of their excellent mechanical properties, versatility in processing, and low cytotoxicity [5,10,43]. Further, the DPs of silk proteins with beta-sheet structures, when exposed to alpha-chymotrypsin, have recently been identified and shown to have no cytotoxicity in vitro neuron cells [11,30].

Based on the beta-sheet structure content and pH of the silk solution, the gelation mechanism of silk solutions was studied, with the results showing that the gelation depends on the beta-sheet content [44,45]. Silk-based hydrogels have also been investigated by Kaplan and coworkers, who found that the gelation of silk solution was induced by pH change, ultrasonication, or vortex [46–50]. Upon the gelation to form physical hydrogel, a water-soluble silk molecule is shown to transform into a beta-sheet structure. Subsequently, the beta-sheet structures aggregated to form cross-linking points, resulting in molecular network structures, namely, hydrogels [48,50]. Human mesenchymal stem cells (hMSCs) grew and proliferated over 21 days in the sonication-induced hydrogel prepared with 4% silk solution, indicating that the silk hydrogel is biocompatible and low-cytotoxic enough to be used in cell encapsulation [48,50].

A facile and quick method was developed to prepare silk hydrogel using ethanol, and also analyzed the gelation behavior, state of water, secondary structure, and mechanical properties of the resulting hydrogel [51]. The highest water content of the silk hydrogels was nearly 99%; this content was observed at a silk concentration was 13 g/L. A series of DSC profiles of the silk hydrogel with various concentrations of silk proteins was recorded (Fig. 5.9). Bound (nonfreezing) and bulk (freezing) water in hydrogels have been investigated by a combination of DSC and the other complementary approaches [31–33]. Based on the previous studies regarding the DSC

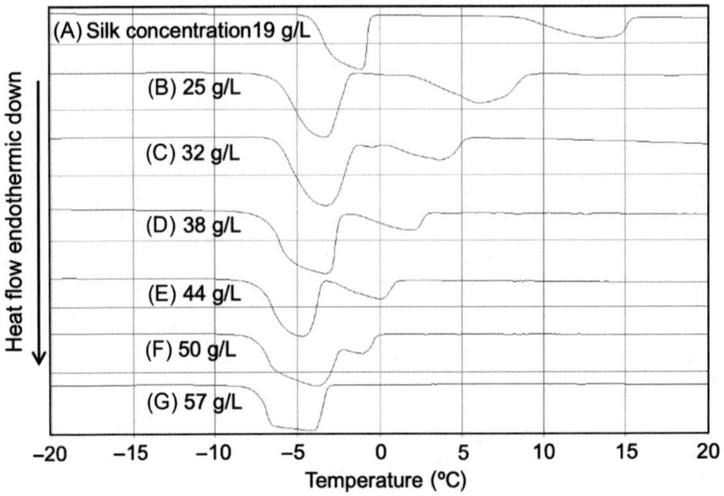

Figure 5.9 DSC endothermic profiles of silk hydrogels prepared at various concentrations of silk proteins [51].

characterizations of hydrogels [33,52–55], the melting peaks of water at a relatively lower temperature, around 0°C, were assigned to bound water, while the melting peaks of water at relatively higher temperature, around 4°C–10°C, were assigned to bulk water. Two major peaks originating from bound (lower temperature) and bulk water (higher temperature) in the silk hydrogels were observed (Fig. 5.9A–F) [51]. The melting temperature (T_m) of the bulk water decreased with an increase in silk concentration, and the peak of the bulk water finally merged with the peak of the bound water around −4°C at a silk concentration of 57 g/L. The T_m and enthalpy of fusion (ΔH) of the bound and bulk water in the silk hydrogels were further characterized quantitatively [51]. The T_m and ΔH of the bulk water in the silk hydrogels significantly decreased with an increase in the concentration of silk proteins, whereas those of bound water were almost invariable throughout the range of concentrations, indicating that the concentration of silk protein in the physical silk hydrogel affects the content and state of water molecules.

The cell viability of hMSC on the silk hydrogels prepared at various silk solution concentrations was evaluated using an MTS assay (Fig. 5.10) [51]. The cell viability of hMSC, namely, the noncytotoxicity of the silk hydrogel, significantly increased with an increase in silk concentration, which may have been due to an increase in the beta-sheet content, elastic modulus, network size, and bound water content. In the previous study

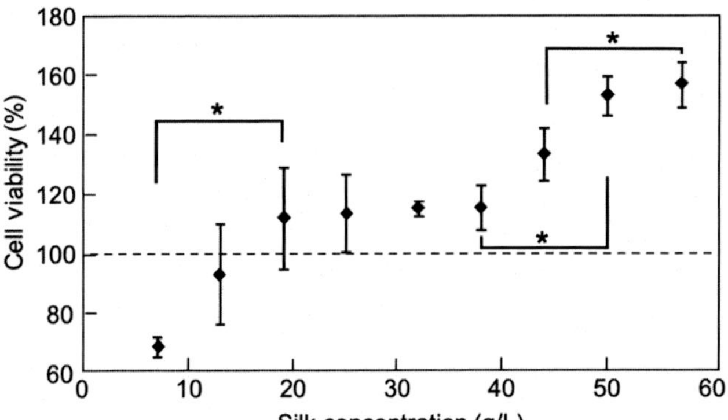

Figure 5.10 Cell viability of hMSC seeded on the silk hydrogels, which was determined from the absorbance at 490 nm measured using the cell cultures after incubation for 48 h. A cell viability of 100% was calculated from a positive control, namely, the cell culture seeded on a cell culture plate after incubation for 48 h. Error bars represent the standard deviation of samples ($n = 3$). *Significant difference between two groups at $P < .05$ [51].

[48], silk hydrogel—based encapsulation with lower beta-sheet content and lower elastic moduli was reported to show higher cell viability. As shown in Fig. 5.10, the cell viability increased monotonically with an increase in silk concentration, whereas the elastic moduli showed no monotonical dependence on silk concentrations. Also, bound water in hydrogels has been reported to play an important role in enzymatic catalysis [39,40,42]. Besides, ionic-surface substrates are considered to be more cell-adhesive materials than hydrophobic-surface substrate, because of the presence of bound water at the surface of the ionic-surface substrates [56,57]. Thus the bound water contents of silk hydrogels may play a more important role in the cytotoxicity of silk hydrogels than the beta-sheet contents or the elastic modulus of silk hydrogels [51].

The DSC data showed that the amount and the mobility of the bulk water in the silk hydrogel decreased with an increase in silk concentrations, whereas the amount of the bulk water decreased, namely, the relative amount of the bound water increased. Considering the studies on the state of water of the silk hydrogel [39,40,42,51,56,57], the cell viability of hMSC on the silk hydrogels was suggested to be under the influence of the ratio of the bulk and bound water. In other words, the improvement of the cell viability of the silk hydrogel implies that the cells and the cell-adhesion proteins in the extracellular matrix preferentially expand and adhere to a substrate containing more bound water. The cell-adhesion proteins also may need bound water to exhibit their functions, similar to the other proteins [39,40,42]. The bound water in the silk hydrogel, therefore, accelerates cell-adhesion proteins, such as fibronectins, in the cellular matrix to interact with the surface of the silk hydrogels, whereas the bulk water would disturb the cell-adhesion proteins to adhere on the surface of the silk hydrogels, due to the relatively higher mobility of water. This new insight into the state of water of hydrogels provides options to design hydrogel-based biomaterials to form cell-interactive biointerfaces.

The bulk water content of the silk hydrogel was found to be readily regulated by the concentration of silk proteins, which helps to investigate the effects of the state of water of polymeric hydrogel on the other properties [51]. The influence of the state of water in the silk hydrogel on the cytotoxicity was recognized by means of DSC and MTS assay. Based on the results, the bound water is considered to support cell-adhesion proteins in the cellular matrix to interact with the surface of the silk hydrogels. On the other hand, the bulk water would disturb the cell-adhesion proteins to adhere on the surface of the silk hydrogels, due to the relatively higher mobility of water.

5.4 Reactive oxygen species response

Reactive oxygen species (ROS) are reactive species that contain oxygen, such as peroxides, superoxides, hydroxyl radicals, and singlet oxygen. In biology, ROS are produced as a by-product of the intracellular metabolism of oxygen. When exposed to environmental stresses (ultraviolet, heat, drought, high salt concentration, etc.), ROS levels rise dramatically and accumulate, to be perceived as oxidative stress. ROS production is strongly influenced by plant stressors response.

As a stimuli-responsive system, we focus on peptide-based systems specifically sensitive to biologically relevant ROS in plant cells. Proline-rich peptides with repeating proline residues are known to be labile in the presence of ROS, leading to a cleavage of peptides at the oligoproline sequence [58–61]. In addition, oligoproline-containing peptides exhibit unique amphiphilic properties, which can be available for nanoassembly formulation and transmembrane unit [62,63]. Nanoplatforms with ROS-responsiveness are promising candidates for controlling the release of cargo molecules, especially in plant cells, since plant tissues produce a tremendous amount of ROS compared to those of animals via their biological activity such as oxidative phosphorylation and/or photosynthesis [64–66]. The generation of ROS from mitochondrion and chloroplast are much higher than that in the cytosol, and these two types of organelles continuously provide fresh ROS. Furthermore, photosynthesis significantly raises the concentration of ROS released from chloroplasts from 250 nM to 0.1–1 mM [66]. For these viewpoints, ROS is an appropriate environmental stimulus for the controlled release with high selectivity and sensitivity in plant cells under light conditions. Oligoproline-containing peptides were used as a ROS sensitive motif. Oligoproline-containing peptides encapsulate hydrophobic molecules and release upon exposure ROS at the physiological condition in plant cells under lights (Fig. 5.11).

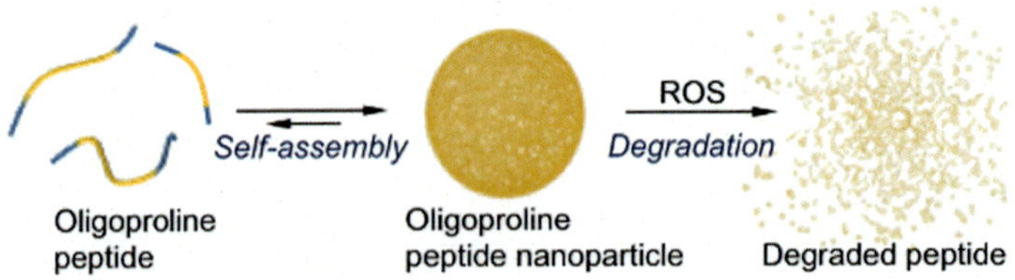

Figure 5.11 Preparation of ROS-triggered degradable nanoparticles using oligoproline-containing peptides.

References

[1] K. Numata, D.L. Kaplan, Biologically Derived Scaffolds, in: D. Farrar (Ed.), Advanced Wound Repair Therapies, Woodhead Publishing, 2011, pp. 524–551.

[2] C. Holland, K. Numata, J. Rnjak-Kovacina, F.P. Seib, The biomedical use of silk: past, present, future, Adv. Healthc. Mater. 8 (1) (2019) 1800465.

[3] C.H. Lee, A. Singla, Y. Lee, Biomedical applications of collagen, Int. J. Pharm. 221 (1-2) (2001) 1–22.

[4] K. Numata, How to define and study structural proteins as biopolymer materials, Polym. J. 52 (9) (2020) 1043–1056.

[5] G.H. Altman, F. Diaz, C. Jakuba, T. Calabro, R.L. Horan, J. Chen, et al., Silk-based biomaterials, Biomaterials 24 (3) (2003) 401–416.

[6] M. Jewell, W. Daunch, B. Bengtson, E. Mortarino, The development of SERI® Surgical Scaffold, an engineered biological scaffold, Ann. N. Y. Acad. Sci. 1358 (2015) 44–55.

[7] Z.Y. Jiao, Y. Song, Y. Jin, C. Zhang, D. Peng, Z.Z. Chen, et al., In vivo characterizations of the immune properties of sericin: an ancient material with emerging value in biomedical applications, Macromol. Biosci. 17 (12) (2017).

[8] S. Franz, S. Rammelt, D. Scharnweber, J.C. Simon, Immune responses to implants—a review of the implications for the design of immunomodulatory biomaterials, Biomaterials 32 (28) (2011) 6692–6709.

[9] A.E. Thurber, F.G. Omenetto, D.L. Kaplan, In vivo bioresponses to silk proteins, Biomaterials 71 (2015) 145–157.

[10] K. Numata, D.L. Kaplan, Silk-based delivery systems of bioactive molecules, Adv. Drug. Deliv. Rev. 62 (15) (2010) 1497–1508.

[11] K. Numata, P. Cebe, D.L. Kaplan, Mechanism of enzymatic degradation of beta-sheet crystals, Biomaterials 31 (10) (2010) 2926–2933.

[12] Y. Wang, D.D. Rudym, A. Walsh, L. Abrahamsen, H.J. Kim, H.S. Kim, et al., In vivo degradation of three-dimensional silk fibroin scaffolds, Biomaterials 29 (24-25) (2008) 3415–3428.

[13] C.A. Ross, M.A. Poirier, What is the role of protein aggregation in neurodegeneration? Nat. Rev. Mol. Cell Bio. 6 (11) (2005) 891–898.

[14] J.D. Sipe, A.S. Cohen, Review: history of the amyloid fibril, J. Struct. Biol. 130 (2-3) (2000) 88–98.

[15] R.D. Terry, An honorable compromise regarding amyloid in Alzheimer disease, Ann. Neurol. 49 (5) (2001) 684.

[16] E.H. Koo, R. Kopan, Potential role of presenilin-regulated signaling pathways in sporadic neurodegeneration, Nat. Med. 10 (7) (2004) S26–S33.

[17] S. Chimon, M.A. Shaibat, C.R. Jones, D.C. Calero, B. Aizezi, Y. Ishii, Evidence of fibril-like beta-sheet structures in a neurotoxic amyloid intermediate of Alzheimer's beta-amyloid, Nat. Struct. Mol. Biol. 14 (12) (2007) 1157–1164.

[18] R.E. Marsh, R.B. Corey, L. Pauling, An investigation of the structure of silk fibroin, Biochim. Biophys. Acta 16 (1) (1955) 1–34.

[19] Y. Takahashi, M. Gehoh, K. Yuzuriha, Structure refinement and diffuse streak scattering of silk (Bombyx mori), Int. J. Biol. Macromol. 24 (2-3) (1999) 127–138.

[20] T. Arai, G. Freddi, R. Innocenti, M. Tsukada, Biodegradation of Bombyx mori silk fibroin fibers and films, J. Appl. Polym. Sci. 91 (4) (2004) 2383–2390.

[21] B. Lotz, A. Gonthiervassal, A. Brack, J. Magoshi, Twisted single-crystals of Bombyx-Mori silk fibroin and related model polypeptides with beta-structure—a correlation with the twist of the beta-sheets in globular-proteins, J. Mol. Biol. 156 (2) (1982) 345–357.

[22] M.Z. Li, M. Ogiso, N. Minoura, Enzymatic degradation behavior of porous silk fibroin sheets, Biomaterials 24 (2) (2003) 357–365.

[23] R.L. Horan, K. Antle, A.L. Collette, Y.Z. Huang, J. Huang, J.E. Moreau, et al., In vitro degradation of silk fibroin, Biomaterials 26 (17) (2005) 3385–3393.
[24] S.W. Ha, H.S. Gracz, A.E. Tonelli, S.M. Hudson, Structural study of irregular amino acid sequences in the heavy chain of Bombyx mori silk fibroin, Biomacromolecules 6 (5) (2005) 2563–2569.
[25] K. Numata, Poly(amino acid)s/polypeptides as potential functional and structural materials, Polym. J. 47 (8) (2015) 537–545.
[26] U. Slotta, S. Hess, K. Spiess, T. Stromer, L. Serpell, T. Scheibel, Spider silk and amyloid fibrils: a structural comparison, Macromol. Biosci. 7 (2) (2007) 183–188.
[27] N.A. Oktaviani, A. Matsugami, A.D. Malay, F. Hayashi, D.L. Kaplan, K. Numata, Conformation and dynamics of soluble repetitive domain elucidates the initial beta-sheet formation of spider silk, Nat. Commun. 9 (2018).
[28] K. Lundmark, G.T. Westermark, A. Olsen, P. Westermark, Protein fibrils in nature can enhance amyloid protein A amyloidosis in mice: Cross-seeding as a disease mechanism, Proc. Natl. Acad. Sci. U S A 102 (17) (2005) 6098–6102.
[29] R. Kisilevsky, L. Lemieux, L. Boudreau, D.S. Yang, P. Fraser, New clothes for amyloid enhancing factor (AEF): silk as AEF, Amyloid 6 (2) (1999) 98–106.
[30] K. Numata, D.L. Kaplan, Differences in cytotoxicity of beta-sheet peptides originated from silk and amyloid beta, Macromol. Biosci. 11 (1) (2011) 60–64.
[31] C.A. Fyfe, A.I. Blazek, Investigation of hydrogel formation from hydroxypropylmethylcellulose (HPMC) by NMR spectroscopy and NMR imaging techniques, Macromolecules 30 (20) (1997) 6230–6237.
[32] H. Hatakeyama, T. Hatakeyama, Interaction between water and hydrophilic polymers, Thermochim. Acta 308 (1-2) (1998) 3–22.
[33] X. Qu, A. Wirsen, A.C. Albertsson, Novel pH-sensitive chitosan hydrogels: swelling behavior and states of water, Polymer 41 (12) (2000) 4589–4598.
[34] M. Carenza, G. Cojazzi, B. Bracci, L. Lendinara, L. Vitali, M. Zincani, et al., The state of water in thermoresponsive poly(acryloyl-L-proline methyl ester) hydrogels observed by DSC and H-1-NMR relaxometry, Radiat. Phys. Chem. 55 (2) (1999) 209–218.
[35] V.J. McBrierty, S.J. Martin, F.E. Karasz, Understanding hydrated polymers: the perspective of NMR, J. Mol. Liq. 80 (2-3) (1999) 179–205.
[36] Y. Ishihara, H. Saito, J. Takano, Differences in the surface membranes and water content between the vegetative cells and spores of Bacillus subtilis, Cell Biochem. Funct. 17 (1) (1999) 9–13.
[37] Y. Ishihara, J. Takano, S. Mashimo, M. Yamamura, Determination of the water necessary for survival of Bacillus-Subtilis vegetative cells and spores, Thermochim. Acta 235 (2) (1994) 153–160.
[38] A.J. Westphal, P.B. Price, T.J. Leighton, K.E. Wheeler, Kinetics of size changes of individual Bacillus thuringiensis spores in response to changes in relative humidity, Proc. Natl. Acad. Sci. U S A 100 (6) (2003) 3461–3466.
[39] G. Careri, A. Giansanti, J.A. Rupley, Proton percolation on hydrated lysozyme powders, Proc. Natl. Acad. Sci. U S A 83 (18) (1986) 6810–6814.
[40] G. Careri, E. Gratton, P.H. Yang, J.A. Rupley, Correlation of IR-spectroscopic, heat-capacity, diamagnetic susceptibility and enzymatic measurements on lysozyme powder, Nature 284 (5756) (1980) 572–573.
[41] H.K. Pessen, T.F. Kumosinski, Measurements of protein hydration by various techniques, in: C.H.W. Hirs, S.N. Timasheff (Eds.), Methods in Enzymology, vol. 117, Academic Press, New York, 1985, p. 219.
[42] A.S. Hoffman, W.R. Gombotz, S. Uenoyama, L.C. Dong, G. Schmer, Immobilization of enzymes and antibodies to radiation grafted polymers for therapeutic and diagnostic applications, Radiat. Phys. Chem. 27 (4) (1986) 265–273.

[43] Y. Wang, H.J. Kim, G. Vunjak-Novakovic, D.L. Kaplan, Stem cell-based tissue engineering with silk biomaterials, Biomaterials 27 (36) (2006) 6064–6082.
[44] Z.H. Ayub, M. Arai, K. Hirabayashi, Quantitative structural-analysis and physical-properties of silk fibroin hydrogels, Polymer 35 (10) (1994) 2197–2200.
[45] Z.H. Ayub, M. Arai, K. Hirabayashi, Mechanism of the gelation of fibroin solution, Biosci. Biotech. Biochem. 57 (11) (1993) 1910–1912.
[46] X.A. Hu, Q.A. Lu, L. Sun, P. Cebe, X.Q. Wang, X.H. Zhang, et al., Biomaterials from ultrasonication-induced silk fibroin-hyaluronic acid hydrogels, Biomacromolecules 11 (11) (2010) 3178–3188.
[47] T. Yucel, P. Cebe, D.L. Kaplan, Vortex-induced injectable silk fibroin hydrogels, Biophys. J. 97 (7) (2009) 2044–2050.
[48] X.Q. Wang, J.A. Kluge, G.G. Leisk, D.L. Kaplan, Sonication-induced gelation of silk fibroin for cell encapsulation, Biomaterials 29 (8) (2008) 1054–1064.
[49] A. Matsumoto, J. Chen, A.L. Collette, U.J. Kim, G.H. Altman, P. Cebe, et al., Mechanisms of silk fibroin sol-gel transitions, J. Phys. Chem. B 110 (43) (2006) 21630–21638.
[50] U.J. Kim, J.Y. Park, C.M. Li, H.J. Jin, R. Valluzzi, D.L. Kaplan, Structure and properties of silk hydrogels, Biomacromolecules 5 (3) (2004) 786–792.
[51] K. Numata, T. Katashima, T. Sakai, State of water, molecular structure, and cytotoxicity of silk hydrogels, Biomacromolecules 12 (6) (2011) 2137–2144.
[52] S. Baumgartner, J. Kristl, N.A. Peppas, Network structure of cellulose ethers used in pharmaceutical applications during swelling and at equilibrium, Pharm. Res. 19 (8) (2002) 1084–1090.
[53] M. Kodama, Y. Kawasaki, H. Aoki, Y. Furukawa, Components and fractions for differently bound water molecules of dipalmitoylphosphatidylcholine-water system as studied by DSC and H-2-NMR spectroscopy, BBA Biomembranes 1667 (1) (2004) 56–66.
[54] J. Ruiz, A. Mantecon, V. Cadiz, States of water in poly(vinyl alcohol) derivative hydrogels, J. Polym. Sci. Pol. Phys. 41 (13) (2003) 1462–1467.
[55] T. Wang, S. Gunasekaran, State of water in chitosan-PVA hydrogel, J. Appl. Polym. Sci. 101 (5) (2006) 3227–3232.
[56] N. Faucheux, R. Schweiss, K. Lutzow, C. Werner, T. Groth, Self-assembled monolayers with different terminating groups as model substrates for cell adhesion studies, Biomaterials 25 (14) (2004) 2721–2730.
[57] S.K. Robertson, A.F. Uhrick, S.G. Bike, TIRM measurements with cells and liposomes, J. Colloid Interf. Sci. 202 (1) (1998) 208–211.
[58] S.H. Lee, T.C. Boire, J.B. Lee, M.K. Gupta, A.L. Zachman, R. Rath, et al., ROS-cleavable proline oligomer crosslinking of polycaprolactone for pro-angiogenic host response, J. Mater. Chem. B 2 (41) (2014) 7109–7113.
[59] S.S. Yu, R.L. Koblin, A.L. Zachman, D.S. Perrien, L.H. Hofmeister, T.D. Giorgio, et al., Physiologically relevant oxidative degradation of oligo(proline) cross-linked polymeric scaffolds, Biomacromolecules 12 (12) (2011) 4357–4366.
[60] E.R. Stadtman, R.L. Levine, Free radical-mediated oxidation of free amino acids and amino acid residues in proteins, Amino Acids 25 (3) (2003) 207–218.
[61] A. Amici, R.L. Levine, L. Tsai, E.R. Stadtman, Conversion of amino acid residues in proteins and amino acid homopolymers to carbonyl derivatives by metal-catalyzed oxidation reactions, J. Biol. Chem. 264 (6) (1989) 3341–3346.
[62] V. Kubyshkin, S.L. Grage, J. Bürck, A.S. Ulrich, N. Budisa, Transmembrane polyproline helix, J. Phys. Chem. Lett. 9 (9) (2018) 2170–2174.
[63] I.W. Hamley, V. Castelletto, A. Dehsorkhi, J. Torras, C. Aleman, I. Portnaya, et al., The conformation and aggregation of proline-rich surfactant-like peptides, J. Phys. Chem. B 122 (6) (2018) 1826–1835.

[64] H.M. Cochemé, C. Quin, S.J. McQuaker, F. Cabreiro, A. Logan, T.A. Prime, et al., Measurement of H_2O_2 within living Drosophila during aging using a ratiometric mass spectrometry probe targeted to the mitochondrial matrix, Cell Metab. 13 (3) (2011) 340–350.
[65] M. Giorgio, M. Trinei, E. Migliaccio, P.G. Pelicci, Hydrogen peroxide: a metabolic by-product or a common mediator of ageing signals? Nat. Rev. Mol. Cell Biol. 8 (2007) 722.
[66] C.H. Foyer, G. Noctor, Redox sensing and signalling associated with reactive oxygen in chloroplasts, peroxisomes and mitochondria, Physiol. Plant. 119 (3) (2003) 355–364.

Questions for this chapter

1. What kinds of factors determine the cytotoxicity of structural proteins?
2. What is the difference between silk and amyloid-beta peptide in terms of cytotoxicity and neurotoxicity?
3. Describe the importance of water molecules with structural proteins for cell adhesion and cytotoxicity.

CHAPTER 6
Stability

6.1 Thermal stability

The studies on thermal stability of structural proteins have not been investigated widely to date. Thermal properties of proteins are usually evaluated in terms of melting temperature, where proteins are denatured, and also glass transition temperature, where protein molecules show transition with water plasticization. Here, with reference to silk protein as a typical structural protein, its thermal stability and degradation are summarized.

6.1.1 Thermal stability of silk

Silk fibroin has been targeted as one of the most promising protein-based biomaterials, owing to its attractive physical and biological properties such as its toughness, lightweight, low cytotoxicity, and biodegradability [1–4]. Besides, silk protein has recently demonstrated versatility in its water- or solvent-mediated processability, and it is able to form a number of materials, such as films, sponges, and hydrogels, as well as being used as a gene/drug carrier for tissue engineering and regenerative medicine [5–7]. Silk-based structural materials on a bulk scale have not yet been realized for practical materials, although natural silk is designed and used as a structural material, such as silkworm cocoons and spider webs/draglines [8]. To exploit silk as a practical structural material on a bulk scale, it is necessary to understand its thermal stability as well as other properties during thermal processing. Although the details of the thermal degradation behaviors are described in Figs. 4.4–4.9, the present section introduces the thermal stability of silk fibroins in various material forms. The sequences of the repetitive fibroin domains from the different samples subjected to the thermal stability tests are listed in Fig. 6.1A [11]. In the case of *Actias aliena* and *Saturnia jonasii* only sequences from the congeneric *Actias selene* and *Saturnia japonica*, respectively, were available in the protein database; these were used for the analysis on the assumption that the composition of the repetitive regions has been conserved at the genus level, as seen among *Antheraea* species [12–14]. The composition of silk tandem repeats is almost conserved within each species on the basis of the current knowledge and available

Biopolymer Science for Proteins and Peptides
DOI: https://doi.org/10.1016/B978-0-12-820555-6.00008-2

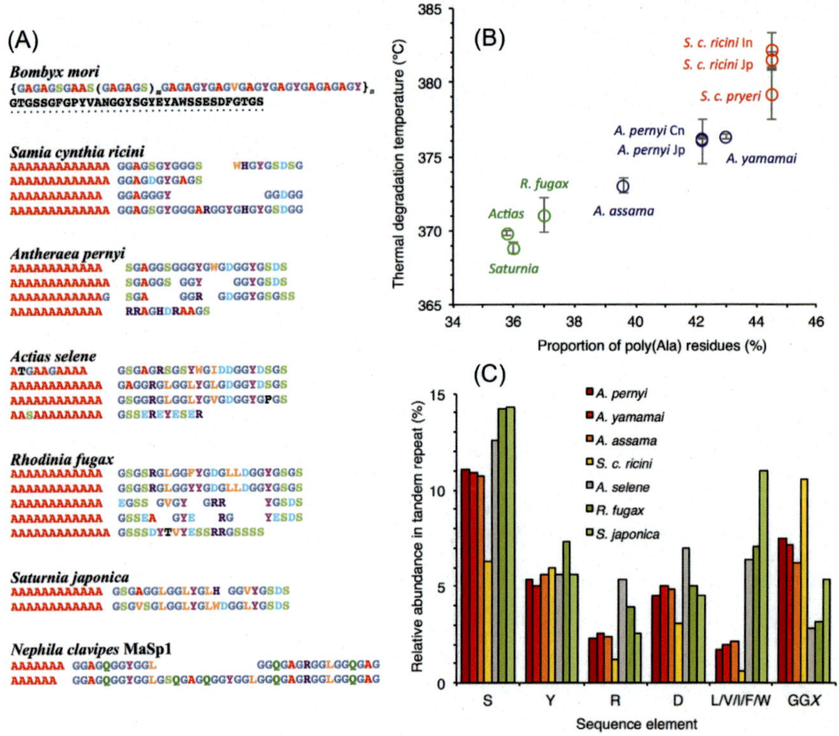

Figure 6.1 Fibroin amino acid sequence analysis. (A) Representative repetitive sequences of the different silk fibroin are shown: *Bombyx mori* heavy chain (GenBank AF226688), demonstrating the hierarchical arrangement of motifs, and with the conserved spacer sequence underlined, *Samia cynthia ricini* (BAQ55621), *Antheraea pernyi* (AAC32606), *Actias selene* (deduced sequence) [9], *Rhodinia fugax* (BAG84270), *Saturnia japonica* (BAH02016), and MaSp1 from spider dragline silk of *Nephila clavipes* [10] (M37137). (B) Relative proportion of poly(Ala) residues within the saturniid fibroin tandem repeats versus the decomposition temperature (T_d) of the different samples derived from TGA, with error bars corresponding to standard deviation values. (C) Average relative abundance of selected residues and motifs within the repetitive regions of the different fibroin samples [11].

information [15–18]. Those amino acid sequences showed a preponderance of Gly, Ala, Ser, and Tyr residues, although the organization of the internal repeats is different between the bombycoid and saturniid fibroins. The *Bombyx mori* sequence features repetitive arrays of $(GA)_nGX$ (where X = S, Y, or V) to form large blocks of varying length and arrangement that are interrupted by ~43-residue spacer sequences [19,20]. In contrast, the saturniid repetitive domains consisted of alternating tandem repeats of poly(Ala) and Gly-rich regions. The poly(Ala) blocks comprised on average

12–13 contiguous Ala residues, whereas the Gly-rich sequences feature various combinations of G*X* and GG*X* (*X* = typically A, S, Y, D, L, or R). Interspecific variations within the Gly-rich sequences were noted. The *Samia cynthia ricini* sequence carried four subtypes bearing GG*X* motifs, while *Antheraea* sequences (*A. pernyi*, *A. yamamai*, and *A. assama*) featured three GG*X*-containing subtypes plus a shorter subtype rich in charged polar residues (e.g., RRAGHDRAA in *A. pernyi*). In *A. aliena*, *Rhodinia fugax*, and *S. jonasii*, the Gly-rich regions contained a relatively low abundance of GG*X* motifs; however, there were a relatively high proportion of bulky, hydrophobic residues.

Quantification of the abundance of poly(Ala) residues revealed that *S. c. ricini* contained the highest level among the saturniid repetitive sequences, at approximately 44.5% of the total length of each tandem repeat, versus 43% for *A. yamamai*, 42.2% for *A. pernyi*, 39.6% for *A. assama*, 37% for *R. fugax*, 36% for *S. japonica*, and 35.8% for *A. selene* [11]. Strikingly, plotting the values of the abundance of poly(Ala) residues among the different species revealed an excellent agreement with the thermal degradation values observed from the thermogravimetric analysis (TGA) (Fig. 6.1B), with the highest stability corresponding to the *Samia* samples, with a degradation temperature (T_d) at approximately 382°C. Conversely, the lowest abundance of poly(Ala) residues, in the *Actias*, *Rhodinia*, and *Saturnia* sequences, exhibited the lowest thermal degradation temperatures, at around 369°C–371°C. These findings support a view that the poly(Ala) runs constitute the β-crystalline component of silk that imparts rigidity to the polymer structure. Interestingly, no such correlation was observed between the average lengths of the poly(Ala) stretches per se and degradation temperature, suggesting that the ratio of crystalline to noncrystalline fractions is the main determinant of thermal stability.

It is useful to compare the sequences of *Antheraea* with *Actias* as they are phylogenetically closely related [21] yet produce silks with divergent material properties; the latter is particularly brittle and exhibits low extensibility, as reported here and elsewhere [22]. Both sequences harbor a similar subset of tandem repeats, although *A. selene* carries a relatively higher proportion of residues with bulky, hydrophobic side chains (notably Leu) compared to *Antheraea*. Plotting the abundance of selected sequence elements within the repetitive regions of the different fibroins gives a clearer view of these differences (Fig. 6.1C) [11]. Comparing the different saturniid species, the remarkable variations were recognized in the relative abundance of large hydrophobic residues, namely Leu, Val, Ile, Phe, and

Trp, and in the abundance of GG*X* motifs. *S. c. ricini* in particular, and the three *Antheraea* species, demonstrated high GG*X* levels but relatively few hydrophobic residues. Conversely, the wild silks of *R. fugax*, *A. selene*, and *S. japonica* all contained much higher levels of the bulky hydrophobic residues but fewer GG*X* motifs [23,24]. These results suggest that the balance between the abundance of large hydrophobic residues and flexible glycine-rich motifs, both situated within the amorphous phase, plays an essential role in determining the overall material properties of the fibers.

6.1.2 Effects of water on thermal degradation

The thermal properties of silks are known to be influenced by water molecules [25–38]. Water molecules in silk materials can primarily be categorized into two types: free water and bound water as explained in Section 3.5.1. Free water is unbound water that behaves like bulk water, whereas bound water strongly interacts with silk molecules and demonstrates different properties from bulk water [33,39–41]. To date, the effects of bound water on the thermal, mechanical, and biological properties of silks have been widely studied [25–38] as introduced in Section 3.1.

The effect of relative humidity (RH) on the thermal stability and degradation of silk films was characterized (see the color chages of silk films in Figs. 6.4–6.12) [42]. To determine the change in molecular weight in more detail, the molecular weight of the silk molecules before and after heat treatment at 240°C was investigated using viscosity measurements with an Ubbelohde viscometer at 40°C. The kinematic viscosities of the silk incubated at 6% RH before and after heating at 240°C were 2.2 ± 0.007 and 2.4 ± 0.020 $mm^2 \cdot s^{-1}$, respectively [42]. The kinematic viscosities at RH 75% before and after heating were 2.2 ± 0.005 and 2.3 ± 0.007 $mm^2 \cdot s^{-1}$, respectively. These results confirmed that the molecular weight of the silk was maintained after the heating process at 240°C. The molecular weight distribution after the heating treatment of silk films was evaluated by SDS-PAGE (sodium dodecyl sulfate polyacrylamide gel electrophoresis) (Fig. 6.2) [42]. As a result, the molecular weight and its distribution did not change even after the heating process at 240°C, regardless of the humidity conditions. In the case of protein/polypeptide, the molecular weight can be measured roughly by gel electrophoresis. Generally, the band position in the gel is affected by the hydrophobicity and electrocharges of protein/polypeptide. Furthermore, disulfide bonds of protein/polypeptide influence the band position in SDS-PAGE.

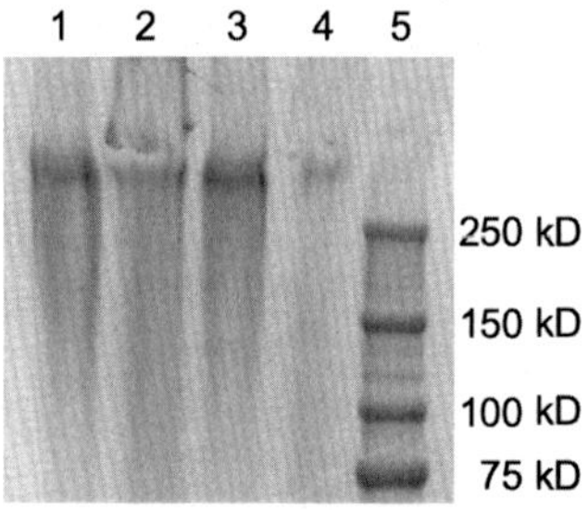

Figure 6.2 SDS-PAGE of silk films before and after heating at 240°C. Silk films incubated under RH 6% at 40°C (lane 1), RH 6% at 240°C (lane 2), RH 75% at 40°C (lane 3), and RH 75% at 240°C (lane 4). Lane 5 denotes a molecular marker. In the case of protein/polypeptide, the molecular weight can be measured roughly by gel electrophoresis [42].

6.1.3 Thermal stability at different heating rates

To study the glass transition and crystallization in silk films, differential scanning calorimetry (DSC) experiments were performed (Fig. 6.3A) [42]. The broad endothermic peak detected between 40 and 200°C was derived from the water removal. The positon of the endothermic peak shifted to lower values as the water content increased, which was the same trend that was observed with the TGA result. In addition, an exothermic peak attributed to beta-sheet crystallization was detected at approximately 220°C, a result in good agreement with previous reports [28,43,44]. The crystallization peak became broader at values over 69% RH until eventually no crystallization was detected at 84% RH.

The T_g derived from the water molecules in the silk films was not detected via conventional DSC analysis at a scanning rate of 20°C/min (Fig. 6.3A); this is because the large endothermic peak derived from the water removal could overlap with the baseline shift that occurred at the T_g. To detect the T_g of silk films, the sample pan should be hermetically sealed to retain the water molecules inside the sample pan, according to a previous study [25]. The goal in this study included identifying the water removal behavior of silk materials; thus the sample pan could not be sealed. Instead, fast-scanning DSC at a scan rate of 6000°C/min can be used to mitigate the water removal peak [45]. The results shown in Fig. 6.3B indicate that T_g decreased as RH increased [42]. The T_g of the dry sample shifted with RH until 23% RH, at which point the T_g shift was saturated; this is because most water molecules are bound at a moderately low RH (i.e., less than 23%), whereas water molecules interact with

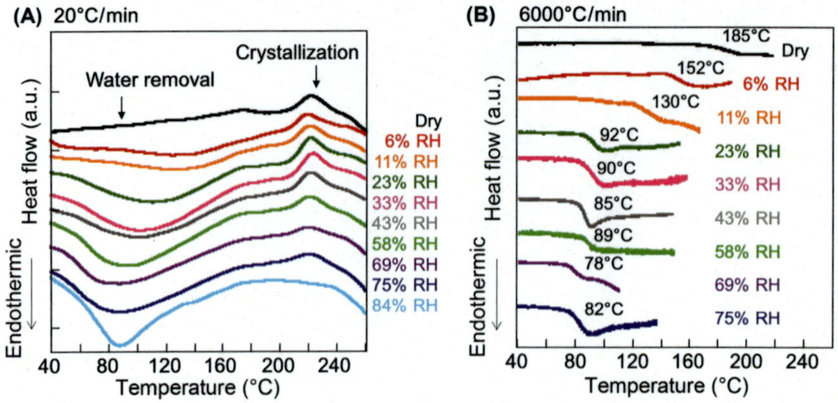

Figure 6.3 DSC measurements of *Bombyx mori* silk films incubated at different RHs: (A) conventional DSC results at the scan rate of 20°C/min; (B) fast-scanning DSC results at the scan rate of 6000°C/min [42].

the silks weakly as free water at higher RH values. The bound water is critical to the water plasticization effect; thus the water molecules that interact with the silk materials at low RH primarily contribute to the T_g shift. After the glass transition, the shape of the silk film changed, and the fast-scanning DSC plots showed noisy and unusual curves.

Torres et al. reported T_g for degummed silk fibers was detected by DSC [46]. They used the silk fibers after three-times degumming to extract the sericin, and hence the silk could contain more amorphous regions to induce T_g. Guan et al. demonstrated that the occurrence of T_g is dependent on the grade of the silk: the low grade of silk with damages showed lower T_g, while the high grade of silk did not show such T_g [27]. We believe that the difference of the degumming procedure could affect the occurrence of T_g. In fact, the T_g was detected in the case of films in Fig. 6.3B, and hence we concluded that no transition was detected in silk raw fibers and degummed fibers by fast-scanning DSC.

6.1.4 Material shape affects thermal properties

It was reported that the effect of water on the crystallization, thermal degradation, water removal of *B. mori* silk cocoon raw fibers, degummed fibers, and films [42]. The thermal stability of silk materials is dependent on their crystallinity and their water content. The DSC and wide-angle X-ray scattering (WAXS) measurements can confirm that water molecules and the thermal treatment induce the crystallization of protein materials due to the plasticization effect of water molecules. Here shows the effects of material shapes on thermal properties.

The surface morphology and retardation of the silk film during the heating process were characterized with the silk films incubated at different relative humidites (RH 6%–RH 75%) under different heating conditions (40°C, 160°C, 180°C, and 240°C), as described in Figs. 4.4–4.12 [42]. The surface morphology and retardation of the silk films can be maintained at humid conditions such as RH 75% up to 240°C, indicating that the excess of water molecules in the silk film was able to prevent its oxidation. The results of the retardation measurements and the optical micrographs of silk films showed that the humidity and thermal treatments did not affect the allignment of crystalline components on the microscale.

TGA analyses of the silk cocoon raw fibers and degummed fibers were performed to elucidate the effect of the material shape and the solid state on the thermal properties of silk [42]. The degradation temperatures that yielded 1 and 5 wt.% losses in the silk materials were similar for different RHs (Fig. 6.4A). The lower thermal stability of the cocoon raw fibers was due to the presence of sericin, and hence the degummed fibers without

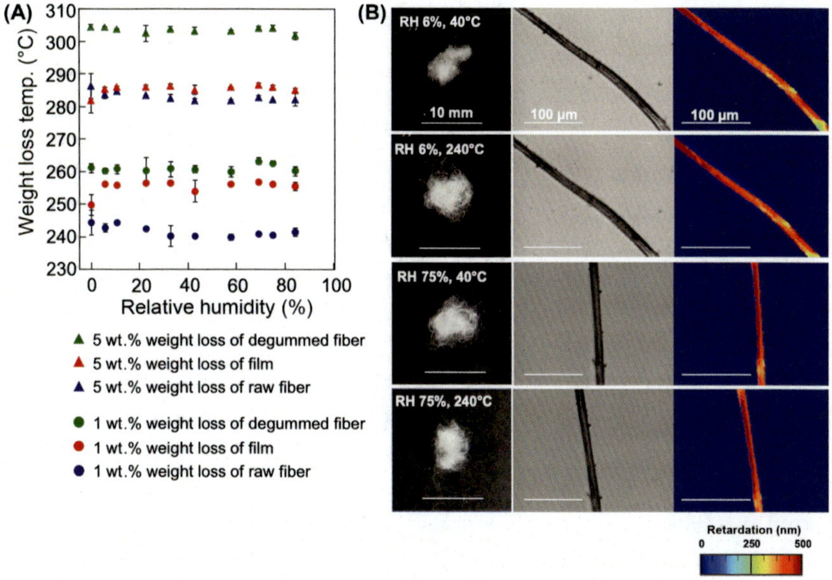

Figure 6.4 Effects of RH on the thremla degradation of silk fibers. (A) 1 wt.% and 5 wt.% loss temperatures. Error bars represent the standard deviation of samples ($n = 3$). (B) Pictures and retardations of silk fibers at RH 6%/RH 75% and 40°C/240°C [42].

sericin exhibited the highest stability. The thermal stability based on the 1 and 5 wt.% losses of the silk films was determined to be between those of the raw and degummed fibers. In terms of the peak of the derivative plot of thermal degradation (Fig. 6.4C), which is the temperature where the most weight loss was detected, both the fiber types showed higher degradation temperatures than those of the films. The thermal stabilities of the cocoon raw fibers and degummed fibers could be explained in terms of their degree of crystallinity. The crystallinity of the films, cocoon raw fibers, and degummed fibers were calculated from the WAXS results at RH 58% and 25°C ($n = 3$) and were found to be equal to 21.3% ± 1.8%, 31.3% ± 2.9%, and 38.2% ± 2.2%, respectively. The lower stability of the silk film was due to the lower amount of crystalline regions contributing to the thermal stability; the degummed fibers contained more crystalline and less amorphous regions, resulting in a higher thermal stability. All of the samples demonstrated a 1 wt.% loss at temperatures greater than 240°C. In addition, the morphology and retardation of the *B. mori* silk cocoon raw fibers and degummed fibers were evaluated at RH 6% and RH 75%, and both properties were shown to remain unaffected during the heating process up to 240°C. Besides, the sericin on the cocoon raw fibers is hydrophilic and thus absorbs water molecules at high RH. Based on Fig. 6.4A, the water molecules bound to the sericin layer destabilized the silk fibroin. The kinematic viscosities of cocoon raw fibers and degummed fibers were measured to investigate the molecular weight changes. However, the molecular weight of silk cocoon raw fibers and degummed fibers were not changed after heat treatment at 240°C. These results suggest that the silk molecule is thermally stable and that the stability can be tuned by the material shape and the crystalline/amorphous states.

Unlike the thermal degradation behaviors of silk materials, the water removal behavior depended on the water content and the RH. The derivative plots of the water removal for silk films, cocoon raw fibers, and degummed fibers were detected via TGA. Water molecules within the silk films evaporated at a higher temperature compared with cocoon raw fibers and degummed fibers, indicating the presence of more bound water in the silk films [42]. The amorphous region of a polymer material contains more bound water than the crystalline region, which was reported for the bound water in cellulose and silk [47–49]. The silk films contained more amorphous regions than both fiber samples; thus more water molecules could exist in the bound state in silk films, which resulted in water removal at a higher temperature than that for the fibers. Thus the water removal behavior of silk materials was influenced by the ratio of bound water to free water, which is strongly related to the crystallinity of silk materials.

Also, WAXS analyses were performed with the silk cocoon raw fibers and degummed fibers, which were prepared at RH 6%, RH 58%, and RH 75% [42]. The WAXS results showed that the beta-sheet structure was predominant in all of the fiber samples before heat treatment and was independent of the water content. The thermal treatment induced the formation of additional beta-sheet structures; however, a clear transition from helical to beta-sheet structures was not detected, which was different from the observations made with the silk films. Thus the silk fibers show more resistance to the water plasticization effect than do the silk films.

The silk fiber was more resistant to the effects of the water molecules, which indicated that the effect of water on silk materials can be prevented by changing the shape and crystalline state of the silk materials [42]. The structural and mechanical characterizations of the silk films demonstrated the silk film prepared at RH 97% showed both crystallinity and ductility simultaneously. This is an exception to the general rule of the material science. The new finding of the water-dependent thermal properties of silk is essential for producing biomedical materials and bulk-scale structural materials with industrial applications, such as car frames, bulletproof jackets, and hydrogen storage tanks.

6.1.5 Disulfide bond

The disulfide bond is one of the most important and dynamic bonds to functionalize polypeptide and protein-based materials. The effect of disulfide bonds on thermal properties was previously studied by using OligoCys with well-defined structures. OligoCys was easily prepared via a proteinase K-catalyzed chemoenzymatic synthesis (Fig. 6.5) [50].

To study changes in the chemical and physical properties of oxidized OligoCys, oxidation was performed by treatment with H_2O_2 solutions of different concentrations under a two-step nucleophilic reaction mechanism [51]. Nucleophilic attack of the thiolate anion on the unionized H_2O_2 to generate cysteine sulfenic acid as an intermediate will yield disulfide or cysteine sulfinic/sulfonic acid. Raman spectra were recorded to show changes in thiol, disulfide, and sulfenic/sulfinic/sulfonic acid by the intensity of the S-H stretch (2563.5 cm^{-1}), S-S stretch (500.1 cm^{-1}), or S-O stretch (1038.3 cm^{-1}), respectively [52]. As shown in Fig. 6.6, 78.0% of free thiol was obtained from freshly synthesized OligoCys, together with 16.1%

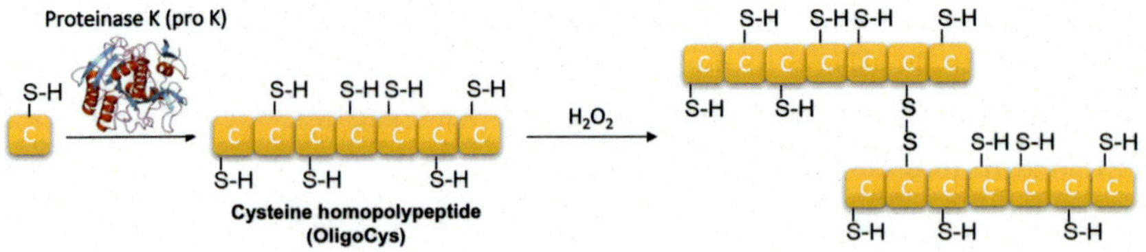

Figure 6.5 Chemoenzymatic synthesis of OligoCys samples with disulfide bonds [50].

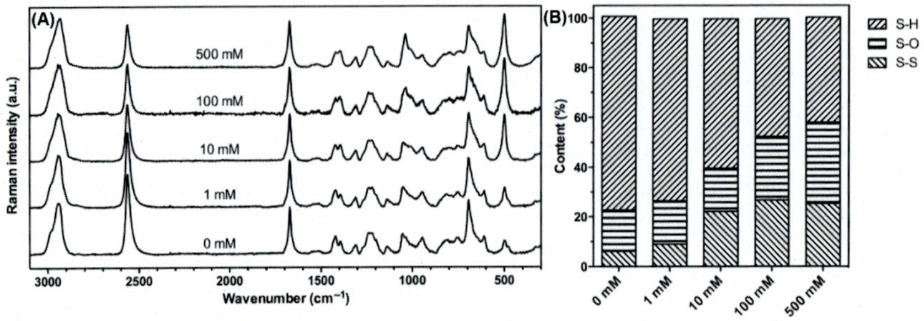

Figure 6.6 Raman spectra (A) and relative contents of S-H, S-O, and S-S (B) in OligoCys treated with different concentration of H_2O_2 solution [50].

sulfenic/sulfinic/sulfonic acid and 5.9% disulfide, suggesting the existence of the cysteine sulfenic acid intermediate in OligoCys and its conversion into disulfides [50]. When treated with H_2O_2, thiol content decreased while sulfenic/sulfinic/sulfonic acid and disulfide increased in OligoCys. For OligoCys treated with 100 or 500 mM H_2O_2, sulfenic/sulfinic/sulfonic acid content increased but disulfides declined due to preferential oxidation into sulfinic/sulfonic acid. Based on the similar mechanism, these results could be applied to mimic the long-term aerobic oxidation of OligoCys [53]. Here, thiol content was chosen to evaluate the degree of oxidation; lower thiol content indicated stronger oxidation.

To evaluate thermal properties of OligoCys with different degrees of oxidation, TGA, DSC and dynamic mechanical analysis (DMA) were applied [50]. Decomposition temperature thresholds of over 200°C were observed (Fig. 6.7A) [50]. When the degree of oxidation increased, temperature of 5% weight loss for OligoCys increased from 225.3°C to 269.6°C, with the decomposition temperature peaking at around 350°C. This suggested that OligoCys oxidation enhanced the thermostability due to the cross-linking of disulfide bonds formed by oxidation. No significant glass transition temperature (T_g), melting temperature (T_m), or decomposition temperature (T_d) was detected from −80°C to 200°C in DSC

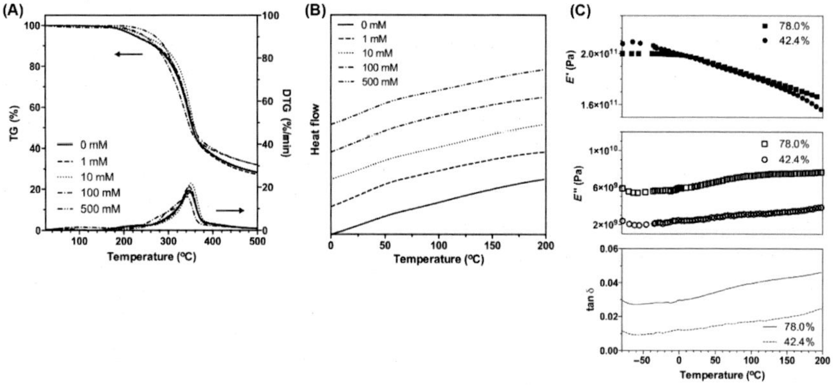

Figure 6.7 Thermograms of TGA (20°C/min, 25°C−500°C) (A) and the DSC first heating run (20°C/min, −80°C−200°C) (B) of the obtained OligoCys with different thiol contents. (C) DMA results of OligoCys. Storage modulus, loss modulus, and phase angle plots of OligoCys containing 78.0% and 42.4% free thiol contents as a function of temperature at 1 Hz [50].

measurements (Fig. 6.7B) [50]. In DMA measurements, powders of OligoCys with different thiol contents were loaded using a material pocket and subjected to oscillating displacement, forcing horizontal shearing of the powder between the two plates of the pocket. Typical experimental parameters were used with frequencies of 0.1, 1, or 10 Hz. However, over the whole detection range, no significant changes for storage modulus or tan δ signal and independence of frequency were observed (Fig. 6.7C), consistent with the DSC results [50]. This observation may be explained by cross-linking via disulfide bond formation, which would fix the local structure of OligoCys chains. Furthermore, the thermal properties of OligoCys without treatment indicate that 6% disulfide bonds may be sufficient for forming the cross-link network structure. Thus the thermal properties of OligoCys were found to be stable in a broad temperature range and even after oxidation, making it possible to use OligoCys preparations as a thermostable bio-based material.

6.1.6 Addition of melting property

For human use of peptides as bulk materials, it would be expected that peptide-based materials would need to be fabricated via a thermal process without any organic solvents [8]. However, peptides/proteins do not have a melting point but instead show decomposition during the heating process because the intermolecular hydrogen bonds in peptides are so strong that the intrachain covalent bonds begin to degrade before the melting transition [54]. To design thermally processable peptide materials, therefore, the copolymerization of peptides and unnatural components has been investigated. As a model amino acid monomer, L-glutamic acid diethyl ester [Glu(Et)$_2$] was selected, because of its relatively high reactivity [55]. Also, four types of commercially available nylon monomers, namely, nylon 1 [ethyl carbamate (nylon 1Et)], nylon 3 [β-alanine methyl ester hydrochloride (nylon 3Me), β-alanine ethyl ester hydrochloride (nylon 3Et)], nylon 4 [methyl 4-aminobutyrate hydrochloride (nylon 4Me), ethyl 4-aminobutyrate hydrochloride (nylon 4Et)], and nylon 6 [methyl 6-aminohexanoate hydrochloride (nylon 6Me)] were selected to copolymerize with Glu(Et)$_2$ [55]. In particular, nylon 4 is biomass-based and biodegradable, and it is considered to be an eco-friendly bioplastic [56]. However, the process window (gap between melting and degradation temperatures) is too small to allow practical industrial processing. Copolymers of the nylon 4 unit and amino acids can be used as an eco-friendly and biomass-based material.

Oligo(GluEt) with a nylon unit, which was synthesized by chemoenzymatic copolymerization, was characterized by ^{1}H NMR (proton nuclear magnetic resonance) and matrix-assisted laser desorption ionization time of flight mass spectrometry to confirm the resultant compositions [55]. The thermal properties of oligo(GluEt) with a nylon unit were also characterized by TGA and DSC to evaluate the effect of the nylon unit on the thermal properties of the peptides. From these results, the relationship between the nylon units and thermal properties was determined, considering the density of hydrogen bonds between the peptide main chains.

To evaluate the effects of the nylon units on the thermal properties of oligo(GluEt), TGA and DSC analyses were performed using the resultant oligo(GluEt) with or without nylon units (Fig. 6.8) [55]. On the basis of the DSC profiles, oligo(GluEt) with or without nylon units showed neither a transition temperature nor a melting temperature but demonstrated an endothermic degradation peak ranging from 230°C to 350°C. The DSC profile of oligo(GluEt) slightly differed from those of the oligo (GluEt) containing nylon units: the degradation peaks of oligo(GluEt-*co*-15 mol% nylon 3Me), oligo(GluEt-*co*-12 mol% nylon 4Me), and oligo (GluEt-*co*-12 mol% nylon 6Me) shifted to higher temperatures in comparison with the profile of oligo(GluEt). On the basis of the weight loss in the TGA data, the degradation temperatures were determined at 1 wt.%,

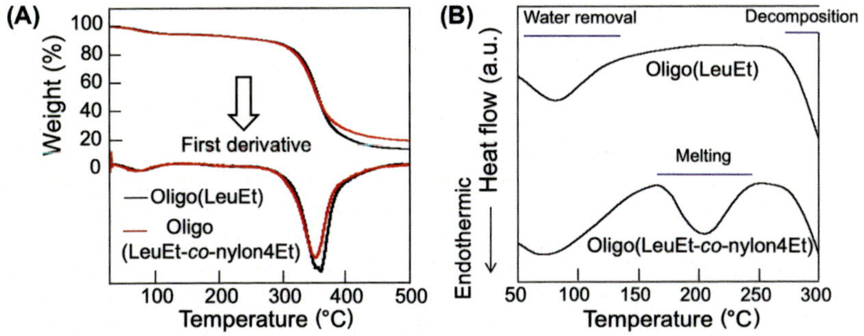

Figure 6.8 (A) TGA and the first derivative of the TGA plot, (B) DSC of oligo(LeuEt-co-nylon4Et) at 10°C/min [57].

5 wt.%, and 10 wt.%. As a result, the oligo(GluEt) with nylon units demonstrated relatively lower thermal stability than oligo (GluEt) did. This result was probably because the intermolecular hydrogen bonds in the oligo(GluEt) with approximately 10 mol% of nylon units are prevented by the presence of the nylon units, and thus the thermal stability of the oligo (GluEt) with nylon units decreased.

The introduction of nylon units into peptides changes the density of hydrogen bonds between the peptide main chains and can lead to new functions such as peptides with unique thermal properties. Nylon 3, 4, and 6 units were introduced into the main chains of Leu peptides through papain-catalyzed chemoenzymatic synthesis [57]. The resultant copolymers demonstrated different thermal properties; thus this result provides a first step toward controlling the thermal properties as well as adding a transition temperature of peptide-based polymers. The peptides containing nylon units are expected to have new functions, such as thermoplasticity, depending on controlling the density of the intermolecular hydrogen bonds [57].

6.2 Stability with water

For the practical use of structural proteins as bulk material in versatile environments, a certain level of water resistance is required. However, a structural protein having some hydrophilic motifs is influenced by water molecules. Further, an artificial structural protein with an optimized amino acid sequence to improve productivity usually has a hydrophilic property as well as a problem with water resistance. As examples, water effects on silk proteins are shown in the following.

6.2.1 Physical properties of spider silk fibers

An intimate relationship links the physical properties of spider silk to the primary structure of the underlying protein components [11,58]. Supercontraction is a typical physical property to demonstrate water sensitivity. A correlation has been found between the abundance of proline residues in major ampullate spidroin 2 (MaSp2) repetitive domains and the degree of elasticity and supercontraction of the corresponding dragline fiber [59,60]. In MaSp2, the proline-rich sections of the Gly-rich regions are thought to adopt β-turn conformations that contribute to the high mobility of the polypeptide chains [61].

Malay et al. performed a systematic analysis of the available MaSp sequences, with the primary aim of elucidating the essential motif elements in MaSp1 and MaSp2 tandem repeat sequences based on empirical criteria [62]. Secondly, they uncovered novel patterns of tandem repeat organization, as well as probed the relationships between conserved motif patterns and other aspects of spider biology. The scope of the sequences included is considerably broader than in previous investigations: all available MaSp (and MaSp-like) sequences were surveyed, taking advantage of the large number of data currently found in the databases, and thus ensuring a wide coverage of spider taxa; in addition, recent insights from spider systematics were incorporated into the analysis [63]. Their motif-based analysis identified a small subset of conserved motif elements associated with MaSp1 and MaSp2 tandem repeats, spanning the Entelegynae clade. In particular, in MaSp2 sequences the prevalence of three motifs, GP, diglutamine (QQ), and GGY, were found to vary as a function of spider web morphology, and are hypothesized to cooperatively modulate the mechanical properties of dragline silk [62]. From the results of sequencing natural spider silk, the presence of QQ, diglutamic acid, or proline enhances the physical property of hypercontraction, which is sensitive to water [62]. This positive correlation between hypercontraction and the presence of diglutamic acid or proline is very useful information when designing amino acid sequences with excellent water resistance.

In addition to supercontraction, the effect of RH on physical/structural properties was also investigated. At different RHs, namely, 0%, 43%, 75%, and 97%, the WAXS measurement of dragline silk fibers was performed [64]. The resultant two-dimensional profiles are displayed in Fig. 6.9A. A circular smearing of the scattering pattern was detected in the WAXS measurement at RH 97% (Fig. 6.9A-iv). The two-dimensional profiles were

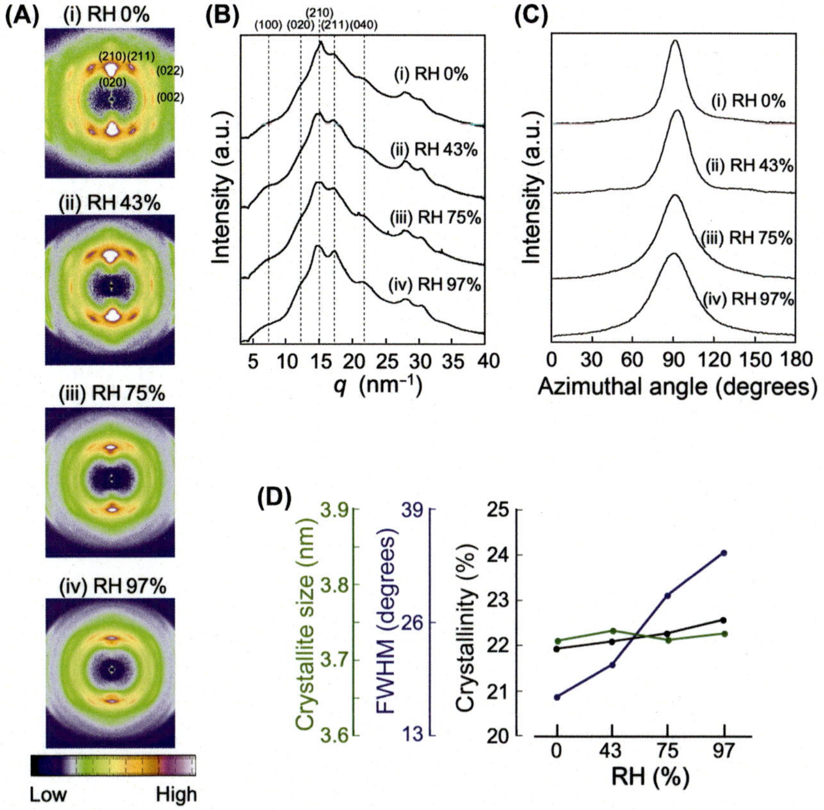

Figure 6.9 WAXS measurements of spider dragline silk fibers under different RHs: (A) two-dimensional profiles; (B) one-dimensional radial integration profiles; (C) azimuthal intensity profiles of the radially integrated (210) peak; and (D) crystallite size, FWHM, and crystallinity of dragline silk fibers incubated at RHs of (i) 0%, (ii) 43%, (iii) 75%, and (iv) 97% [64].

converted to one-dimensional profiles through a radial integration of the WAXS data (Fig. 6.9B). The *d*-spacing values of each plane remained constant upon changing the RHs based on the one-dimensional WAXS profiles, indicating that the crystalline beta-sheet structure was not changed by the RH conditions, consistent with the TGA results. The two-dimensional WAXS pattern varied depending on the experimental RH conditions, leading to observable changes in the one-dimensional radial integration profiles, namely, the (211) peak became more conspicuous under higher RH conditions. In addition, the azimuthal intensity profiles of the radially integrated (210) peak are plotted in Fig. 6.9C. As the RH increased, the full width at half maximum (FWHM) value increased, suggesting a decrease in the orientation of crystalline beta-sheets. To estimate the degree of crystallinity, which was calculated from the area of crystalline peaks divided by the total area of crystalline peaks and the amorphous halo, the one-dimensional radial integration profiles were deconvoluted to multiple peaks assigned to the crystalline beta-sheets and amorphous halo. The degree of crystallinity was calculated to be constant at approximately 22% under all RH values investigated. In contrast, changes in FWHM values were detected as a function of RH (Fig. 6.9D). The FWHM value was 37.5 degrees at RH 97% and decreased to 14.5 degrees at RH 0%, suggesting that the degree of orientation was higher under lower RH conditions. The crystallite size was calculated by Scherrer's equation using the FWHM of the (210) peak (Fig. 6.9D). Thus, crystallite sizes were constant and close to those identified in a previous report [65]. As a result, the high RH condition reduced the degree of orientation of the crystalline beta-sheets but did not affect the other crystalline state.

To evaluate the amorphous region in addition to the crystalline beta-sheets, birefringence measurements have been used for various polymeric materials, including spider silks [66]. The retardance values of dragline silk fibers under different RHs were measured, from which the birefringence values were calculated in combination with fiber diameter measurements, obtained via scanning electron microscopy (SEM) (Fig. 6.10A) [64]. The birefringence values were found to be approximately constant at around 41.5×10^{-3} between RH 0% and RH 75% (Fig. 6.10B). In contrast, a relatively large decrease in birefringence was observed at RH 97% (Fig. 6.10B). The decrease in birefringence at RH 97% suggests a more random molecular arrangement of the constituent protein chains, and agrees with the WAXS results. Water molecules are known to function as plasticizers and induce reorientation of silk molecules in the amorphous region, as observed in the supercontraction of the dragline silk fibers [67–69].

6.2.2 Relationship between strain rates and humidity

Dragline silk fibers are used by spiders as a lifeline to prevent unexpected falling and to capture prey, such as flying insects. Strain rates greater than $1\ s^{-1}$ are thought to be applied to the dragline silk fibers in the natural state [70]. This natural range of strain rates is in agreement with the range of strain rates selected, namely, from $3.3 \times 10^{-5}\ s^{-1}$ to $3.3\ s^{-1}$ [64]. In these strain rates, the toughness of the dragline silk at RH 43% and RH

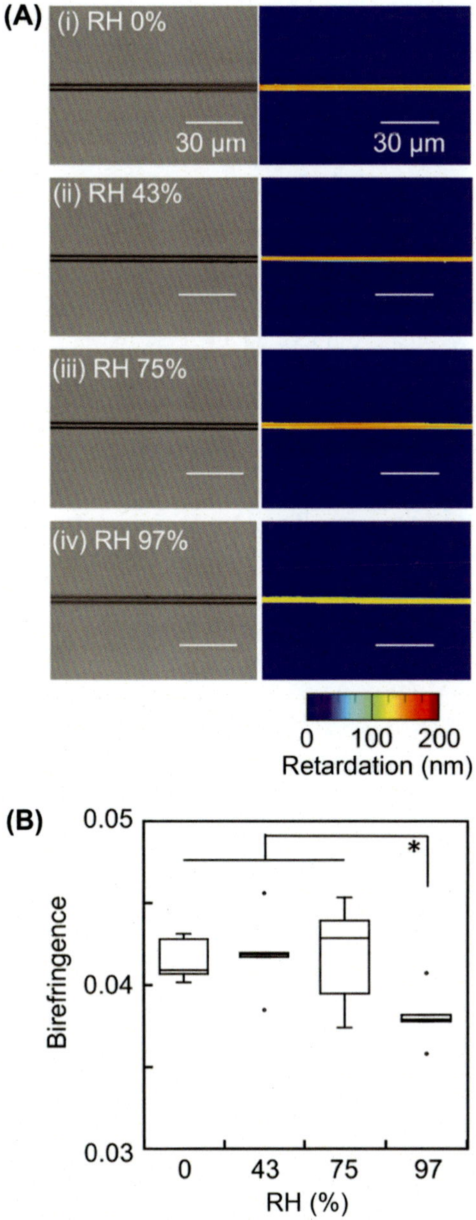

Figure 6.10 Retardation and birefringence of spider dragline silk fibers. (A) Optical micrographs and retardation color mapping of spider dragline silk fibers incubated at (i) RH 0%, (ii) RH 43%, (iii) RH 75%, and (iv) RH 97%. Each scale bar denotes 30 μm. The retardation values ranged from 0 to 200 nm, as shown in the color scale. (B) Birefringence value as a function of RH. *Significant differences between groups at $P < .05$ [64].

75% was higher than that at RH 0% and RH 97% [64]. The environmental humidity in the natural habitat of spiders could be moderate. Based on the effect of RH on toughness at a strain rate of 3.3 s^{-1}, the spider dragline silk fibers might evolve to exhibit toughness at a moderate RH, namely, a naturally humid environment and the relatively high deformation rates that occur in natural spider behaviors.

Tensile tests of the dragline silk fibers were conducted at different strain rates (3.3×10^{-5} s^{-1} to 3.3 s^{-1}) and RHs (0%, 75%, and 97%) [64]. The resultant stress–strain curves are displayed in Fig. 4.4. The tensile strength and elongation at break tended to increase with increasing strain rate. Mechanical properties at different strain rates and RH conditions were determined from the stress–strain curves. The tensile strength, elongation at break, and Young's modulus increased linearly as the strain rate increased, regardless of the experimental RH conditions. The tensile strength increased approximately 1.4-fold, 2.3-fold, and 1.5-fold as the strain rate increased from 3.3×10^{-5} s^{-1} to 3.3 s^{-1} at RH 0%, RH 43%, and RH 97%, respectively [64]. The tensile strength decreased as the RH increased [71]. The elongation at break increased as the RH increased under all the strain rates utilized in this study. Regardless of the applied strain rates, the elongation at break at RH 97% was approximately 2-fold higher than that at RH 0%. The Young's modulus linearly decreased as the RH increased. As discussed in the previous paragraph, the decrease in Young's modulus originated from the plasticization effect of water molecules at high RH conditions [67,69,72–74]. The toughness of the dragline silk was not significantly changed by the RH conditions at strain rates ranging from 3.3×10^{-5} s^{-1} to 3.3×10^{-1} s^{-1}. In contrast, at a strain rate of 3.3 s^{-1}, the toughness was highest at RH43%, followed by the toughness at RH 75%. Overall, at RH 0%, the toughness at high deformation rates originated from the increase in yield stress, whereas the dragline at RH 75% and RH 97% exhibited rubber-like behavior because of the increase in rubbery components induced by water plasticization [64].

The two-dimensional WAXS data were also converted to one-dimensional radial integration profiles under different RHs. The crystallite sizes as a function of strain were calculated from the FWHM of (210) peaks of the one-dimensional profiles. The (210) peak corresponds to the direction along the hydrogen bonds of the crystalline beta-sheets and is denoted here as the *b*-axis of the crystallite [75]. The crystallite sizes were slightly decreased upon extension, regardless of the RHs and strain rates. Because the mobility of silk molecular chains could be enhanced by a plasticization effect of water molecules at high RH condition, it would be expected that the amorphous entropy elastic stretching preferentially happens, instead of unfolding and breaking at the crystalline region. However, the crystallite size decreased even at low deformation range at high RH [64]. Glišović et al. also reported the crystallite size decrease upon stretching using the *Nephila* dragline silk which was immersed in water [76]. The decrease in crystallite size can be originated from a partial splitting of the crystallites along the noncovalent bonds or an unfolding of silk chains at the interface of the amorphous region [76]. Glišović et al. calculated the total energy, which is necessary to break the hydrogen bonds, in the crystallite and compared with the typical stretching energy needed in the tensile tests. The result indicates that the total energy which is necessary to break the hydrogen bonds is in the same range as the total energy for stretching in the tensile test data [76]. The crystallinity did not change significantly under the measured RHs and strain rates. Thus crystal regions tend to be deformed upon extension at low RH, and both crystalline and amorphous regions are thought to be deformed upon extension at high RH. This phenomenon occurs because when the amorphous motility is high, the silk behaves as a rubber-like elastomer and molecular reorientation easily occurs. On the other hand, in the dry state, the silk behaves like a plastic material, and the efficiency of reorientation is lower. In the natural condition for Japanese *Trichonephila* (formerly known as *Nephila*) spiders, namely, at moderate humidity, the spider dragline is a rubber-like elastomer with high strength originating from oriented crystalline beta-sheets.

The effect of strain rates on the mechanical and structural properties of spider dragline silk fibers was studied under different RH conditions [64]. The tensile strength improved as the strain rate increased. Although the increase in tensile strength with increasing strain rate was smaller at RH 97%, the high elongation at break at RH 97% contributed to maintaining the toughness value of the dragline silk fibers. The fracture surfaces of the dragline fibers were observed via SEM, demonstrating that the fracture surfaces were relatively smooth at lower deformation rates and rough at higher deformation rates [64]. The resultant fracture surfaces indicated that the dragline silk fibers were broken at macroscopic structural defects at lower deformation rates, whereas the fibers were broken at microfibrils at higher deformation rates. Water molecules could be substituted with the intermolecular hydrogen bonding between silk molecular chains, inducing easy flow under the application of load. At high RH condition, the dragline silk has a highly disordered structure, which was confirmed by the lower birefringence and degree of orientation in the present study. Thus the creep and stress relaxation behavior of spider dragline silks are expected to be higher at high RH condition and it is speculated that the fracture surfaces would be detected at the structural defect under higher RH condition. However, the SEM observation of the fracture surface after the tensile test demonstrated that the fracture surfaces of the dragline silks tended to be smooth at relatively low strain rates, whereas the fracture surfaces were rough under relatively high strain rates, regardless of the experimental RH. The rough fracture surfaces of the dragline silk fibers suggest that the fibers were broken at the microfibril level at faster deformation rate. Even though the mobility of silk molecular chain is facilitated under high RH condition, the tensile force was applied to both crystal and amorphous region, which was confirmed by the crystallite size decrease based on the simultaneous WAXS-tensile test. Hence, the load is expected to be applied to the microfibrils of spider dragline silks during stretching at higher deformation rate before the stress relaxation completed. Accordingly, the dragline silk could be broken at the microfibril level rather than at a macroscopic structural defect at faster deformation rate. Combined with the SEM observation and simultaneous WAXS-tensile test, microstructural changes of spider dragline silk fibers could be depicted as shown in Fig. 6.11 [64].

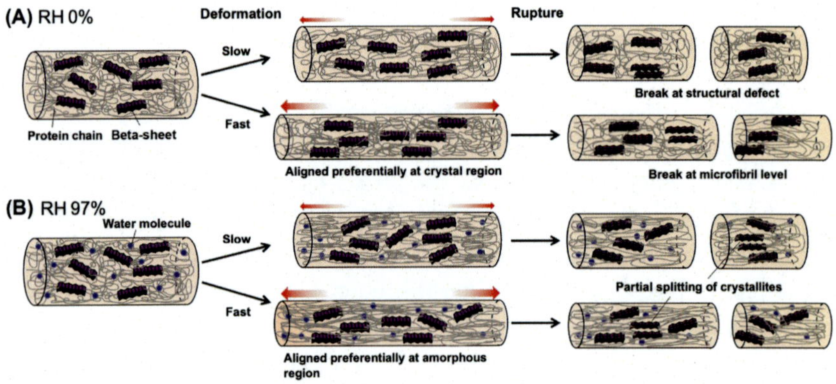

Figure 6.11 A schematic model demonstrating microstructural changes of spider dragline silk fibers at different strain rates and humidity. Silk molecular chains in the crystal region deform and align along the fiber axis at dry condition (A), whereas water molecules bind to amorphous regions and facilitate the molecular movement at wet condition (B). With further stretching, the crystallites start to split. The fracture surfaces would be detected at the structural defect at slower deformation rate, while the dragline silk could be broken at the microfibril level rather than at a macroscopic structural defect at faster deformation rate [64].

6.2.3 Silk materials

To evaluate the potential of silks as structural material in industry, the mechanical properties of silk films were characterized at different RHs. The stress–strain curves obtained from the tensile tests are shown in Figs. 3.12 and 4.3 [42]. The stress–strain curves of the silk films incubated at RH 97% were different from those obtained from silk films incubated up to RH 84%. Based on the stress–strain curves, four types of mechanical properties, namely, tensile strength, Young's modulus, elongation at break, and toughness, were determined (Fig. 4.3). Those mechanical properties of the silk films did not change significantly up to RH 84%. On the other hand, the mechanical properties of films changed drastically in the case of films incubated at RH 97%. The elongation at break and toughness of the silk films incubated at RH 97% increased 15- and 7-fold higher than those incubated at RH 84%. To clear the mechanism behind the significant change in mechanical properties, the crystal structure of the silk films at different RH was characterized by WAXS (Fig. 3.13) [42]. The crystallinity maintained at approximately 30% up to RH 75%. Meanwhile, the crystallinity increased to approximately 40% at over RH 84%. The FWHM of the peaks corresponding to beta-sheet crystal structure at RH 84% and 97% was significantly decreased, which means that larger crystallites were formed at very humid conditions according to the Scherrer's theory. Based on the structural and mechanical characterizations, the silk films prepared at RH 97% demonstrated high crystallinity and elongation at break. Although there is generally a trade-off relationship between crystallinity and ductility in material science, the silk film at RH 97% can realize both the properties simultaneously. This is mainly because of the plasticizing effect of water molecules on amorphous silk molecules. At RH 97%, amorphous regions of silk films were plasticized, resulting in the increase of elasticity of silk films. Additionally, water molecules induced the beta-sheet formation via helix–helix interactions.

6.3 Biological stability

6.3.1 Stability with proteases

Protein materials are biodegradable in natural environments as well as inside animal bodies. The major degradation reactions are protease-mediated digestion and chemical hydrolysis. Here, the stability of the protein materials with various proteases is summarized.

B. mori silkworm silk has been studied widely to clarify its crystalline structures, a particularly beta-sheet structure composed of the highly repetitive Gly-Ala-Gly-Ala-Gly-Ser (GAGAGS) domain, due to the remarkable mechanical properties that originate from a combination of crystalline, semicrystalline, and amorphous domains. The antiparallel beta-sheet structures are composed of alanine-glycine or alanine repeats. An increase in the fraction of beta-sheets also reduces the enzymatic degradability of silk fibroin [77]. Several proteolytic enzymes have been used to digest silk fibroin, with protease XIV considered to show high activity towards beta-sheet structures in silk-based fibers, films, and scaffolds, whereas alpha-chymotrypsin can digest the less crystalline regions of the assembled silk structures but does not degrade the beta-sheet crystals [78–81].

Abnormal aggregations or deposition of misfolded proteins, such as amyloid fibrils with beta-sheet structures, have been recognized as the molecular pathogenesis of Alzheimer's disease, Parkinson's disease, and Huntington's disease. Amyloids are generally described as stacked beta-sheet structures aligned perpendicular to the fibril axis, known as cross beta-sheets, and are often based on glutamine-rich or hydrophobic interactions. The amyloids, in contrast to the silks that contain antiparallel beta-pleated sheets, do not degrade at reasonable rates in vivo and also lead to significant physiological complications due to toxicity. Amyloid-beta (Aβ) peptides in Alzheimer's disease have been studied to clarify the mechanism of neurotoxicity by Aβ fibrils. Monomers, intermediates (oligomers), and fibrils of Aβ peptides with different molecular weights have been investigated for neurotoxicity, and recently a spherical amyloid intermediate of 15–35 nm diameter, which had predominantly beta-sheet structures, demonstrated higher toxicity than Aβ monomers and fibrils [82].

Protease XIV and alpha-chymotrypsin, previously reported to degrade silk fibroin, were used as proteolytic enzymes [77–80]. Alpha-chymotrypsin degrades only the noncrystalline (non antiparallel beta-pleated sheets) in silk, while protease XIV degrades the entire structure. The silk crystals exposed to the protease XIV for 12 h showed the crystals were degraded initially into fibril-like fragments (nanofibrils) that were around 80 nm wide and 4 nm in height (Fig. 5.1) [81]. The degradation of the nanofilaments was the slowest aspect of the degradation process, because of the tight molecular packing of the nanofilaments in comparison with the other crystalline regions containing loose chain-packing hydrophobic blocks of silk fibroin.

To determine the secondary structure of soluble silk molecules after enzymatic degradation, supernatant of degradation products after the enzymatic degradation of silk crystals by protease XIV or alpha-chymotrypsin for 24 h at 37°C were characterized by circular dichroism (CD) [81]. The CD spectrum of the soluble degradation products by alpha-chymotrypsin showed random-coil (unordered) structures with a negative peak at 203 nm, whereas protease XIV demonstrated beta-sheet structure with a negative at 216 nm shoulder (Fig. 5.2) [81]. Estimation of the secondary structure contents of the soluble fragments yielded 37% beta-strand and 50% unordered structures for the soluble degradation products by protease XIV. In the case of the soluble fragments from alpha-chymotrypsin, their secondary structures were estimated to be 5% beta-strand and 57% unordered. Comparison of these secondary structures suggests that only protease XIV produces soluble beta-sheet fragments from the enzymatic degradation of silk crystals. On the basis of the digestion pattern of silk molecules by protease XIV described above, the soluble degradation products contain several hydrophobic domains, so that these hydrophobic sequences, $(GAGAGS)_n$, retained beta-sheet structure in a buffer solution. The cytotoxicity of the silk degradation products is discussed in Section 5.2.3.

6.3.2 Environmental degradation

Polymer materials such as plastics are used in various material fields because of their excellent physical properties and processability. On the other hand, it has been reported that degradability during or after the use of polymer materials causes serious problems in various usage environments. In recent years, it has become clear that, in marine pollution by plastic products, microplastics that remain as minute fragments without being decomposed have long-term adverse effects on various marine organisms. In order to solve these problems, attempts have been made to manufacture disposable plastic bags and straws from marine-degradable polymers. In the white paper from Chemical Sciences and Society Summit (CS3), London, 2019 (Science to enable sustainable plastics; https://www.rsc.org/news-events/articles/2020/jun/science-to-enable-sustainable-plastics/), structural proteins such as spider silk are also introduced as environmentally degradable polymers. The examples of using plastics that utilize biodegradability, especially the examples of utilizing on-demand biodegradation that induces degradation after use, are very limited. Precision control of polymer degradation is very important not only for ensuring the properties of degradable polymers, but also for extending the long-term stability of polymer materials. In current polymer science, it is not easy to suppress deterioration and decomposition of polymer materials and improve the stability of mechanical properties, and it is a scientific factor that cannot take the position of iron as a major structural material. It is also possible to tackle various problems of stability and degradability of polymer materials by understanding the decomposition of polymers from a macro level to a molecular level in a multilevel manner and systematically academically.

The "biodegradability" of macromolecules is widely used as one of the functions and physical properties of substances. However, the concept of biodegradability varies slightly depending on the country/region or research field, and is not unique. The fact that the polymeric material becomes invisible does not necessarily mean that it has been decomposed, but it is necessary to show that the decomposed product is converted into a substance existing in the natural environment and reduced in a concentration range that does not affect the environment. It is expected to be decomposed in some form because it dissolves in seawater or freshwater, but reduction to the environment always requires scientific grounds.

Environmental degradability is the property where low-molecular-weight substances generated by the decomposition of macromolecules are converted into substances that originally exist in nature by further chemical decomposition and biological metabolism. It is not enough that the degradation products are invisible and have no short-term effect on the organism. In addition, even if the decomposed product is a natural compound, if the rate or concentration of its occurrence is significantly different from the natural environment, it is necessary to investigate the effect on the environment scientifically.

6.3.3 Biological/environmental stability of keratin

Keratin is the main structural protein that forms the hair, wool, feathers, nails, and horns of many types of animals [83]. The protein has high contents of cysteine (7%–20% of the total amino acid residues), which is known to form intramolecular and intermolecular disulfide bonds [84,85]. It has been reported that α-keratin, which has a helical structure, can be reduced to form β-keratin upon stretching treatment, and this structural change affects the mechanical, thermal, and chemical properties of the material [86]. Keratin shows biological stability even though it is a natural protein, and hence some mummies still maintain their hairs and nails [87]. This is because a keratin-degrading enzyme is not everywhere. In the field where a keratin-degrading enzyme is not present, keratin shows long-term stability even in the natural environments. The biological stability of keratin should be applicable for a wide variety of material design and purpose.

6.3.4 Environmental stability of silk

Silks are an attractive biopolymeric alternative to petroleum-based high-strength polymeric materials, because silks exhibit excellent mechanical properties in addition to high biodegradability and biocompatibility [88]. One of the great advantages of using silks is that silk can be formed into various types of materials, including fibers, films, sponges, gels, and tubes, by an established solubilization technique [6]. The shape versatility of regenerated silk materials can broaden the range of possible applications of biopolymers. In addition, the good processability enables facile composite formation using the regenerated silk with other polymeric components.

The biodegradation of silk materials in the natural environments has been studied widely [88,89]. Silk resins are considered to show biodegradability and to be unstable in natural environments, because silk is one of the natural and biological polymers [88]. To verify their biodegradability, the biochemical oxygen demand (BOD) test was performed with silk resins (see the experimental details of BOD tests in Chapter 10, Experimental Details). Biodegradability was determined from the BOD and theoretical oxygen demand (ThOD) from incubating the silk resins under active sludge conditions for 30 days (Fig. 6.12) [89]. The dehydration treatment (water content) did not disturb the biodegradability of the silk resins in natural environments. Furthermore, at the relatively high

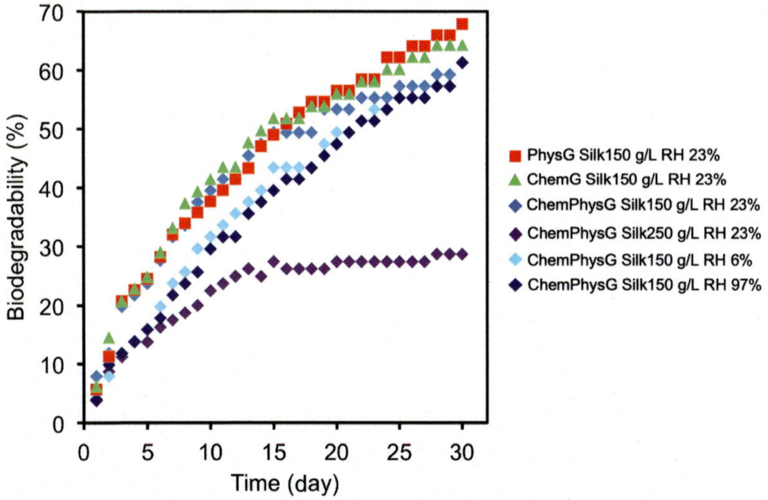

Figure 6.12 Biochemical oxygen demand (BOD) curves of silk resins in active sludge at 25°C [89].

concentrations of silk molecules, it was necessary to incorporate β-sheet cross-links, chemical dityrosine links, and entanglements and assembly via hydrophobic interactions of silk molecules to design and develop silk-based materials with high strength and toughness. The control of the network structure is an easier method than mimicking the native hierarchical structures to prepare tough silk-based materials. These tough and tunable silk-based resins have the potential to replace biodegradable plastics and natural biopolymers in various applications such as car parts.

Polypeptides and proteins composed of natural amino acids are considered to show biodegradability in natural environments; however, polypeptides containing unnatural units such as poly(alanine-nylon-alanine) [poly(AlaNyl*X*Ala), $X = 3-11$] need to be studied for biodegradability [90]. To confirm the biodegradation of the copolymer of alanine and nylon units, poly(AlaNyl*X*Ala), a BOD test was performed using poly (AlaNyl*X*Ala) powder samples under seawater conditions at 25°C (Fig. 6.13) [90]. The biodegradability of the sample was calculated from the ThOD and experimental BOD values as described in the experimental section. As a result, the significant biodegradability of poly(AlaNyl*X*Ala) was confirmed, even though poly(Ala) showed higher biodegradability. The biodegradability based on the BOD tests did not indicate significant differences among the poly(AlaNyl*X*Ala) samples.

The environmental toxicity of the samples was tested according to the Organisation for Economic Co-operation and Development (OECD) test guideline 202, acute immobilization test (OECD), with slight modification

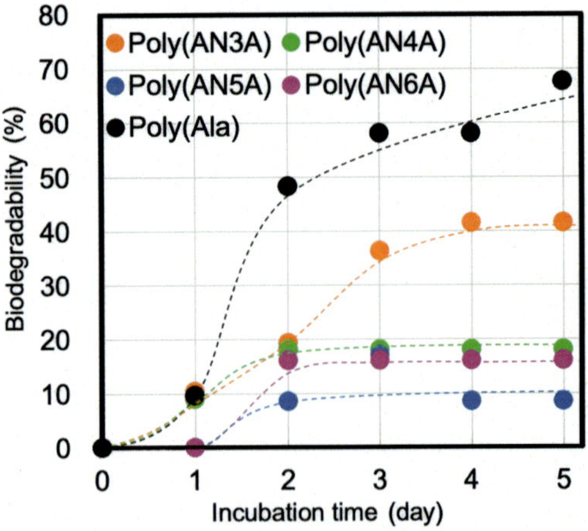

Figure 6.13 BOD tests of poly(AlaNylXAla) and poly(Ala) in seawater at 25°C [90].

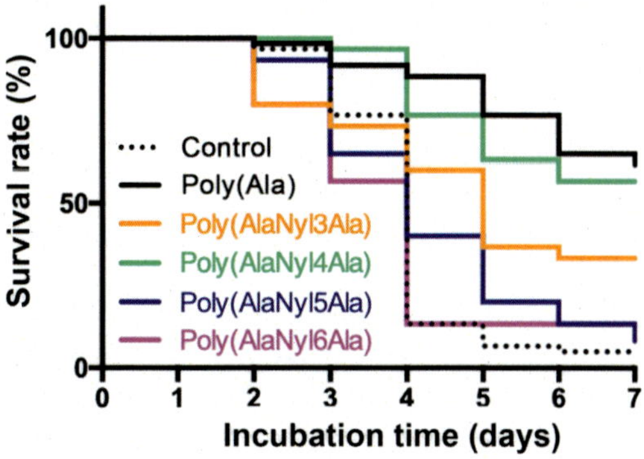

Figure 6.14 Kaplan–Meier survival curves to evaluate the effects of poly(AlaNyl*X*Ala) and poly(Ala) on the aquatic model animal *Daphnia magna* at 20°C. Control denotes the background values without any test samples [90].

[91,92] (see more details in Chapter 10, Experimental Details). Briefly, 30 *Daphnia magna*, which were not more than 24 h old at the beginning of the test, were used for each test sample and control. The final concentration of the samples was set at 100 mg/L, which was the higher limit provided by the test guidelines [90]. Survival curves of each treatment were produced based on daily observations during the test period. The effects of the samples on their survival were evaluated by comparing the Kaplan–Meier survival curves using the log-rank test followed by Bonferroni correction [90]. The results were that no reduction in survival was observed in any of the conditions compared to the control (Fig. 6.14). In contrast, poly(Ala) significantly improved their survival compared to the control, in which most animals experienced death from starvation during the test period, suggesting that poly(Ala) may act as a nutrient rather than a toxicant for daphnids. These results with *D. magna* demonstrate that poly(AlaNylXAla) and its degradation products did not have toxic effects on the natural model animal. Considering that the types of nylon units in poly(AlaNylXAla) showed different thermal and structural properties, the biodegradability and environmental toxicity also show different behaviors among the different poly (AlaNylXAla) samples. In particular, similar to poly(Ala), poly(AlaNyl4Ala) showed higher survival rate in average.

6.3.5 Marine stability of silk

Although seawater generally shows a weak ability to degrade bio-based polymers due to lower concentrations of organisms in ocean, amide bonds

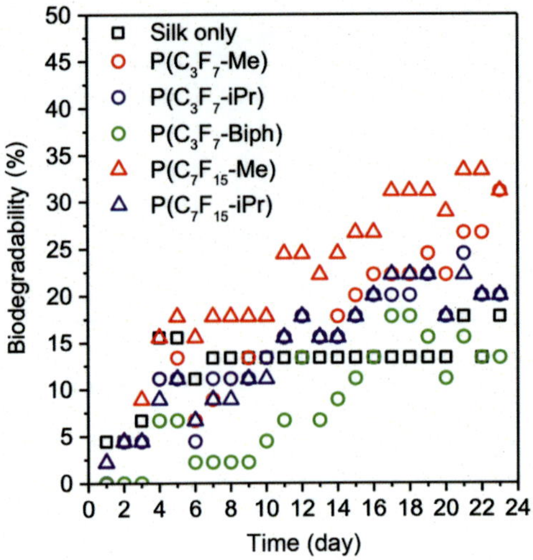

Figure 6.15 Biodegradability of the silk films coated with 1 wt.% fluorinated polypeptides in seawater [93].

can be degraded by the microbes in seawater. Even a synthetic bio-based polyamide, Nylon 4, showed good degradability in seawater [56]. Besides, silk resins also exhibit excellent microbial biodegradability [89]. To investigate the biodegradability of the silk films, a BOD assay of the silk films was performed using seawater collected at Yokohama, Japan. The silk film was coated with the fluorinated polypeptides using ethanol solutions, and the BOD values and the residual film weights were monitored over 3 weeks of immersion in seawater (Fig. 6.15) [93]. The uncoated silk film was gradually degraded by the seawater, and the final biodegradability after 3 weeks reached 30%. The silk films coated with fluorinated polypeptides showed degradation behaviors similar to that of the uncoated silk film. The final biodegradabilities of the polypeptide-coated silk films after 3 weeks were in the range of 15%–33% when the 1 wt.% solution was used for coating.

The appearance of the films during the BOD tests was monitored by SEM as shown in Fig. 6.16 [93]. Before the BOD test, all the films were transparent and had smooth surfaces, indicating that the polypeptide was homogeneously coated on the surface even using the 1 wt.% solutions. The side view of the films revealed that the polypeptide layer coated on the silk film was very thin relative to the thickness of the silk. As the silk films degraded in the seawater, the film surfaces became noticeably rougher and turbid. The side views of some of the silk films after 3 weeks

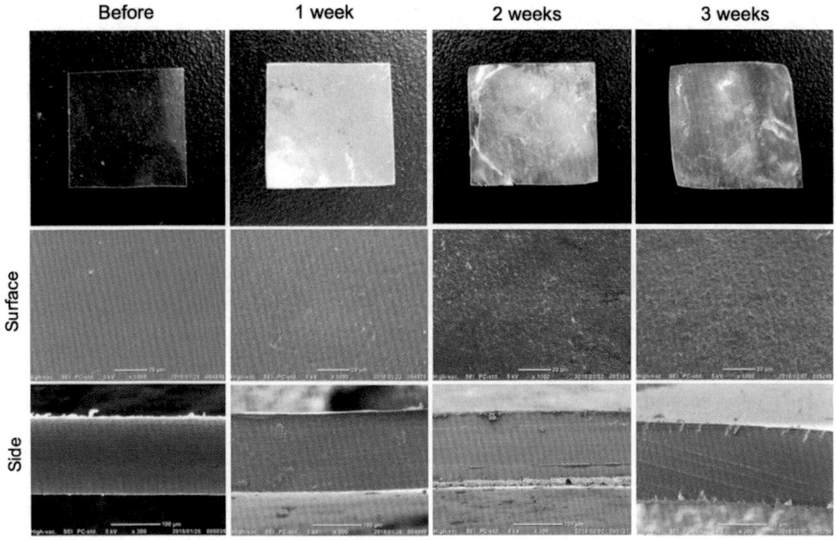

Figure 6.16 Appearance and SEM images of the surface and side of the silk films before the BOD test and after 3 weeks in a seawater [93].

showed cracks in the surface, indicating that the seawater penetrated into the bulk of the silk films. These changes in the appearance of the films also support that the polypeptide-coated silk films had good biodegradability in the seawater.

The film degradation was also investigated by measuring the remaining weight of the silk films. All the silk films showed continuous decreases in residual weight regardless of the fluorinated polypeptide used, even though the BOD values (biodegradability) varied slightly among the polypeptide-coated silk films [93]. The residual weights of the silk films reached approximately 85% of their initial weights by the end of the 3-week BOD test. The degraded weights were larger than BOD values by approximately 10%–15%. The discrepancy between weight loss and BOD is assumed to be due to the bound water absorbed in the degraded silk films during the BOD test.

6.3.6 In vitro stability of silk as biomedical materials

When silk is used as a biomedical material, we need to consider the in vitro stability of silk materials. One of the traditional biomedical applications of silk is silk sutures. Silk sutures are classified by regulatory agencies as nonbiodegradable because regulatory guidelines expect a loss of most tensile strength within 60 days post implantation [94]. Silk sutures can maintain their mechanical properties over the timescale, as they require more extended time frames in vitro [95]. In patients undergoing two-stage breast reconstruction, histological evidence of breast tissue samples taken from 60 patients at stage 2 (median 152 days after initial scaffold implantation, please see more details in the Ref. [96]) showed consistent degradation of one of the commercial silk products, SERI Surgical Scaffold. Many studies indicated that silk-based materials, such as films, gel, microparticles, tubes, and porous scaffolds, degrade at various degradation rates [97,98]. The implantation of SERI Surgical Scaffold in a sheep model of two-stage breast reconstruction showed progressive degradation and vascularization of the silk mesh [94]. At 1 month postimplantation, tissue ingrowth and marked vascularization were evident. At 4 months after the implantation, the mesh was no longer felt through the skin. After 12 months, the mesh was degraded and vascularization was recognized but with substantial silk loss that precluded mechanical testing of the remaining silk material [94,99]. In other cases with silk-based material, clinical hernia repair in a horse showed incomplete SERI Surgical Scaffold degradation at 2 years postimplantation, but no hernia relapse [94,100].

The in vivo stability of silk-based materials for implantation is dependent on several factors, including (1) the amount of material, (2) gross morphology, (3) silk secondary structure, (4) silk treatment history, (5) mechanical environment, and (6) implantation site (or final destination) [94]. The implantation site directly impacts the type of proteolytic enzyme encountered by the silk, because these enzymes vary between tissues, cells, and subcellular location. As described in Chapter 5, Biological Properties With Cells, the silk protein is known to degrade in vitro and in vivo in response to proteolytic enzymes [101]. Silk fibroin sequence alignment indicates susceptibility to several types of proteases (e.g., protease XIV, α-chymotrypsin, proteinase K, papain, matrix metalloproteinases, collagenase, etc.) [102,103].

The secondary structure and crystalline state is also important to determine the degradation behaviors. For *B. mori* silk, degradation begins with the 11 hydrophilic amorphous segments in the silk heavy chain, as well as the *C*- and *N*-terminal and the silk light chain, which consist of completely nonrepeating amino acid sequences; this is then followed by degradation of the more crystalline sequences [3,102]. The tightly packed crystalline domains are degraded last (Fig. 5.3). Furthermore, the silk format is a critical factor in determining degradation rates, as in vivo studies in rodent models indicated faster degradation for open silk structures than for tightly packed monolithic silk fibroin films, namely, hydrogel > silk scaffold > monolithic film [101].

References

[1] D.L. Kaplan, C.M. Mello, S. Arcidiacono, S. Fossey, K. Senecal, W. Muller, Silk, Protein-Based Materials, Springer, 1997, pp. 103–131.

[2] E.M. Pritchard, D.L. Kaplan, Silk fibroin biomaterials for controlled release drug delivery, Expert. Opin. Drug. Deliv. 8 (6) (2011) 797–811.

[3] K. Numata, P. Cebe, D.L. Kaplan, Mechanism of enzymatic degradation of beta-sheet crystals, Biomaterials 31 (10) (2010) 2926–2933.

[4] F. Vollrath, D. Porter, Spider silk as a model biomaterial, Appl. Phys. A Mater 82 (2) (2006) 205–212.

[5] G.H. Altman, F. Diaz, C. Jakuba, T. Calabro, R.L. Horan, J. Chen, et al., Silk-based biomaterials, Biomaterials 24 (3) (2003) 401–416.

[6] K. Numata, D.L. Kaplan, Silk-based delivery systems of bioactive molecules, Adv. Drug. Deliv. Rev. 62 (15) (2010) 1497–1508.

[7] D.N. Rockwood, R.C. Preda, T. Yucel, X. Wang, M.L. Lovett, D.L. Kaplan, Materials fabrication from Bombyx mori silk fibroin, Nat. Protoc. 6 (10) (2011) 1612–1631.

[8] K. Numata, Poly(amino acid)s/polypeptides as potential functional and structural materials, Polym. J. 47 (8) (2015) 537–545.

[9] Y. Dong, F. Dai, Y. Ren, H. Liu, L. Chen, P. Yang, et al., Comparative transcriptome analyses on silk glands of six silkmoths imply the genetic basis of silk structure and coloration, BMC Genomics 16 (2015) 203.

[10] M. Xu, R.V. Lewis, Structure of a protein superfiber: spider dragline silk, Proc. Natl. Acad. Sci. U S A 87 (18) (1990) 7120–7124.

[11] A.D. Malay, R. Sato, K. Yazawa, H. Watanabe, N. Ifuku, H. Masunaga, et al., Relationships between physical properties and sequence in silkworm silks, Sci. Rep. 6 (2016).

[12] K.A. Gupta, K. Mita, K.P. Arunkumar, J. Nagaraju, Molecular architecture of silk fibroin of Indian golden silkmoth, Antheraea assama, Sci. Rep. 5 (2015) 12706.

[13] J.S. Hwang, J.S. Lee, T.W. Goo, E.Y. Yun, K.S. Lee, Y.S. Kim, et al., Cloning of the fibroin gene from the oak silkworm, Antheraea yamamai and its complete sequence, Biotechnol. Lett. 23 (16) (2001) 1321–1326.

[14] H. Sezutsu, K. Yukuhiro, Dynamic rearrangement within the Antheraea pernyi silk fibroin gene is associated with four types of repetitive units, J. Mol. Evol. 51 (4) (2000) 329–338.

[15] R.F. Manning, L.P. Gage, Internal structure of the silk fibroin gene of Bombyx mori. II. Remarkable polymorphism of the organization of crystalline and amorphous coding sequences, J. Biol. Chem. 255 (19) (1980) 9451–9457.

[16] K. Mita, S. Ichimura, T.C. James, Highly repetitive structure and its organization of the silk fibroin gene, J. Mol. Evol. 38 (6) (1994) 583–592.

[17] H. Sezutsu, K. Uchino, I. Kobayashi, T. Tamura, K. Yukuhiro, Extensive sequence rearrangements and length polymorphism in fibroin genes in the wild silkmoth, Antheraea yamamai (Lepidoptera, Saturniidae), Int. J. Wild Silkmoth Silk 15 (2010) 35–50.

[18] A. Chinali, W. Vater, B. Rudakoff, A. Sponner, E. Unger, F. Grosse, et al., Containment of extended length polymorphisms in silk proteins, J. Mol. Evol. 70 (4) (2010) 325–338.

[19] F. Sehnal, M. Zurovec, Construction of silk fiber core in Lepidoptera, Biomacromolecules 5 (3) (2004) 666–674.
[20] C.Z. Zhou, F. Confalonieri, N. Medina, Y. Zivanovic, C. Esnault, T. Yang, et al., Fine organization of Bombyx mori fibroin heavy chain gene, Nucleic Acids Res. 28 (12) (2000) 2413–2419.
[21] Y.H. Sima, M. Chen, R. Yao, Y.P. Li, T. Liu, X. Jin, et al., The complete mitochondrial genome of the Ailanthus silkmoth, Samia cynthia cynthia (Lepidoptera: Saturniidae), Gene 526 (2) (2013) 309–317.
[22] F. Chen, D. Porter, F. Vollrath, Structure and physical properties of silkworm cocoons, J. R. Soc. Interface 9 (74) (2012) 2299–2308.
[23] H. Sezutsu, T. Tamura, K. Yukihiro, Uniform size of leucine-rich repeats in a wild silk moth Saturnia japonica (Lepidoptera Saturniidae) fibroin, Int. J. Wild Silkmoth Silk 13 (2008) 53–60.
[24] H. Sezutsu, T. Tamura, K. Yukuhiro, Leucine-rich fibroin gene of the Japanese wild silkmoth, Rhodinia fugax (Lepidoptera: Saturniidae), Eur. J. Entomol. 105 (4) (2008) 561–566.
[25] N. Agarwal, D.A. Hoagland, R.J. Farris, Effect of moisture absorption on the thermal properties of Bombyx mori silk fibroin films, J. Appl. Polym. Sci. 63 (3) (1997) 401–410.
[26] C. Fu, D. Porter, Z. Shao, Moisture effects on Antheraea pernyi silk's mechanical property, Macromolecules 42 (20) (2009) 7877–7880.
[27] J. Guan, D. Porter, F. Vollrath, Thermally induced changes in dynamic mechanical properties of native silks, Biomacromolecules 14 (3) (2013) 930–937.
[28] X. Hu, D. Kaplan, P. Cebe, Effect of water on the thermal properties of silk fibroin, Thermochim. Acta 461 (1) (2007) 137–144.
[29] X. Hu, K. Shmelev, L. Sun, E.S. Gil, S.H. Park, P. Cebe, et al., Regulation of silk material structure by temperature-controlled water vapor annealing, Biomacromolecules 12 (5) (2011) 1686–1696.
[30] W. Huang, S. Krishnaji, O.R. Tokareva, D. Kaplan, P. Cebe, Influence of water on protein transitions: morphology and secondary structure, Macromolecules 47 (22) (2014) 8107–8114.
[31] W. Huang, S. Krishnaji, O.R. Tokareva, D. Kaplan, P. Cebe, Influence of water on protein transitions: thermal analysis, Macromolecules 47 (22) (2014) 8098–8106.
[32] C. Mo, P. Wu, X. Chen, Z. Shao, The effect of water on the conformation transition of Bombyx mori silk fibroin, Vib. Spectrosc. 51 (1) (2009) 105–109.
[33] K. Numata, T. Katashima, T. Sakai, State of water, molecular structure, and cytotoxicity of silk hydrogels, Biomacromolecules 12 (6) (2011) 2137–2144.
[34] G.R. Plaza, G.V. Guinea, J. Pérez-Rigueiro, M. Elices, Thermo-hygro-mechanical behavior of spider dragline silk: glassy and rubbery states, J. Polym. Sci. B Polym. Phys. 44 (6) (2006) 994–999.
[35] Q. Yuan, J. Yao, L. Huang, X. Chen, Z. Shao, Correlation between structural and dynamic mechanical transitions of regenerated silk fibroin, Polymer 51 (26) (2010) 6278–6283.
[36] T. Asakura, M. Demura, Y. Watanabe, K. Sato, H-1 Pulsed NMR-study of Bombyx-Mori silk fibroin—dynamics of fibroin and of absorbed water, J. Polym. Sci. Pol. Phys. 30 (7) (1992) 693–699.
[37] M. Ishida, T. Asakura, M. Yokoi, H. Saito, Solvent-induced and mechanical-treatment-induced conformational transition of silk fibroins studied by high-resolution solid-state C-13 NMR-spectroscopy, Macromolecules 23 (1) (1990) 88–94.
[38] H. Yoshimizu, T. Asakura, The structure of Bombyx-Mori Silk fibroin membrane swollen by water studied with ESR, C-13-NMR, and FT-IR spectroscopies, J. Appl. Polym. Sci. 40 (9-10) (1990) 1745–1756.

[39] Y.S. Kim, L. Dong, M.A. Hickner, T.E. Glass, V. Webb, J.E. McGrath, State of water in disulfonated poly(arylene ether sulfone) copolymers and a perfluorosulfonic acid copolymer (Nafion) and its effect on physical and electrochemical properties, Macromolecules 36 (17) (2003) 6281–6285.
[40] I.D. Kuntz, Hydration of macromolecules. III. Hydration of polypeptides, J. Am. Chem. Soc. 93 (2) (1971) 514–516.
[41] K.Y. Lee, W.S. Ha, DSC studies on bound water in silk fibroin/S-carboxymethyl kerateine blend films, Polymer 40 (14) (1999) 4131–4134.
[42] K. Yazawa, K. Ishida, H. Masunaga, T. Hikima, K. Numata, Influence of water content on the beta-sheet formation, thermal stability, water removal, and mechanical properties of silk materials, Biomacromolecules 17 (3) (2016) 1057–1066.
[43] X. Hu, D. Kaplan, P. Cebe, Dynamic protein – water relationships during β-sheet formation, Macromolecules 41 (11) (2008) 3939–3948.
[44] A. Motta, L. Fambri, C. Migliaresi, Regenerated silk fibroin films: thermal and dynamic mechanical analysis, Macromol. Chem. Phys. 203 (10-11) (2002) 1658–1665.
[45] P. Cebe, B.P. Partlow, D.L. Kaplan, A. Wurm, E. Zhuravlev, C. Schick, Using flash DSC for determining the liquid state heat capacity of silk fibroin, Thermochim. Acta 615 (2015) 8–14.
[46] F.G. Torres, O.P. Troncoso, C. Torres, W. Cabrejos, An experimental confirmation of thermal transitions in native and regenerated spider silks, Mat. Sci. Eng. C Mater 33 (3) (2013) 1432–1437.
[47] T. Asakura, K. Isobe, A. Aoki, S. Kametani, Conformation of crystalline and noncrystalline domains of [3-13C]Ala-, [3-13C]Ser-, and [3-13C]Tyr-Bombyx mori silk fibroin in a hydrated state studied with 13C DD/MAS NMR, Macromolecules 48 (22) (2015) 8062–8069.
[48] K. Nakamura, T. Hatakeyama, H. Hatakeyama, Effect of bound water on tensile properties of native cellulose, Text. Res. J. 53 (11) (1983) 682–688.
[49] S. Vyas, S. Pradhan, N. Pavaskar, A. Lachke, Differential thermal and thermogravimetric analyses of bound water content in cellulosic substrates and its significance during cellulose hydrolysis by alkaline active fungal cellulases, Appl. Biochem. Biotechnol. 118 (1-3) (2004) 177–188.
[50] Y.A. Ma, R. Sato, Z.B. Li, K. Numata, Chemoenzymatic synthesis of oligo(L-cysteine) for use as a thermostable bio-based material, Macromol. Biosci. 16 (1) (2016) 151–159.
[51] D. Luo, S.W. Smith, B.D. Anderson, Kinetics and mechanism of the reaction of cysteine and hydrogen peroxide in aqueous solution, J. Pharm. Sci. 94 (2) (2005) 304–316.
[52] H.E. Van Wart, H.A. Scheraga, Raman spectra of strained disulfides. Effect of rotation about sulfur-sulfur bonds on sulfur-sulfur stretching frequencies, J. Phys. Chem. 80 (16) (1976) 1823–1832.
[53] F.J.R. Hird, J.R. Yates, The oxidation of cysteine, glutathione and thioglycollate by iodate, bromate, persulphate and air, J. Sci. Food Agric. 12 (2) (1961) 89–95.
[54] P. Cebe, X. Hu, D.L. Kaplan, E. Zhuravlev, A. Wurm, D. Arbeiter, et al., Beating the heat—fast scanning melts silk beta sheet crystals, Sci. Rep. 3 (2013) 1130.
[55] K. Yazawa, K. Numata, Papain-catalyzed synthesis of polyglutamate containing a nylon monomer unit, Polymers (Basel) 8 (5) (2016) 194.
[56] K. Tachibana, Y. Urano, K. Numata, Biodegradability of nylon 4 film in a marine environment, Polym. Degrad. Stabil. 98 (9) (2013) 1847–1851.
[57] K. Yazawa, J. Gimenez-Dejoz, H. Masunaga, T. Hikima, K. Numata, Chemoenzymatic synthesis of a peptide containing nylon monomer units for thermally processable peptide material application, Polym. Chem. 8 (29) (2017) 4172–4176.
[58] J. Gatesy, C. Hayashi, D. Motriuk, J. Woods, R. Lewis, Extreme diversity, conservation, and convergence of spider silk fibroin sequences, Science 291 (5513) (2001) 2603–2605.

[59] T.A. Blackledge, J. Perez-Rigueiro, G.R. Plaza, B. Perea, A. Navarro, G.V. Guinea, et al., Sequential origin in the high performance properties of orb spider dragline silk, Sci. Rep. 2 (2012).
[60] Y. Liu, Z.Z. Shao, F. Vollrath, Relationships between supercontraction and mechanical properties of spider silk, Nat. Mater. 4 (12) (2005) 901–905.
[61] J.E. Jenkins, M.S. Creager, E.B. Butler, R.V. Lewis, J.L. Yarger, G.P. Holland, Solid-state NMR evidence for elastin-like beta-turn structure in spider dragline silk, Chem. Commun. 46 (36) (2010) 6714–6716.
[62] A.D. Malay, K. Arakawa, K. Numata, Analysis of repetitive amino acid motifs reveals the essential features of spider dragline silk proteins, PLoS One 12 (8) (2017) e0183397.
[63] J.E. Bond, N.L. Garrison, C.A. Hamilton, R.L. Godwin, M. Hedin, I. Agnarsson, Phylogenomics resolves a spider backbone phylogeny and rejects a prevailing paradigm for orb web evolution, Curr. Biol. 24 (15) (2014) 1765–1771.
[64] K. Yazawa, A. Malay, H. Masunaga, Y. Norma-Rashid, K. Numata, Simultaneous effect of strain rate and humidity on the structure and mechanical behavior of spider silk, Commun. Mater. (1)(2020) 10.
[65] D.T. Grubb, L.W. Jelinski, Fiber morphology of spider silk: the effects of tensile deformation, Macromolecules 30 (10) (1997) 2860–2867.
[66] K. Numata, R. Sato, K. Yazawa, T. Hikima, H. Masunaga, Crystal structure and physical properties of Antheraea yamamai silk fibers: long poly(alanine) sequences are partially in the crystalline region, Polymer 77 (2015) 87–94.
[67] C. Boutry, T.A. Blackledge, Wet webs work better: humidity, supercontraction and the performance of spider orb webs, J. Exp. Biol. 216 (19) (2013) 3606–3610.
[68] T. Giesa, R. Schuetz, P. Fratzl, M.J. Buehler, A. Masic, Unraveling the molecular requirements for macroscopic silk supercontraction, ACS Nano 11 (10) (2017) 9750–9758.
[69] G.V. Guinea, M. Elices, J. Perez-Rigueiro, G.R. Plaza, Stretching of supercontracted fibers: a link between spinning and the variability of spider silk, J. Exp. Biol. 208 (Pt 1) (2005) 25–30.
[70] J.M. Gosline, P.A. Guerette, C.S. Ortlepp, K.N. Savage, The mechanical design of spider silks: from fibroin sequence to mechanical function, J. Exp. Biol. 202 (23) (1999) 3295–3303.
[71] T. Vehoff, A. Glisovic, H. Schollmeyer, A. Zippelius, T. Salditt, Mechanical properties of spider dragline silk: humidity, hysteresis, and relaxation, Biophys. J. 93 (12) (2007) 4425–4432.
[72] T.A. Blackledge, C. Boutry, S.-C. Wong, A. Baji, A. Dhinojwala, V. Sahni, et al., How super is supercontraction? Persistent versus cyclic responses to humidity in spider dragline silk, J. Exp. Biol. 212 (13) (2009) 1981–1989.
[73] M. Elices, G.R. Plaza, J. Perez-Rigueiro, G.V. Guinea, The hidden link between supercontraction and mechanical behavior of spider silks, J. Mech. Behav. Biomed. Mater. 4 (5) (2011) 658–669.
[74] Y. Liu, Z. Shao, F. Vollrath, Relationships between supercontraction and mechanical properties of spider silk, Nat. Mater. 4 (2005) 901.
[75] S. Sampath, T. Isdebski, J.E. Jenkins, J.V. Ayon, R.W. Henning, J.P.R.O. Orgel, et al., X-ray diffraction study of nanocrystalline and amorphous structure within major and minor ampullate dragline spider silks, Soft Matter 8 (25) (2012) 6713–6722.
[76] A. Glišović, T. Vehoff, R.J. Davies, T. Salditt, Strain dependent structural changes of spider dragline silk, Macromolecules 41 (2) (2008) 390–398.
[77] T. Arai, G. Freddi, R. Innocenti, M. Tsukada, Biodegradation of Bombyx mori silk fibroin fibers and films, J. Appl. Polym. Sci. 91 (4) (2004) 2383–2390.

[78] B. Lotz, A. Gonthiervassal, A. Brack, J. Magoshi, Twisted single-crystals of Bombyx-Mori silk fibroin and related model polypeptides with beta-structure—a correlation with the twist of the beta-sheets in globular-proteins, J. Mol. Biol. 156 (2) (1982) 345−357.
[79] M.Z. Li, M. Ogiso, N. Minoura, Enzymatic degradation behavior of porous silk fibroin sheets, Biomaterials 24 (2) (2003) 357−365.
[80] R.L. Horan, K. Antle, A.L. Collette, Y.Z. Huang, J. Huang, J.E. Moreau, et al., In vitro degradation of silk fibroin, Biomaterials 26 (17) (2005) 3385−3393.
[81] K. Numata, D.L. Kaplan, Mechanisms of enzymatic degradation of amyloid beta microfibrils generating nanofilaments and nanospheres related to cytotoxicity, Biochemistry 49 (15) (2010) 3254−3260.
[82] S. Chimon, M.A. Shaibat, C.R. Jones, D.C. Calero, B. Aizezi, Y. Ishii, Evidence of fibril-like beta-sheet structures in a neurotoxic amyloid intermediate of Alzheimer's beta-amyloid, Nat. Struct. Mol. Biol. 14 (12) (2007) 1157−1164.
[83] K. Numata, How to define and study structural proteins as biopolymer materials, Polym. J. 52 (9) (2020) 1043−1056.
[84] L.M. Dowling, W.G. Crewther, D.A. Parry, Secondary structure of component 8c-1 of alpha-keratin. An analysis of the amino acid sequence, Biochem. J. 236 (3) (1986) 705−712.
[85] L.M. Dowling, W.G. Crewther, A.S. Inglis, The primary structure of component 8c-1, a subunit protein of intermediate filaments in wool keratin. Relationships with proteins from other intermediate filaments, Biochem. J. 236 (3) (1986) 695−703.
[86] L. Pauling, R.B. Corey, Compound helical configurations of polypeptide chains: structure of proteins of the alpha-keratin type, Nature 171 (4341) (1953) 59−61.
[87] K. Numata, D.L. Kaplan, Biologically Derived Scaffolds, Woodhead Publishing Materials Series 2011, pp. 524−551.
[88] K. Tsuchiya, N. Ifuku, Y. Koyama, K. Numata, Development of regenerated silk films coated with fluorinated polypeptides to achieve high water repellency and biodegradability in seawater, Polym. Degrad. Stabil. 160 (2019) 96−101.
[89] K. Numata, N. Ifuku, H. Masunaga, T. Hikima, T. Sakai, Silk resin with hydrated dual chemical-physical cross-links achieves high strength and toughness, Biomacromolecules 18 (6) (2017) 1937−1946.
[90] P.G. Gudeangadi, K. Uchida, A. Tateishi, K. Terada, H. Masunaga, K. Tsuchiya, et al., Poly(alanine-nylon-alanine) as a bioplastic: chemoenzymatic synthesis, thermal properties and biological degradation effects, Polym. Chem. 11 (30) (2020) 4920−4927.
[91] R. Abe, K. Toyota, H. Miyakawa, H. Watanabe, T. Oka, S. Miyagawa, et al., Diofenolan induces male offspring production through binding to the juvenile hormone receptor in Daphnia magna, Aquat. Toxicol. 159 (2015) 44−51.
[92] B.P. Elendt, W.R. Bias, Trace nutrient deficiency in Daphnia-magna cultured in standard medium for toxicity testing—effects of the optimization of culture conditions on life-history parameters of Daphnia-magna, Water Res. 24 (9) (1990) 1157−1167.
[93] K. Numata, N. Ifuku, A. Isogai, Silk composite with a fluoropolymer as a water-resistant protein-based material, Polymers (Basel) 10 (4) (2018).
[94] C. Holland, K. Numata, J. Rnjak-Kovacina, F.P. Seib, The biomedical use of silk: past, present, future, Adv. Healthc. Mater. 8 (1) (2019) e1800465.
[95] G.H. Altman, F. Diaz, C. Jakuba, T. Calabro, R.L. Horan, J.S. Chen, et al., Silk-based biomaterials, Biomaterials 24 (3) (2003) 401−416.
[96] N.A. Fine, M. Lehfeldt, J.E. Gross, S. Downey, G.M. Kind, G. Duda, et al., SERI surgical scaffold, prospective clinical trial of a silk-derived biological scaffold in two-stage breast reconstruction: 1-year data, Plast. Reconstr. Surg. 135 (2) (2015) 339−351.

[97] L. Meinel, S. Hofmann, V. Karageorgiou, C. Kirker-Head, J. McCool, G. Gronowicz, et al., The inflammatory responses to silk films in vitro and in vivo, Biomaterials 26 (2) (2005) 147–155.
[98] Y. Wang, D.D. Rudym, A. Walsh, L. Abrahamsen, H.J. Kim, H.S. Kim, et al., In vivo degradation of three-dimensional silk fibroin scaffolds, Biomaterials 29 (24-25) (2008) 3415–3428.
[99] J.E. Gross, R.L. Horan, M. Gaylord, R.E. Olsen, L.D. McGill, J.M. Garcia-Lopez, et al., An evaluation of SERI surgical scaffold for soft-tissue support and repair in an ovine model of two-stage breast reconstruction, Plast. Reconstr. Surg. 134 (5) (2014) 700e–704e.
[100] J. Haupt, J.M. Garcia-Lopez, K. Chope, Use of a novel silk mesh for ventral midline hernioplasty in a mare, BMC Vet. Res. 11 (2015) 58.
[101] A.E. Thurber, F.G. Omenetto, D.L. Kaplan, In vivo bioresponses to silk proteins, Biomaterials 71 (2015) 145–157.
[102] T. Wongpinyochit, B.F. Johnston, F.P. Seib, Degradation behavior of silk nanoparticles-enzyme responsiveness, ACS Biomater. Sci. Eng. 4 (3) (2018) 942–951.
[103] J. Brown, C.L. Lu, J. Coburn, D.L. Kaplan, Impact of silk biomaterial structure on proteolysis, Acta Biomater. 11 (2015) 212–221.

Questions for this chapter

1. Explain why protein materials might show different physical properties at different places/environments.
2. To evaluate the environmental degradability of protein material, what kinds of experiments should be performed?
3. Why does the crystalline region show slower biodegradation in polymeric materials?

CHAPTER 7

Structural proteins in nature

Generally, proteins are classified into fibrillar and globular types. Structural proteins introduced in the book are mainly fibrillar proteins. Fig. 7.1 shows some examples of structural proteins and related information [1,2]. In this chapter, structural proteins and also nonglobular proteins, which play a structural role in nature, are summarized.

7.1 Spider dragline silk

Silk proteins can be synthesized by various organisms, including spiders, silkworms, and bagworms (Fig. 7.2) [1]. A wide variety of spiders are known to produce multiple types of fibers, and some spin up to seven types (Fig. 7.3) [3–5]. The thread spun from the major ampullate gland is called dragline and is used as a framework for reticular spider webs and as lifelines. As can be inferred from their role in nature, spider draglines exhibit excellent mechanical properties and are known as typical "tough

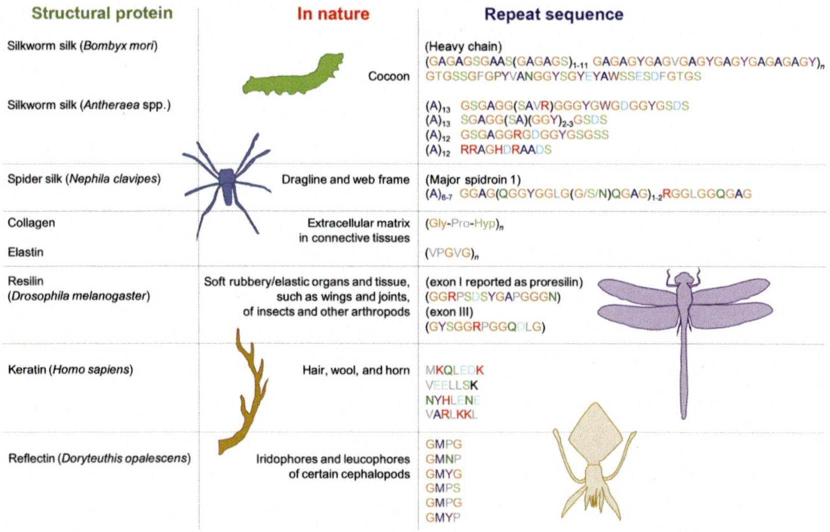

Figure 7.1 Examples of structural proteins in nature. Their roles/functions in nature and typical repeat sequences are listed [1].

Biopolymer Science for Proteins and Peptides
DOI: https://doi.org/10.1016/B978-0-12-820555-6.00004-5

Figure 7.2 Pictures of silk producers and silk threads. (A) *Trichonephila clavata* spider and threads; (B) *Nephila pilipes* spider on the author's hand. Other silk producers, silkworms (C), and bagworm (D); (E) spider egg case; (F) and silkworm cocoons and silk threads [1].

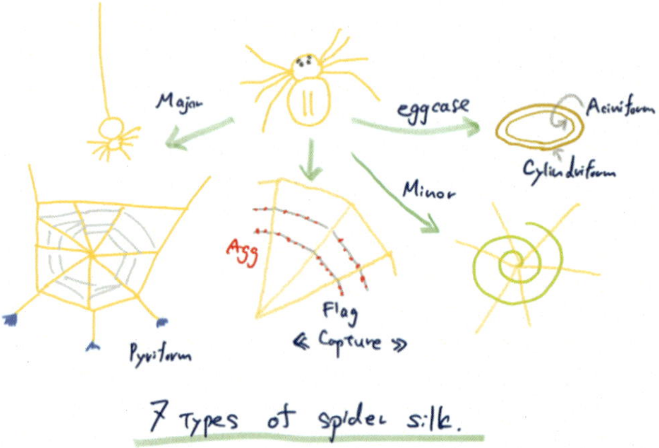

Figure 7.3 Various spider threads. Tow yarn, warp, and weft forming a circular net, egg case.

spider silk fibers." In particular, the toughness of these fibers in terms of resistance to breakage due to stretching is significantly superior to that of synthetic polymers and other natural materials due to their relatively high strength and elongation at break [6,7]. Therefore spider silk fibers are expected to be applicable in lightweight and tough structural materials, such as automotive parts. The mechanical and physical properties of the spider dragline are significantly influenced by deformation rate and humidity [8–10]. Unlike silkworm silk fibers, the physical properties of spider dragline, namely, its supercontraction, are influenced by interactions with water molecules. Recently, proline and diglutamine (QQ) in the amino acid sequences have been proposed to contribute to supercontraction based on comparative studies of the supercontraction of different spider silks [8].

Although the formation mechanism of silk fibers has been studied since 1900, the molecular mechanism by which the yarn is formed and the formation process of the hierarchical structure have not been clarified yet [11–14]. The surface of the yarn has a skin layer consisting of glycoproteins and lipids. Although this skin layer does not affect mechanical properties, it has been reported to impart resistance to biodegradation. The protein fibers that form the center of the yarn are formed in bundles of microfibers whose main component is silk protein. In addition, microfibers are formed so as to connect granular structures by atomic force microscopy, birefringence analysis, small-angle X-ray scattering (SAXS)/wide-angle X-ray scattering (WAXS) analysis, etc. (Fig. 7.4).

The mechanism of spinning inside the granule structure, that is, at the molecular level, has been studied in various countries around the world. The different domains of silk proteins respond to the coordinated changes during the spinning process [5]. Notably, the *N*-terminal domains dimerize in response to the decreasing pH in the spinning duct to form continuous polypeptide chains. On the other hand, the shear forces that arise during spinning induce self-assembly of intermolecular beta-sheet nanocrystals, oriented parallel to the fiber axis, embedded in a "soft" amorphous protein matrix, giving rise to the characteristic viscoelasticity of silk fibers [15,16]. Recent advances in structural analysis technology have revealed that the structures of the *N*- and *C*-termini of silk proteins that make up spider silk contribute to the self-assembly of the spinning process (Fig. 7.5). In particular, as the pH decreases during the spinning process, the *N*-terminal is dimerized to form an intermolecular network. This dimerization of the

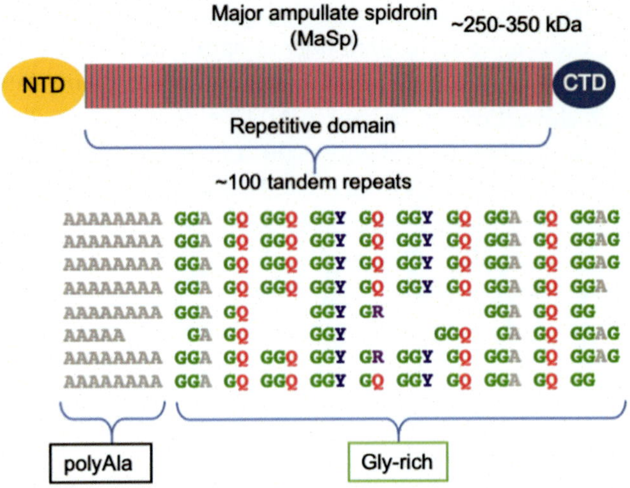

Figure 7.4 Schematic illustration of spider silk formation. MaSp, one of the major components of spider silk, forms liquid–liquid phase separation via *C*-terminal domain (CTD) interactions and subsequently forms a microfibrillar network via *N*-terminal domain (NTD) dimerization. The microfibrillar network is sheared into a silk fiber that is a bundle of microfibrils.

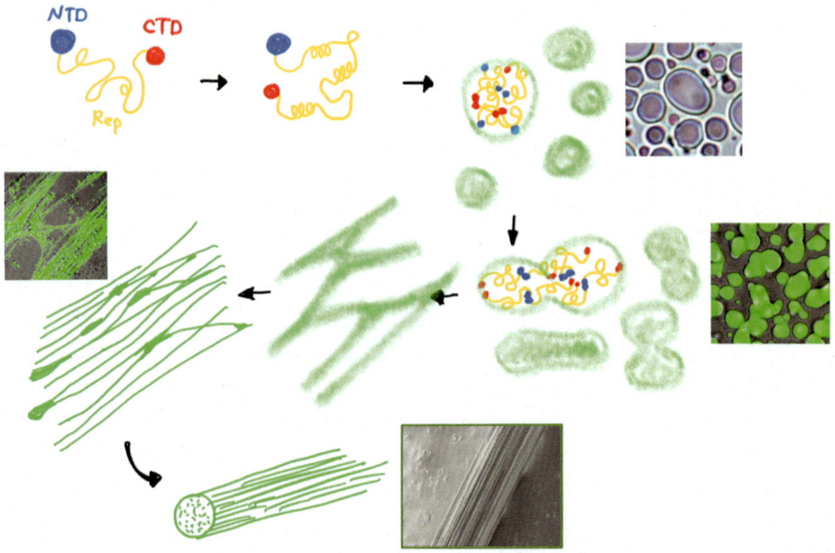

Figure 7.5 A schematic diagram of the *N*-terminus, *C*-terminus, and repeating sequence of silk proteins. The repeating sequence has crystal regions and amorphous regions repeatedly present.

N-terminal domains induces beta-sheet structure formation as a crystal component and, at the same time, achieves a specific mechanical and physical property by simulating a network-like molecular structure by the interaction of molecular chain ends.

Following the *N*- and *C*-termini, the influence of the central repeat sequence on the initial process of spinning has also been studied. Spider dragline silk has a tandem sequence of polyalanine (polyAla)-rich and glycine-rich sequences. The extended repeating domains include polyAla stretches (6–13 alanine residues) alternating with glycine-rich regions consisting of concatenated G*X* and GG*X* motifs, where *X* represents a small subset of residues [8,17]. polyAla is a potentially crystalline motif, whereas the glycine-rich sequences form amorphous regions. This characteristic repeating motif allows the formation of higher-order structures in which microcrystals comprising beta-sheet structures are highly oriented along the fiber axis [18–20]. Also with regard to the structure of silk protein before spinning, the analysis of nuclear magnetic resonance (NMR) and circular dichroism shows that the noncrystalline arrangement containing a large amount of glycine is polyproline II helix due to the existence of the repeated arrangement of silk twice or more in the solution [21]. In addition, it has also been reported that pH change does not affect the repeat sequence during the spinning process, but affects only the structure of *N*-terminus and *C*-terminus.

Spider silk has multiple structural components. In other words, silk, in its natural state, is not a single-component material but rather a composite material [5]. Spider dragline is typically composed of more than two major ampullate spidroin (MaSp) proteins (MaSp1, MaSp2, and others) that have similar overall architectures but different amino acid motif compositions (e.g., proline residues and QQ motifs in MaSp2, which are absent in MaSp1) [8]. QQ motif can be seen in sticky proteins such as glutenin in bread dough (Fig. 7.6). Differences in the MaSp1/MaSp2 ratio are thought to influence mechanical properties, such as tensile strength versus extensibility [22]. Although the hierarchical interactions between MaSp1 and MaSp2 remain poorly characterized, they are considered to cause phenomena such as supercontraction in response to water [22]; interestingly, evidence suggests that MaSp1 and MaSp2 occupy distinct spatial domains within dragline silk fibers [23]. Polymerized MaSp protein, which forms the core of dragline silk, is further surrounded by outer layers, whose composition includes glycoproteins and lipids, and these layers are thought to provide protective functions [24,25].

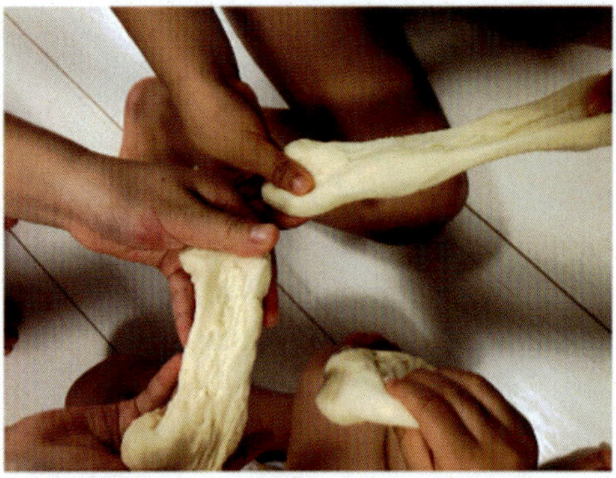

Figure 7.6 Pictures of bread dough containing a sticky protein, glutenin. This sticky property is partially originated from the sequences containing proline and QQ motifs.

7.2 Spider viscid silk

Viscid silk and glues are noncrystalline and can only be crystallized partially at low temperatures, according to a study by Craig [26]. The synchrotron SAXS and WAXS structural analyses of spider capture silk during stretching deformation were also reported [18]. Its structure is predominantly amorphous and is not converted to a crystalline state even when the silk is stretched by 500% (Fig. 7.7) [18]. The amino acid sequence and composition of the capture silk of *Trichonephila clavata* differ from those of *T. clavata* dragline silk and *Bombyx mori* silk. The amino acid sequences for the beta-sheet formations found in *T. clavata* dragline silk and *B. mori* silk are AAAAAA and GAGAGS, respectively; however, these are not found in *T. clavata* capture silk, which mostly contains glycine- and proline-rich β-spiral structures [27,28]. The capture spiral silk does not exhibit any crystalline components and nor does it undergo crystallization before and during the stretching deformation. This allows it to retain its elastic and optical properties during deformation. Spider webs made of capture silk are known to display multiple colors during the reflection, refraction, and transmission of sunlight [29,30]. The results of the structural analysis of capture silk demonstrate that this silk is perfectly amorphous during stretching. It is for this reason that crystallization does not have an effect on the optical properties of spider webs of this silk. These attributes of capture silk, which is one of the main components of spider webs, ensure that it is better suited than other types of silks for catching prey.

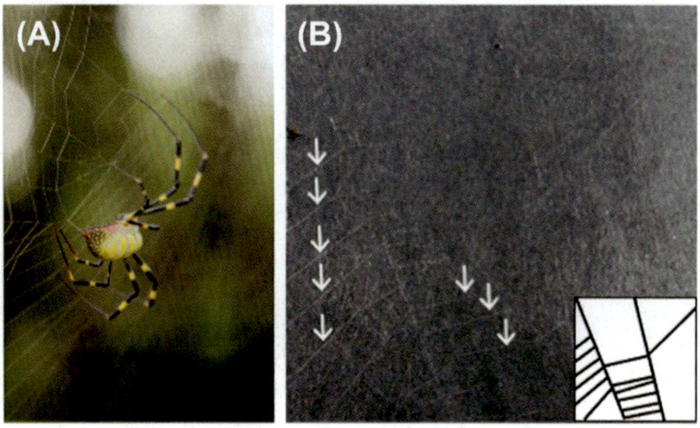

Figure 7.7 Pictures of orb-web (A) and viscid/capture silk (B) [18].

7.3 Silkworm silk

Silk protein is produced not only by spiders but also by animals such as silkworms and scorpions. The silkworm fibers are then enveloped by layers of sericin, an amorphous protein that serves as a gumming agent, to maintain the structural integrity of the cocoon [16], and many studies focusing on the material properties of sericin have been reported [31]. The major component of silkworm silk, the fibroin heavy chain, forms complexes with the fibroin light chain and P25 glycoprotein [16]. The fibroin heavy chain (>100 kDa) has architecture featuring conserved terminal domains that flank an extended middle section consisting of numerous repeating low-complexity modules [15]. In silkworm silk, this middle section mainly consists of extended $(GAGA)_nG(Y/S)$ arrays (where G = glycine, A = alanine, Y = tyrosine, and S = serine) of heterogeneous length interspersed with conserved linker regions [32,33]. The presence of beta-sheet structures as crystal components is similar, but the amino acid sequences and physical properties obviously vary among the different species. A consortium of Asian countries has reevaluated silk as a polymer material and reported the variety of physical properties offered by silkworm silk thread [32] (Fig. 7.8). Differences in mechanical and thermal properties can occur due to the variations in the hierarchical structure resulting from the amino acid sequence of the silk protein.

The beta-sheet structure of silkworm silk is formed in the silkworm cocoons, according to their amino acid sequences (Fig. 7.9) [18]. The beta-sheet crystals of silkworm silk must be formed by silkworm's spinning and dehydration of spun silk fibers. The silkworm cocoon silk must

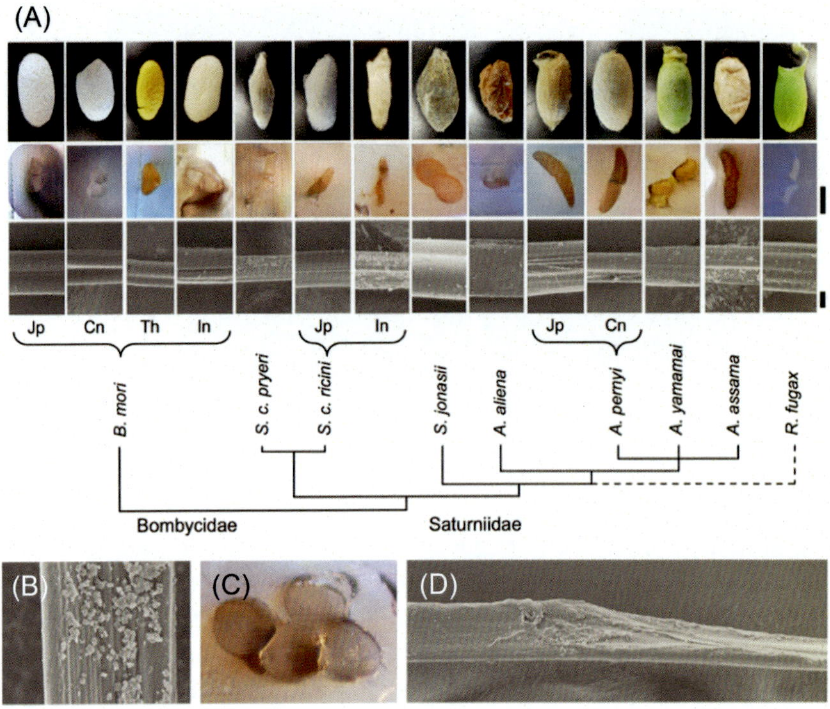

Figure 7.8 Morphologies of the different silks used in this study. (A) First row, representative photographs of cocoons (not to scale); second row, representative photomicrographs showing cross sections of silk fibers embedded in cyanoacrylate glue (scale bar = 20 μm); third row, representative scanning electron microscopic (SEM) images of fiber surfaces (scale bar = 10 μm). The phylogenetic assignment is based on Sima et al. [34]. Branch lengths are arbitrary, and the position of *Rhodinia fugax* is tentatively assigned. Japan (Jp), China (Cn), Thailand (Th), and India (In) denote the country where the silk was sampled. (B) SEM image showing the presence of calcium oxalate crystals on the surface of *Samia cynthia pryeri* silk fiber. (C) Cross-sectional photomicrograph showing the common occurrence of silk fibers stuck together with gum material in *Saturnia jonasii* cocoons. (D) SEM image showing an example of abrupt changes in the morphology and fiber diameter in *Actias aliena. Source: Reprinted with permission from A.D. Malay, R. Sato, K. Yazawa, et al., Relationships between physical properties and sequence in silkworm silks, Sci. Rep. 6 (2016) 27573 [32].*

contain a sufficient number of crystalline regions to improve the toughness and strength of cocoons and to protect the chrysalis from external physical impact and predators. It has thus not evolved to respond to stretching deformation. This conclusion is supported by the mechanical properties exhibited by the two types of silk fibers investigated, namely, the fact that *B. mori* silkworm silk breaks after elongation by ~15% and is

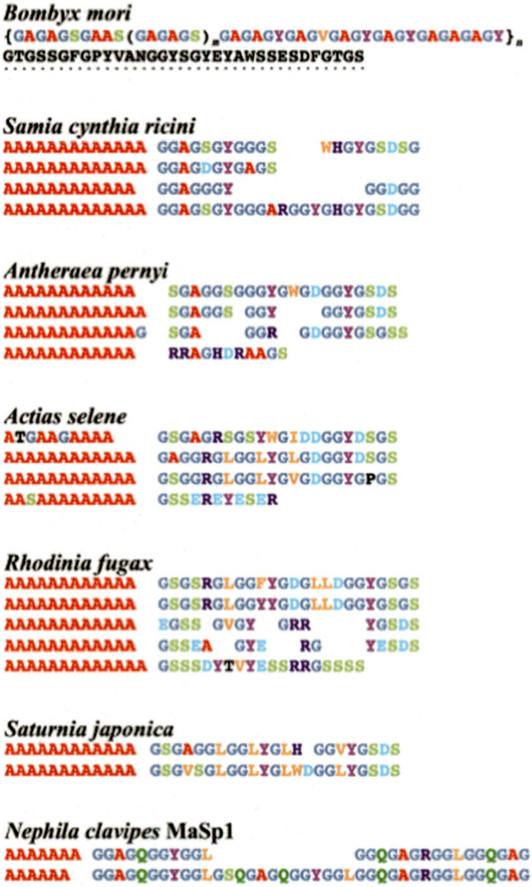

Figure 7.9 Representative repetitive sequences of the different silk fibroin are shown: *Bombyx mori* heavy chain (GenBank AF226688), demonstrating the hierarchical arrangement of motifs, and with the conserved spacer sequence underlined, *Samia cynthia ricini* (BAQ55621), *Antheraea pernyi* (AAC32606), *Actias selene* (deduced sequence) [35], *Rhodinia fugax* (BAG84270), *Saturnia japonica* (BAH02016), and MaSp1 from spider dragline silk of *Nephila clavipes* [20] (M37137). *Sources: Modified with permission from A.D. Malay, R. Sato, K. Yazawa, et al., Relationships between physical properties and sequence in silkworm silks, Sci. Rep. 6 (2016) 27573 [32] and J.M. Ageitos, K. Yazawa, A. Tateishi, K. Tsuchiya, K. Numata, The benzyl ester group of amino acid monomers enhances substrate affinity and broadens the substrate specificity of the enzyme catalyst in chemoenzymatic copolymerization, Biomacromolecules 17 (1) (2016) 314–323 [36] © 2017 American Chemical Society.*

less elastic [37], in contrast to *Trichonephila edulis* (formerly known as *Nephila edulis*) spider dragline silk, whose average breaking elongation is approximately 40% [38]. In addition, spinning speed exerts a significant effect on the mechanical properties of silkworm silk [39]. Thus the silkworm cocoon silk has a high crystallinity and low elasticity; these properties are consistent with its role as a tough outer component of the chrysalis [18].

Removal of sericin from silkworm silk is necessary to prepare nonallergic and noncytotoxic silk-based materials. Recently, methods to extract and regenerate silk fibroin have been developed, and several silk-based biomaterials, such as silk porous scaffolds, silk films, hydrogels, coatings, and electrospun nanofibers, have been processed from silk solutions [40–45] As silk fibers have been used as sutures, silk is attractive as a biomedical material for regenerative medicine, tissue engineering, and drug delivery [40,46], because of excellent biocompatibility, biodegradability, and less cytotoxicity [47,48]. In addition, some silks (e.g., *Antheraea yamamai*) have an arginine–glycine–aspartic acid sequence, which contributes to cell adhesion, making them useful for biomaterials [49–51]. The degradation products of silk fibroin proteins with beta-sheet structures from the action of proteases, such as alpha-chymotrypsin, have recently been reported, and no cytotoxicity was observed to neurons in vitro [47].

7.4 Bagworm silk

The bagworm is one of the insects that produces and uses silk fibers throughout its life stages. The silkworm and the saturniid are closely related species; however, the bagworm has a slight difference in the use of silk [52]. The bagworm family (Psychidae) includes over 1000 species, and all of their larval development is conducted within a self-enclosing bag [53]. This self-enclosing bag is different from a cocoon produced by silkworm since the bag is combined with plant materials [54,55].

Eumeta variegata is a bagworm moth (Lepidoptera, Psychidae) that uses silken thread throughout all life stages (Fig. 7.10). Notably, the bagworm-specific uses of silk include larval development in a bag coated with silk and plant materials and the use of silk attachments to hang pupae on plant leaves. The first bagworm genome, including a silk fibroin gene was reported by Kono et al. [52]. The structural feature of the fibroin gene is shared among other lepidopterans, the obtained repetitive motif varied. The bagworm fibroin gene combines both polyAla and an alternating glycine-alanine motif in a repetitive domain [32]. The repetitive motif in bagworm fibroin contains a characteristic motif combining the properties of both the silkworm and the saturniid. The mechanical properties of the

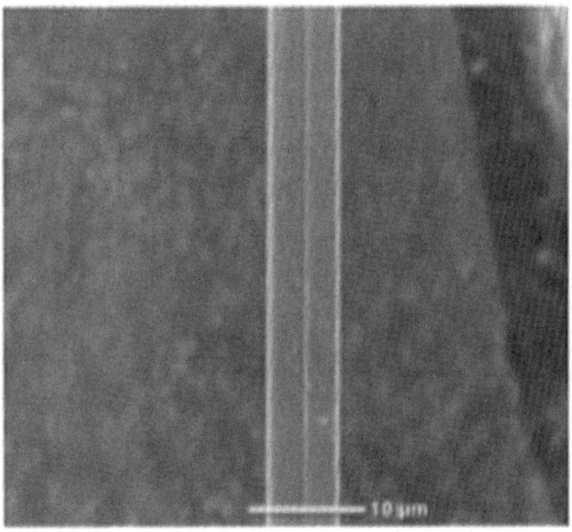

Figure 7.10 Silk fibers of *Eumeta variegata* bagworm moth [52].

bagworm showed a tensile strength of 636 ± 55 MPa, an extensibility of 19.5% ± 4.5%, a Young's modulus of 5.67 ± 0.66 GPa, and a toughness of 70.3 ± 66.00 MJ/m^3. The wide-angle X-ray diffraction pattern showed the orientation, and the analysis indicated that the degree of crystallinity was 13.4%. Upon comparison, the bagworm silk showed a significantly higher tensile strength than the Japanese silkworm and all saturniids [52]. On the other hand, the extensibility of the bagworm silk was significantly less than that of the silkworm *A. yamamai*.

7.5 Other silks from bees, ants, and hornets

In addition to spiders and silkworms, the larvae of bees, ants, and hornets produce silk [56]. The characteristic molecular structure of these silks is α-helical proteins assembled into tetrameric coiled coils, which are different from the beta-sheet structure predominant spider silks [57,58]. In nature, the roles of the coiled coil silk are different from spider silks and it is used for various purposes, for instance, the elaborate nests and the cocoons of bulldog ant pupae and bumblebees [56]. The patterns from honeybee silks show a four-stranded coiled coil structure [59]. The core of the coiled coil structure is composed of hydrophobic residues (alanine, leucine, and isoleucine) [60]. The ant silk was reported to contain both the coiled coil and also beta-sheet structures [56,61], whereas the bee silk is dominated by the coiled coil signal. Similarly, both honeybee and hornet silk contain an α-helix and beta-sheet structure, according to NMR characterizations [62,63].

7.6 Sericin

Sericin is known as a glue protein for silkworm silk fibers and has traditionally been linked to the adverse effects reported for virgin silk [46,64]. Silk sericin is a serine-rich protein and hence is biodegradable and hot water soluble. Over the past decades, sericin has emerged as an attractive and functional biopolymer [65]. The biocompatibility studies on sericin showed encouraging results in relation to the allergenic and immunogenic profile of sericin [66]. An increasing number of studies report the biomedical use of the biopolymer sericin. For example, the development of composite sericin/silicone nerve guides or sericin/polyacrylamide hydrogels is proposed for dermal repair [67,68]. However, there is still the possibility for an excess amount of sericin on silk fibers to be an allergen.

7.7 Collagen

Collagen is the most abundant protein in mammals and the main structural component of the extracellular matrix, with $(\text{Gly-X-Y})_n$ repeating units longer than 1400 amino acid residues and with three residues per one helical turn structure [69]. The most common tripeptide unit of collagen is $(\text{Gly-Pro-Hyp})_n$. Hyp is a characteristic amino acid to indicate collagen due to the posttranslational modification of proline in mammal cells. As a structural protein, collagen has excellent biocompatibility and cell adhesion and can promote cell proliferation and differentiation. In addition to mammals, the larvae of a sawfly species, fruit fly *Drosophila melanogaster*, locust, and cockroach produce silk from three small collagen proteins [70–74]. Insect collagens are apparently homologous to molecules found in mammals.

Collagen is also one of the most studied biologically derived polymer scaffolds because of its biocompatibility, cell adhesion, growth, and differentiation promoting properties [75,76]. Collagen along with glycosaminoglycans are important factors for cell attachment, proliferation, and differentiation [77]. Collagen-glycosaminoglycan scaffolds have been studied for skin regeneration [78], porous scaffolds for the attachment and proliferation of fibroblasts, chondrocytes, and neurons, as well as for supporting osteogenesis of rat mesenchymal stem cells, mouse osteogenic cell line MC3T3, and human osteoblast cell line hFOB [75,79–83]. The disadvantage of collagen-based biomaterials is the rapid degradation rate and lower hydrostability, which results in the rapid loss of mechanical properties in vivo [84]. Several methods to stabilize collagen scaffolds with biologically derived polymers have been reported. Elastin and glycoaminoglycan enhance the stability of collagen scaffolds when mixed for composite materials [85]. Cross-linking can also be used to stabilize collagen by forming molecular networks [86]. Porous scaffolds of collagen cross-linked with hyaluronate or calcium phosphate (CaP) demonstrated good biocompatibility and potential as osteochondral implants [87,88]. To form scaffolds with adequate mechanical properties, cross-linking elastin-like polymer with collagen has been developed with enzymatically resistant covalent bonds between collagens and elastins to increase mechanical strength in a dose-dependent manner without significantly affecting the porosity or thermal properties of the original scaffolds [89].

7.8 Elastin

Elastin is a famous structural protein that can contribute to the elasticity and stretchability of tissues. It consists mainly of glycine, valine, alanine, and proline residues and possesses a molecular weight of approximately 66 kDa. Elastin exists in connective, vascular, and load-bearing tissues and has very elastic mechanical properties [1,2]. Elastin plays a mechanical role at small stress with small deformations in rat skin [90]. The elastic modulus, strength, extensibility, toughness, and resilience are 0.0011 GPa, 0.002 GPa, 150%, 1.6 MJ/m^3, and 90%, respectively [91]. Therefore elastin is considered an important structural protein and material for scaffolds that require elastic physical properties and cell adhesiveness in cell culture. However, elastin is poorly soluble in aqueous solutions, making it difficult to process, and this is a common problem among structural proteins. Besides, elastin is difficult to obtain, synthesize, and produce in large quantities, and hence studies on biomaterials composed of elastin alone are very limited compared to reports on scaffolds composed of other structural proteins. A scaffold of elastin derived from bovine ligament has been reported, and its mechanical properties have been studied, focusing on its elastic modulus [92]. The elastin's elasticity makes it an effective additive for improving collagen-based tissue engineering vessels, showing that elastin hybrid composites offer unique mechanical properties such as significant elasticity, indicating that elastin is a useful structural protein for preparing elastic structural materials [93].

Elastin, which is mainly composed of Gly, Val, Ala, and Pro with a molecular weight of approximately 66 kDa, is present in connective, vascular, and load-bearing tissues, and has highly elastic mechanical properties. Elastin is attractive as a biomaterial as scaffolds for tissue engineering. However, elastin is insoluble and difficult to process, as well as less available in terms of quantities, therefore there are few reports about elastin as biomaterials scaffolds in comparison to the other protein-based scaffolds.

An elastin tissue scaffold derived from bovine nuchal ligament was reported and characterized for mechanical properties [92]. The elastic modulus was $1.2 \times 10^6 \pm 1 \times 10^5$ N/m^2 (parallel to fiber orientation) [92]. The elasticity of elastin is useful to improve collagen-based tissue-engineered blood vessels, which do not have sufficient mechanical properties for bypass grafts. Hybrid tissue-engineered blood vessels of type I collagen and elastin with either human dermal fibroblasts (HDFs) or rat smooth muscle cells (RASMs) exhibited increased tensile strength (11-fold in HDFs; 7.5-fold in RASMs) and linear stiffness moduli (4-fold in HDFs; 1.8-fold in RASMs) compared with collagen control constructs with no exogenous elastin scaffold. These data indicate that the elastin hybrid constructs show useful elastic solid mechanical properties [93].

The low ultimate tensile strength of elastin has limited its use in biomaterials. Scaffolds consisting of a purified elastin tubular conduits were strengthened with fibrin bonded layers of acellular small intestinal submucosa for potential use as small diameter vascular grafts [94]. The addition of acellular small intestinal submucosa increased the ultimate tensile strength of the elastin conduits 9-fold. Burst pressures for the elastin composite vascular scaffold (1396 ± 309 mmHg) were significantly higher than pure elastin conduits (162 ± 36 mmHg) and comparable to native saphenous veins. The average suture pullout strength of the elastin composite vascular scaffolds is 14.612 ± 3.677 N, significantly higher than the pure elastin conduit (0.402 ± 0.098 N), but comparable to native porcine carotid arteries (13.994 ± 4.344 N) [94]. In vivo cellular repopulation of a tissue-derived tubular elastin scaffold was also reported. Elastin tubes filled with agarose gel containing basic fibroblast growth factor for sustained release of the growth factor showed significantly more cell infiltration at 28 days compared to those without growth factor [95]. Elastin scaffolds formed from cross-linked elastin-like polypeptide hydrogels were investigated to identify relationships between scaffold formulation parameters (cross-link density, molecular weight, and concentration) and properties including mechanical, matrix accumulation, metabolite use and production, and histological appearance [96]. Cross-link density was the strongest predictor of most outcomes related to neuron functions, followed by elastin-like polymer concentration [96].

Recombinant elastin and tropoelastin have recently been reviewed [97–99]. Elastin fibers are predominantly composed of the secreted monomer tropoelastin. The important role of cell interactions with recombinant human tropoelastin includes integrin alpha(V)beta(3) as the major fibroblast cell surface receptor for human tropoelastin [100]. The *C*-terminal GRKRK motif of tropoelastin can bind to cells in a divalent cation-dependent manner [100]. Assemblies of the elastin generated by recombinant DNA means, permitted the construction of elastic sponges via chemical cross-linking with bis(sulfosuccinimidyl) suberate. These matrices exhibited a Young's modulus from 220 to 280 kPa with linearity of extension to at least 150% [101]. Synthetic elastin is extensible by 200%–370% [102]. The constructs behaved as hydrogels and displayed stimuli-responsive characteristics toward temperature and salt concentrations. Further, the elastin scaffolds have shown in vitro growth and proliferation of cells and have been well tolerated in vivo [102,103]. Recombinant human tropoelastin and α-elastin as biopolymeric materials have been used to fabricate tissue-engineered scaffolds by electrospinning and with different cross-linking methods. Cell culture studies confirmed that the electrospun protein scaffolds supported the attachment and growth of human embryonic palatal mesenchymal cells [104,105].

7.9 Resilin

Resilin, an entropic elastomer (rubber)-like protein found within structures where energy storage and long-range elasticity are needed, shows an elongation to break of 300%–400%, low solubility, and thermal stability up to 140°C [1,106]. The resilience of resilin is approximately 92% due to the covalent cross-links between tyrosine residues [107–109]. The mechanical properties of resilin are similar to those of elastin, but resilin shows higher resilience and lower modulus. Resilin can be found in various biological joints, such as the vein joints of dragonfly wings [110]; however, natural resilin is not widely available for biological and physical characterization (Fig. 7.11).

Cloning and expression of the first exon of *Drosophila* CG15920 gene, which was identified as encoding a resilin-like protein, showed that this recombinant protein can be cast into a rubber-like biomaterial by rapid photochemical cross-linking [111]. Artificial elastomeric proteins that

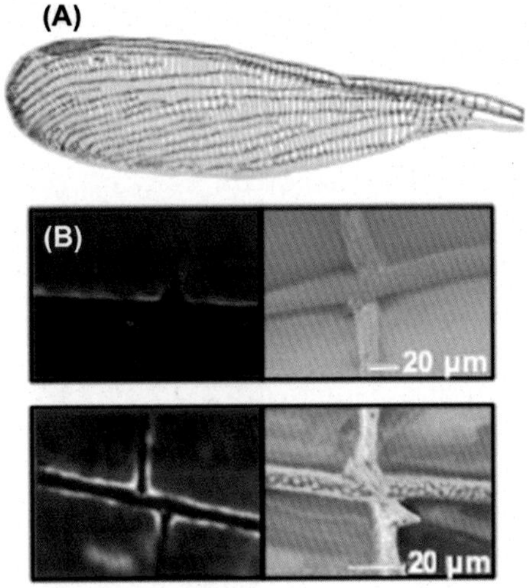

Figure 7.11 Distribution of resilin in the vein joints of the *Rhinocypha fenestrella* wing. (A) Resilin present on the dorsal side of the wing. The purple dots indicate resilin. (B) Fluorescence microscopy (left panels) and scanning electron microscopy images (right panels) of the selected joints. *Adopted from Ref [110].*

mimic the molecular architecture of titin have been characterized through the combination of well-characterized protein domains GB1 and resilin [112,113]. The elastomeric proteins containing resilin can be photochemically cross-linked and processed into solid biomaterials. This cross-linking was achieved by peroxidase, which catalyzes dityrosine formation. The recombinant resilin proteins were expressed and evaluated as biomaterials [114–116]. These resilin-based biomaterials are rubber-like materials with a high resilience at low strain, and also represent a new muscle-mimetic biomaterial [1]. The mechanical properties of resilin-based materials can be tuned by adjusting the composition of the elastomeric proteins, and hence there is the opportunity to develop relining-based biomaterials that mimic different types of muscles for application in tissue engineering scaffolds as well as bulk scale chemicals [113].

7.10 Keratin

Keratin is the major structural fibrous protein to form hair, wool, feathers, nails, and horns of many kinds of animals (Fig. 7.12). Keratin demonstrates excellent biodegradation resistance in some natural environments, because of the lack of keratin-degrading enzymes [1,2]. Also, keratin has a

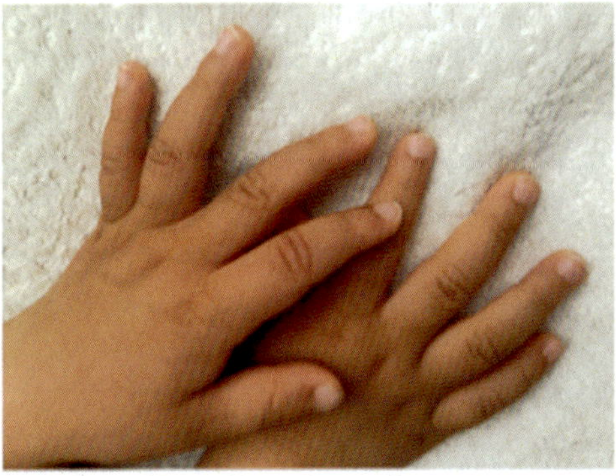

Figure 7.12 Skins and nails of a human.

high concentration of cysteine, 7%–20% of the total amino acid residues, that form inter- and intramolecular disulfide bonds, which enhances its physical and biological stability [117]. α-Keratin with helical structures declines and β-keratin appears upon stretching elastin, which affects mechanical, thermal, and chemical properties [118,119]. The α-helices are arranged parallel to their length, with all the *N*-terminal residues present at the same end. Three α-helices form a left-handed supercoil, leading to a protofibril. Eleven protofibrils form a microfibril. Skin is stretchable because it has fewer cross-links, whereas nails are inflexible and tough because they have many more cross-links.

A stable aqueous solution of reduced keratins can be prepared by extracting the proteins from wool with a mixture of urea, mercaptanol, sodium dodecyl sulfate as a surfactant, and water at 40°C–60°C [120]. A clear film from the keratin solution with glycerol can be prepared, is insoluble in water and organic solvents including dimethyl sulfoxide, and is degradable in vitro and in vivo [120]. Comparative culture assay on keratins, collagen, and glass revealed that the keratins are more adhesive for cells and more supportive of cell proliferation than collagen and glass [121]. Keratin sponge scaffolds, with a homogeneous porous microstructure with pore sizes of 100 μm, have been developed by lyophilization of an aqueous wool keratin solution after controlled freezing [122]. Rapid cell growth of mouse fibroblast cells (L929) on the sponge (doubling time 29 h) for at least 7 days, as well as maximum cell number of 7.4 million, or approximately 37 times higher than on the cell culture dishes, was reported. These data indicate that wool keratin sponges are useful scaffolds for long-term and high-density cell cultivation [122]. Chemically modified keratin sponge scaffolds with functional groups and basic bioactive proteins such as CaP precipitation, carboxyl groups, hydroxyapatite particles, thiol (SH) groups, lysozymes, and bone morphogenetic protein 2 were also reported [123–125]. Additions of chitosan or glycerol into keratin films provide strong and flexible film. The composite as well as keratin and chitosan films demonstrate high fibroblast attachment and proliferation for mammalian cell culture [126]. Films were compression molded at 120°C, were insoluble and slightly swelled in water, and demonstrated maximum strength of 27.8 ± 2.9 and Young's modulus of 1218 ± 80 MPa [127]. A compression-molding/particulate-leaching method was reported for the fabrication of keratin sponges with controlled pore size (<100, 100–300, and 300–500 μm) and porosity (more than 90%) [128]. In vivo degradation of human hair keratin scaffolds was due to extracellular enzymatic degradation by ubiquitin systems, generating particles. Subsequently the satellite cells grown on the scaffolds were activated to proliferate and eventually fused into generated muscle fibers [129]. Chicken calamus keratin conduits as a tissue-engineered scaffold material and its degradation products demonstrated low toxicity based on skin sensitization in rats, intracutaneous stimulation in rabbits, acute systemic toxicity in mice, and cytotoxicity to L929 cells [130].

7.11 Reflectin

Reflectin is not a structural protein but forms layered structures to induce a specific optical properties of tissue or proteins [1]. Reflectin has been found in certain cephalopods (especially squid), including *Euprymna scolopes* and *Doryteuthis opalescens* [131]. Squid are capable of changing their body color to camouflage themselves, namely, mimic their surroundings, with reflectin's function. However, the studies on reflectin protein itself are very limited, as is the scientific understanding of the molecular mechanism. Reflectin is speculated to have random coil and beta barrels as the main motifs [132], but the three-dimensional structure has not been fully elucidated. The amino acid sequence of reflectin is rich in aromatic and sulfur-containing amino acids, and these motifs are used to refract incident light in certain environments in certain cephalopods; at the same time, protein families that exist in cephalopod pigment cells (rainbow cells, a type of pigment cell that changes the wavelength of reflected light) and white pigment vesicles and that contribute to camouflage function have been recently reported [133,134]. However, not all the molecular mechanisms have been fully clarified to date. *Aliivibrio fischeri*, a marine luminescent bacterium, possesses the reflectin gene of cephalopods due to horizontal gene transmission [135]. Recently, several material scientists have been interested in reflectin as a new functional biopolymer and have reported novel optical biomaterials containing recombinant reflectin [1,136].

References

[1] K. Numata, How to define and study structural proteins as biopolymer materials, Polym. J. 52 (2020) 1043–1056.

[2] K. Numata, D.L. Kaplan, Biologically derived scaffolds, Advanced Wound Repair Therapies, Woodhead Publishing Series in Biomaterials, 2011, pp. 524–551.

[3] P.L. Babb, N.F. Lahens, S.M. Correa-Garhwal, et al., The *Nephila clavipes* genome highlights the diversity of spider silk genes and their complex expression, Nat. Genet. 49 (6) (2017) 895–903.

[4] F. Vollrath, D. Porter, C. Holland, There are many more lessons still to be learned from spider silks, Soft Matter 7 (20) (2011) 9595–9600.

[5] T. Katashima, A.D. Malay, K. Numata, Chemical modification and biosynthesis of silk-like polymers, Curr. Opin. Chem. Eng. 24 (2019) 61–68.

[6] J.M. Gosline, M.W. Denny, M.E. DeMont, Spider silk as rubber, Nature 309 (5968) (1984) 551–552.

[7] J.M. Gosline, P.A. Guerette, C.S. Ortlepp, K.N. Savage, The mechanical design of spider silks: from fibroin sequence to mechanical function, J. Exp. Biol. 202 (Pt 23) (1999) 3295–3303.

[8] A.D. Malay, K. Arakawa, K. Numata, Analysis of repetitive amino acid motifs reveals the essential features of spider dragline silk proteins, PLoS One 12 (8) (2017) e0183397.

[9] K. Yazawa, K. Ishida, H. Masunaga, T. Hikima, K. Numata, Influence of water content on the beta-sheet formation, thermal stability, water removal, and mechanical properties of silk materials, Biomacromolecules 17 (3) (2016) 1057–1066.

[10] K. Yazawa, A.D. Malay, H. Masunaga, K. Numata, Simultaneous effect of strain rate and humidity on the structure and mechanical behavior of spider silk, Commun. Mater. 1 (2020) 10.

[11] H.-J. Jin, D.L. Kaplan, Mechanism of silk processing in insects and spiders, Nature 424 (2003) 1057.

[12] T.Y. Lin, H. Masunaga, R. Sato, et al., Liquid crystalline granules align in a hierarchical structure to produce spider dragline microfibrils, Biomacromolecules 18 (4) (2017) 1350–1355.

[13] L.R. Parent, D. Onofrei, D. Xu, et al., Hierarchical spidroin micellar nanoparticles as the fundamental precursors of spider silks, Proc. Natl. Acad. Sci. U S A 115 (45) (2018) 11507.

[14] F. Vollrath, D.P. Knight, Liquid crystalline spinning of spider silk, Nature 410 (6828) (2001) 541–548.

[15] M. Andersson, J. Johansson, A. Rising, Silk spinning in silkworms and spiders, Int. J. Mol. Sci. 17 (8) (2016).

[16] C. Fu, Z. Shao, V. Fritz, Animal silks: their structures, properties and artificial production, Chem. Commun. 43 (2009) 6515–6529.

[17] J. Gatesy, C. Hayashi, D. Motriuk, J. Woods, R. Lewis, Extreme diversity, conservation, and convergence of spider silk fibroin sequences, Science 291 (5513) (2001) 2603–2605.
[18] K. Numata, H. Masunaga, T. Hikima, S. Sasaki, K. Sekiyama, M. Takata, Use of extension-deformation-based crystallisation of silk fibres to differentiate their functions in nature, Soft Matter 11 (31) (2015) 6335–6342.
[19] J.D. van Beek, S. Hess, F. Vollrath, B.H. Meier, The molecular structure of spider dragline silk: folding and orientation of the protein backbone, Proc. Natl. Acad. Sci. U S A 99 (16) (2002) 10266.
[20] M. Xu, R.V. Lewis, Structure of a protein superfiber: spider dragline silk, Proc. Natl Acad. Sci. U S A 87 (18) (1990) 7120–7124.
[21] N.A. Oktaviani, A. Matsugami, A.D. Malay, F. Hayashi, D.L. Kaplan, K. Numata, Conformation and dynamics of soluble repetitive domain elucidates the initial beta-sheet formation of spider silk, Nat. Commun. 9 (2018) 2121.
[22] Y. Liu, A. Sponner, D. Porter, F. Vollrath, Proline and processing of spider silks, Biomacromolecules 9 (1) (2008) 116–121.
[23] A. Sponner, E. Unger, F. Grosse, K. Weisshart, Differential polymerization of the two main protein components of dragline silk during fibre spinning, Nat. Mater. 4 (10) (2005) 772–775.
[24] A. Sponner, W. Vater, S. Monajembashi, E. Unger, F. Grosse, K. Weisshart, Composition and hierarchical organisation of a spider silk, PLoS One 2 (10) (2007) e998.
[25] K. Yazawa, A.D. Malay, H. Masunaga, K. Numata, Role of skin layers on mechanical properties and supercontraction of spider dragline silk fiber, Macromol. Biosci. 19 (3) (2018) e1800220.
[26] C.L. Craig, Spiderwebs and silk: tracing evolution from molecules to genes to phenotypes, Oxford University Press, Oxford England; New York, 2003.
[27] N. Becker, E. Oroudjev, S. Mutz, et al., Molecular nanosprings in spider capture-silk threads, Nat. Mater. 2 (4) (2003) 278–283.
[28] C.Y. Hayashi, R.V. Lewis, Molecular architecture and evolution of a modular spider silk protein gene, Science 287 (5457) (2000) 1477–1479.
[29] G.R.S. Deb, M. Kane, N. Naidoo, D.J. Little, M.E. Herberstein, Optics of spider "sticky" orb webs, in: Proceedings of SPIE Smart Structures and Materials + Nondestructive Evaluation and Health Monitoring, Vol. 7975, San Diego, CA, 2011, 13 p.
[30] C.L. Craig, Alternative foraging modes of orb web weaving spiders, Biotropica 21 (3) (1989) 257–264.
[31] Y.Q. Zhang, Applications of natural silk protein sericin in biomaterials, Biotechnol. Adv. 20 (2) (2002) 91–100.
[32] A.D. Malay, R. Sato, K. Yazawa, et al., Relationships between physical properties and sequence in silkworm silks, Sci. Rep. 6 (2016) 27573.
[33] C.Z. Zhou, F. Confalonieri, M. Jacquet, R. Perasso, Z.G. Li, J. Janin, Silk fibroin: structural implications of a remarkable amino acid sequence, Proteins 44 (2) (2001) 119–122.
[34] Y.H. Sima, M. Chen, R. Yao, et al., The complete mitochondrial genome of the Ailanthus silkmoth, Samia cynthia cynthia (Lepidoptera: Saturniidae), Gene 526 (2) (2013) 309–317.
[35] Y. Dong, F. Dai, Y. Ren, et al., Comparative transcriptome analyses on silk glands of six silkmoths imply the genetic basis of silk structure and coloration, BMC Genom. 16 (2015) 203.
[36] J.M. Ageitos, K. Yazawa, A. Tateishi, K. Tsuchiya, K. Numata, The benzyl ester group of amino acid monomers enhances substrate affinity and broadens the substrate

specificity of the enzyme catalyst in chemoenzymatic copolymerization, Biomacromolecules 17 (1) (2016) 314–323.
[37] M.A. Wilding, J.W.S. Hearle, Fibre structure, in: J.C. Salamone (Ed.), Polymeric Materials Encyclopedia, CRC Press Inc., Boca Raton, Florida, 1996, pp. 8307–8322.
[38] F. Vollrath, B. Madsen, Z.Z. Shao, The effect of spinning conditions on the mechanics of a spider's dragline silk, Proc. R. Soc. B Biol. Sci. 268 (1483) (2001) 2339–2346.
[39] Z. Shao, F. Vollrath, Surprising strength of silkworm silk, Nature 418 (6899) (2002) 741.
[40] K. Numata, D.L. Kaplan, Silk-based delivery systems of bioactive molecules, Adv. Drug Deliv. Rev. 62 (15) (2010) 1497–1508.
[41] K. Makaya, S. Terada, K. Ohgo, T. Asakura, Comparative study of silk fibroin porous scaffolds derived from salt/water and sucrose/hexafluoroisopropanol in cartilage formation, J. Biosci. Bioeng. 108 (1) (2009) 68–75.
[42] Y. Tamada, New process to form a silk fibroin porous 3-D structure, Biomacromolecules 6 (6) (2005) 3100–3106.
[43] V. Karageorgiou, D. Kaplan, Porosity of 3D biomaterial scaffolds and osteogenesis, Biomaterials 26 (27) (2005) 5474–5491.
[44] R. Nazarov, H.J. Jin, D.L. Kaplan, Porous 3-D scaffolds from regenerated silk fibroin, Biomacromolecules 5 (3) (2004) 718–726.
[45] S. Sofia, M.B. McCarthy, G. Gronowicz, D.L. Kaplan, Functionalized silk-based biomaterials for bone formation, J. Biomed. Mater. Res. 54 (1) (2001) 139–148.
[46] C. Holland, K. Numata, J. Rnjak-Kovacina, F.P. Seib, The biomedical use of silk: past, present, future, Adv. Healthc. Mater. 8 (1) (2019) e1800465.
[47] K. Numata, P. Cebe, D.L. Kaplan, Mechanism of enzymatic degradation of beta-sheet crystals, Biomaterials 31 (10) (2010) 2926–2933.
[48] K. Numata, T. Katashima, T. Sakai, State of water, molecular structure, and cytotoxicity of silk hydrogels, Biomacromolecules 12 (6) (2011) 2137–2144.
[49] P. Gupta, M. Kumar, N. Bhardwaj, et al., Mimicking form and function of native small diameter vascular conduits using mulberry and non-mulberry patterned silk films, ACS Appl. Mater. Inter. 8 (25) (2016) 15874–15888.
[50] K. Numata, R. Sato, K. Yazawa, T. Hikima, H. Masunaga, Crystal structure and physical properties of *Antheraea yamamai* silk fibers: long poly(alanine) sequences are partially in the crystalline region, Polymer 77 (2015) 87–94.
[51] B.B. Mandal, S. Das, K. Choudhury, S.C. Kundu, Implication of silk film RGD availability and surface roughness on cytoskeletal organization and proliferation of primary rat bone marrow cells, Tissue Eng. Part A 16 (7) (2010) 2391–2403.
[52] N. Kono, H. Nakamura, R. Ohtoshi, M. Tomita, K. Numata, K. Arakawa, The bagworm genome reveals a unique fibroin gene that provides high tensile strength, Commun. Biol. 2 (2019) 148.
[53] M. Rhainds, D.R. Davis, P.W. Price, Bionomics of bagworms (Lepidoptera: Psychidae), Annu. Rev. Entomol. 54 (2009) 209–226.
[54] D.L. Cox, D.L. Potter, Aerial dispersal behavior of larval bagworms, Thyridopteryx ephemeraeformis (Lepidoptera: Psychidae), Can. Entomol. 118 (6) (1986) 525–535.
[55] S. Sugiura, Bagworm bags as portable armour against invertebrate predators, PeerJ 4 (2016) e1686.
[56] T.D. Sutherland, S. Weisman, A.A. Walker, S.T. Mudie, Invited review the coiled coil silk of bees, ants, and hornets, Biopolymers 97 (6) (2012) 446–454.
[57] R.H. Hedley, K.M. Rudall, Extracellular silk fibers in Stannophyllum (Rhizopodea-Protozoa), Cell Tissue Res. 150 (1) (1974) 107–111.
[58] K.D. Parker, K.M. Rudall, Structure of the silk of Chrysopa egg-stalks, Nature 179 (4566) (1957) 905–906.

[59] E.D.T. Atkins, A four-strand coiled-coil model for some insect fibrous proteins, J. Mol. Biol. 24 (1) (1967) 139–140.

[60] T.D. Sutherland, S. Weisman, H.E. Trueman, A. Sriskantha, J.W.H. Trueman, V.S. Haritos, Conservation of essential design features in coiled coil silks, Mol. Biol. Evol. 24 (11) (2007) 2424–2432.

[61] R.M. Crewe, P.R. Thompson, Oecophylla silk—functional adaptation in a biopolymer, Die Naturwissenschaften 66 (1) (1979) 57–58.

[62] T. Kameda, K. Kojima, M. Miyazawa, S. Fujiwara, Film formation and structural characterization of silk of the hornet Vespa simillima xanthoptera Cameron, Z. Naturforsch C. 60 (11-12) (2005) 906–914.

[63] T. Kameda, Y. Tamada, Variable-temperature C-13 solid-state NMR study of the molecular structure of honeybee wax and silk, Int. J. Biol. Macromol. 44 (1) (2009) 64–69.

[64] C. Pecquet, Allergic reactions to insect secretions, Eur. J. Dermatol. 23 (6) (2013) 767–773.

[65] R.I. Kunz, R.M.C. Brancalhao, L.D.C. Ribeiro, M.R.M. Natali, Silkworm sericin: properties and biomedical applications, Biomed. Res. Int. 2016 (2016).

[66] Z.Y. Jiao, Y. Song, Y. Jin, et al., In vivo characterizations of the immune properties of sericin: an ancient material with emerging value in biomedical applications, Macromol. Biosci. 17 (12) (2017).

[67] H.J. Xie, W. Yang, J.H. Chen, et al., Sericin/silicone nerve guidance conduit promotes regeneration of a transected sciatic nerve, Adv. Healthc. Mater. 4 (15) (2015) 2195–2205.

[68] B. Kundu, S.C. Kundu, Silk sericin/polyacrylamide in situ forming hydrogels for dermal reconstruction, Biomaterials 33 (30) (2012) 7456–7467.

[69] C.H. Lee, A. Singla, Y. Lee, Biomedical applications of collagen, Int. J. Pharm. 221 (1-2) (2001) 1–22.

[70] T.D. Sutherland, Y.Y. Peng, H.E. Trueman, et al., A new class of animal collagen masquerading as an insect silk, Sci. Rep. 3 (2013).

[71] S. Yasothornsrikul, J.D. Wendy, G. Cramer, D.A. Kimbrell, C.R. Dearolf, *viking*: identification and characterization of a second type IV collagen in *Drosophila*, Gene 198 (1-2) (1997) 17–25.

[72] S.J. Fowler, S. Jose, X.M. Zhang, R. Deutzmann, M.P. Sarras, R.P. Boot-Handford, Characterization of hydra type IV collagen—type IV collagen is essential for head regeneration and its expression is up-regulated upon exposure to glucose, J. Biol. Chem. 275 (50) (2000) 39589–39599.

[73] D.E. Ashhurst, A.J. Bailey, Locust collagen—morphological and biochemical-characterization, Eur. J. Biochem. 103 (1) (1980) 75–83.

[74] J. Francois, D. Herbage, S. Junqua, Cockroach collagen—isolation, biochemical and biophysical characterization, Eur. J. Biochem. 112 (2) (1980) 389–396.

[75] T.M. Freyman, I.V. Yannas, L.J. Gibson, Cellular materials as porous scaffolds for tissue engineering, Prog. Mater. Sci. 46 (3-4) (2001) 273–282.

[76] C.B. Weinberg, E. Bell, A blood vessel model constructed from collagen and cultured vascular cells, Science 231 (4736) (1986) 397–400.

[77] J.S. Pieper, A. Oosterhof, P.J. Dijkstra, J.H. Veerkamp, T.H. van Kuppevelt, Preparation and characterization of porous crosslinked collagenous matrices containing bioavailable chondroitin sulphate, Biomaterials 20 (9) (1999) 847–858.

[78] I.V. Yannas, E. Lee, D.P. Orgill, E.M. Skrabut, G.F. Murphy, Synthesis and characterization of a model extracellular matrix that induces partial regeneration of adult mammalian skin, Proc. Natl. Acad. Sci. U S A 86 (3) (1989) 933–937.

[79] E.M. Byrne, E. Farrell, L.A. McMahon, et al., Gene expression by marrow stromal cells in a porous collagen-glycosaminoglycan scaffold is affected by pore size and mechanical stimulation, J. Mater. Sci. Mater Med. 19 (11) (2008) 3455–3463.

[80] M.J. Jaasma, F.J. O'Brien, Mechanical stimulation of osteoblasts using steady and dynamic fluid flow, Tissue Eng. Part A 14 (7) (2008) 1213–1223.
[81] M.J. Jaasma, N.A. Plunkett, F.J. O'Brien, Design and validation of a dynamic flow perfusion bioreactor for use with compliant tissue engineering scaffolds, J. Biotechnol. 133 (4) (2008) 490–496.
[82] M.B. Keogh, F.J. O'Brien, J.S. Daly, A novel collagen scaffold supports human osteogenesis—applications for bone tissue engineering, Cell Tissue Res. 340 (1) (2010) 169–177.
[83] L.A. McMahon, A.J. Reid, V.A. Campbell, P.J. Prendergast, Regulatory effects of mechanical strain on the chondrogenic differentiation of MSCs in a collagen-GAG scaffold: experimental and computational analysis, Ann. Biomed. Eng. 36 (2) (2008) 185–194.
[84] P. Angele, J. Abke, R. Kujat, et al., Influence of different collagen species on physico-chemical properties of crosslinked collagen matrices, Biomaterials 25 (14) (2004) 2831–2841.
[85] W.F. Daamen, H.T. van Moerkerk, T. Hafmans, et al., Preparation and evaluation of molecularly-defined collagen-elastin-glycosaminoglycan scaffolds for tissue engineering, Biomaterials 24 (22) (2003) 4001–4009.
[86] J. Glowacki, S. Mizuno, Collagen scaffolds for tissue engineering, Biopolymers 89 (5) (2008) 338–344.
[87] L.S. Liu, A.Y. Thompson, M.A. Heidaran, J.W. Poser, R.C. Spiro, An osteoconductive collagen/hyaluronate matrix for bone regeneration, Biomaterials 20 (12) (1999) 1097–1108.
[88] M.B. Yaylaoglu, C. Yildiz, F. Korkusuz, V. Hasirci, A novel osteochondral implant, Biomaterials 20 (16) (1999) 1513–1520.
[89] Y. Garcia, N. Hemantkumar, R. Collighan, M. Griffin, J.C. Rodriguez-Cabello, A. Pandit, In vitro characterization of a collagen scaffold enzymatically cross-linked with a tailored elastin-like polymer, Tissue Eng. Part A 15 (4) (2009) 887–899.
[90] H. Oxlund, J. Manschot, A. Viidik, The role of elastin in the mechanical-properties of skin, J. Biomech. 21 (3) (1988) 213–218.
[91] B.B. Aaron, J.M. Gosline, Elastin as a random-network elastomer—a mechanical and optical analysis of single elastin fibers, Biopolymers 20 (6) (1981) 1247–1260.
[92] S.J. Kirkpatrick, M.T. Hinds, D.D. Duncan, Acousto-optical characterization of the viscoelastic nature of a nuchal elastin tissue scaffold, Tissue Eng 9 (4) (2003) 645–656.
[93] J.D. Berglund, R.M. Nerem, A. Sambanis, Incorporation of intact elastin scaffolds in tissue-engineered collagen-based vascular grafts, Tissue Eng 10 (9-10) (2004) 1526–1535.
[94] M.T. Hinds, R.C. Rowe, Z. Ren, et al., Development of a reinforced porcine elastin composite vascular scaffold, J. Biomed. Mater. Res. A 77 (3) (2006) 458–469.
[95] A. Kurane, D.T. Simionescu, N.R. Vyavahare, In vivo cellular repopulation of tubular elastin scaffolds mediated by basic fibroblast growth factor, Biomaterials 28 (18) (2007) 2830–2838.
[96] D.L. Nettles, M.A. Haider, A. Chilkoti, L.A. Setton, Neural network analysis identifies scaffold properties necessary for in vitro chondrogenesis in elastin-like polypeptide biopolymer scaffolds, Tissue Eng. Part A 16 (1) (2010) 11–20.
[97] J.F. Almine, D.V. Bax, S.M. Mithieux, et al., Elastin-based materials, Chem. Soc. Rev. 39 (9) (2010) 3371–3379.
[98] S.M. Mithieux, A.S. Weiss, Elastin, Adv. Protein Chem. 70 (2005) 437–461.
[99] S.G. Wise, A.S. Weiss, Tropoelastin, Int. J. Biochem. Cell Biol. 41 (3) (2009) 494–497.

[100] D.V. Bax, U.R. Rodgers, M.M. Bilek, A.S. Weiss, Cell adhesion to tropoelastin is mediated via the C-terminal GRKRK motif and integrin alphaVbeta3, J. Biol. Chem. 284 (42) (2009) 28616–28623.
[101] W.J. Wu, B. Vrhovski, A.S. Weiss, Glycosaminoglycans mediate the coacervation of human tropoelastin through dominant charge interactions involving lysine side chains, J. Biol. Chem. 274 (31) (1999) 21719–21724.
[102] S.M. Mithieux, J.E. Rasko, A.S. Weiss, Synthetic elastin hydrogels derived from massive elastic assemblies of self-organized human protein monomers, Biomaterials 25 (20) (2004) 4921–4927.
[103] Y. Tu, S.G. Wise, A.S. Weiss, Stages in tropoelastin coalescence during synthetic elastin hydrogel formation, Micron 41 (3) (2010) 268–272.
[104] M.Y. Li, M.J. Mondrinos, M.R. Gandhi, F.K. Ko, A.S. Weiss, P.I. Lelkes, Electrospun protein fibers as matrices for tissue engineering, Biomaterials 26 (30) (2005) 5999–6008.
[105] L. Nivison-Smith, J. Rnjak, A.S. Weiss, Synthetic human elastin microfibers: stable cross-linked tropoelastin and cell interactive constructs for tissue engineering applications, Acta Biomater. 6 (2) (2010) 354–359.
[106] A.S. Tatham, P.R. Shewry, Comparative structures and properties of elastic proteins, Philos. Trans. R. Soc. Lond. B Biol. Sci. 357 (1418) (2002) 229–234.
[107] S.O. Andersen, The cross-links in resilin identified as dityrosine and trityrosine, Biochim. Biophys. Acta 93 (1964) 213–215.
[108] J. Gosline, M. Lillie, E. Carrington, P. Guerette, C. Ortlepp, K. Savage, Elastic proteins: biological roles and mechanical properties, Philos. Trans. R. Soc. Lond. B Biol. Sci. 357 (1418) (2002) 121–132.
[109] R.E. Lyons, E. Lesieur, M. Kim, et al., Design and facile production of recombinant resilin-like polypeptides: gene construction and a rapid protein purification method, Protein Eng. Des. Sel. 20 (1) (2007) 25–32.
[110] N. Mamat, K. Yazawa, K. Numata, Y. Norma-Rashid, Morphological and mechanical properties of flexible resilin joints on damselfly wings (Rhinocypha spp.), PLoS One 13 (3) (2018) e0193147.
[111] C.M. Elvin, A.G. Carr, M.G. Huson, et al., Synthesis and properties of crosslinked recombinant pro-resilin, Nature 437 (7061) (2005) 999–1002.
[112] Y. Cao, H. Li, Polyprotein of GB1 is an ideal artificial elastomeric protein, Nat. Mater. 6 (2) (2007) 109–114.
[113] S. Lv, D.M. Dudek, Y. Cao, M.M. Balamurali, J. Gosline, H. Li, Designed biomaterials to mimic the mechanical properties of muscles, Nature 465 (7294) (2010) 69–73.
[114] G.K. Qin, S. Lapidot, K. Numata, et al., Expression, cross-linking, and characterization of recombinant chitin binding resilin, Biomacromolecules 10 (12) (2009) 3227–3234.
[115] G.K. Qin, A. Rivkin, S. Lapidot, et al., Recombinant exon-encoded resilins for elastomeric biomaterials, Biomaterials 32 (35) (2011) 9231–9243.
[116] G.K. Qin, X. Hu, P. Cebe, D.L. Kaplan, Mechanism of resilin elasticity, Nat. Commun. 3 (2012).
[117] L.M. Dowling, W.G. Crewther, D.A. Parry, Secondary structure of component 8c-1 of alpha-keratin. An analysis of the amino acid sequence, Biochem. J. 236 (3) (1986) 705–712.
[118] L. Pauling, R.B. Corey, Configuration of polypeptide chains, Nature 168 (4274) (1951) 550–551.
[119] L. Pauling, R.B. Corey, Compound helical configurations of polypeptide chains: structure of proteins of the alpha-keratin type, Nature 171 (4341) (1953) 59–61.

[120] K. Yamauchi, A. Yamauchi, T. Kusunoki, A. Kohda, Y. Konishi, Preparation of stable aqueous solution of keratins, and physiochemical and biodegradational properties of films, J. Biomed. Mater. Res. 31 (4) (1996) 439–444.
[121] K. Yamauchi, M. Maniwa, T. Mori, Cultivation of fibroblast cells on keratin-coated substrata, J. Biomater. Sci. Polym. Ed. 9 (3) (1998) 259–270.
[122] A. Tachibana, Y. Furuta, H. Takeshima, T. Tanabe, K. Yamauchi, Fabrication of wool keratin sponge scaffolds for long-term cell cultivation, J. Biotechnol. 93 (2) (2002) 165–170.
[123] A. Kurimoto, T. Tanabe, A. Tachibana, K. Yamauchi, Keratin sponge: immobilization of lysozyme, J. Biosci. Bioeng. 96 (3) (2003) 307–309.
[124] A. Tachibana, S. Kaneko, T. Tanabe, K. Yamauchi, Rapid fabrication of keratin-hydroxyapatite hybrid sponges toward osteoblast cultivation and differentiation, Biomaterials 26 (3) (2005) 297–302.
[125] A. Tachibana, Y. Nishikawa, M. Nishino, S. Kaneko, T. Tanabe, K. Yamauchi, Modified keratin sponge: binding of bone morphogenetic protein-2 and osteoblast differentiation, J. Biosci. Bioeng. 102 (5) (2006) 425–429.
[126] T. Tanabe, N. Okitsu, A. Tachibana, K. Yamauchi, Preparation and characterization of keratin-chitosan composite film, Biomaterials 23 (3) (2002) 817–825.
[127] K. Katoh, M. Shibayama, T. Tanabe, K. Yamauchi, Preparation and physicochemical properties of compression-molded keratin films, Biomaterials 25 (12) (2004) 2265–2272.
[128] K. Katoh, T. Tanabe, K. Yamauchi, Novel approach to fabricate keratin sponge scaffolds with controlled pore size and porosity, Biomaterials 25 (18) (2004) 4255–4262.
[129] D.F. Qiao, Y.M. Lu, W.Y. Fu, Y.J. Piao, Degradation of human hair keratin scaffold implanted for repairing injured skeletal muscles, Di Yi Jun Yi Da Xue Xue Bao 22 (10) (2002) 902–904.
[130] W.R. Dong, B.L. Zhao, Y.Q. Xiao, X.X. Qiu, Y.H. Chen, Z.Z. Zou, Toxicity evaluation of chicken calamus keratin conduit as a tissue-engineering scaffold biomaterial, Nan Fang Yi Ke Da Xue Xue Bao 27 (7) (2007) 931–935.
[131] W.J. Crookes, L.L. Ding, Q.L. Huang, J.R. Kimbell, J. Horwitz, M.J. McFall-Ngai, Reflectins: the unusual proteins of squid reflective tissues, Science 303 (5655) (2004) 235–238.
[132] R.M. Kramer, W.J. Crookes-Goodson, R.R. Naik, The self-organizing properties of squid reflectin protein, Nat. Mater. 6 (7) (2007) 533–538.
[133] D.G. DeMartini, M. Izumi, A.T. Weaver, E. Pandolfi, D.E. Morse, Structures, Organization, and function of reflectin proteins in dynamically tunable reflective cells, J. Biol. Chem. 290 (24) (2015) 15238–15249.
[134] K.L. Naughton, L. Phan, E.M. Leung, et al., Self-assembly of the cephalopod protein reflectin, Adv. Mater. 28 (38) (2016) 8405–8412.
[135] Z. Guan, T.T. Cai, Z.M. Liu, et al., Origin of the reflectin gene and hierarchical assembly of its protein, Curr. Biol. 27 (18) (2017) 2833–2842.
[136] G.K. Qin, P.B. Dennis, Y.J. Zhang, et al., Recombinant reflectin-based optical materials, J. Polym. Sci. Pol. Phys. 51 (4) (2013) 254–264.

Questions for this chapter

1. To compare different structural proteins in mechanical properties, we need to consider the experimental conditions and sample morphologies/shapes/types. Explain why we need to compare them carefully.

2. How do you design and develop rubber-like materials using structural proteins? Explain your molecular design.
3. Reflectin is not classified into structural protein based on its role in nature. Explain why reflectin is more of a functional protein rather than structural protein.

CHAPTER 8

Biopolymer material and composite

Structural proteins play roles in constructing the frames and structures of organisms and hence have been used as biological materials. Silk fibroin (SF) is one of the most widely studied structural proteins as a biomaterial. Silkworm silk has been used as biomedical sutures because of its biocompatibility and mechanical strength. Once the sericin is properly removed from SFs, there is a minimal response from the core fibroin structural proteins, as described in Chapter 5, Biological Properties With Cells. Methods to extract and regenerate SF have been developed [1]. According to this primary method, silk solution is prepared at a wide range of concentrations. From the silk solution after lyophilization, silk sponges can be obtained. The silk sponges (or silk powder) can be dissolved in 1,1,1,3,3,3-hexafluoro-2-propanol (HFIP), which is one of the solvents that can dissolve silk powders. Various silk-based materials can be processed from silk solutions (Fig. 8.1). For example, aqueous-derived and HFIP-derived silk porous scaffolds are prepared using salt leaching, gas forming, or freeze-drying method [4–7]. Silk films are prepared by cast or layer-by-layer deposition of silk aqueous or HFIP solution with various silk concentrations [3,8,9]. The film thickness can be tuned easily by the silk solution concentrations. This character of silk film allows us to use silk for thin film and coating applications. Hydrogel of silk protein is formed via sol–gel transitions by sonication, vortexing, or the presence of alcohol (poor solvent), acid, and/or ions [10–13]. Nanofibers of SF can be prepared by electrospinning or liquid–liquid phase separation (LLPS)–mediated method [14,15]. For almost all silk-based materials including nanomaterial to bulk materials, methanol, heat, and water vapor treatments can induce beta-sheet structure (crystalline phase), which makes them water-insoluble and more slowly biodegradable [8]. Alternatively, water annealing has also been developed for such transitions, avoiding the use of any organic solvents.

Biopolymer Science for Proteins and Peptides
DOI: https://doi.org/10.1016/B978-0-12-820555-6.00009-4

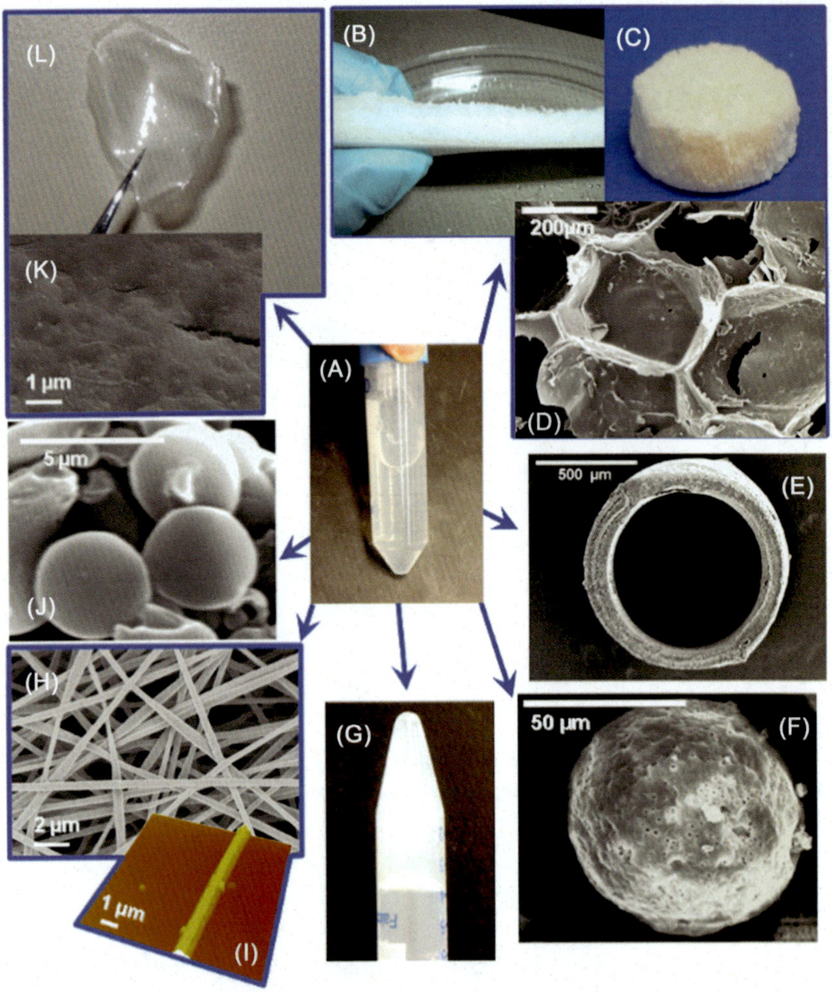

Figure 8.1 Various silk-based biomaterials prepared from silk solutions [2]. (A) Silk solution; (B and C) silk scaffolds, (D) scanning electron microscopic (SEM) image of porous structure of scaffold; (E) SEM image of silk tube; (F) SEM image of polymeric microsphere coated with silk layers; (G) silk hydrogel; (H) SEM image of silk electrospun fibers; (I) atomic force microscopic (AFM) image of single electrospun fibers of silk; (J) SEM image of silk-based microspheres; (K) SEM image of the surface of silk films; (L) silk film. *Sources: Reproduced with permission from K. Numata, D.L. Kaplan, Silk-based delivery systems of bioactive molecules, Adv. Drug Delivery Rev. 62(15) (2010) 1497–1508, published by Elsevier [2] and X. Wang, E. Wenk, X. Hu, G.R. Castro, L. Meinel, X. Wang, et al., Silk coatings on PLGA and alginate microspheres for protein delivery, Biomaterials 28 (28) (2007) 4161–4169 published by Elsevier [3].*

8.1 Nanofibers and fibers

8.1.1 Aqueous system

Based on the natural spinning system, many engineered materials have been developed and reviewed by many groups [16–19]. Recently, there has been an increasing focus on exploring "greener" alternatives for producing silk materials in aqueous environments, thereby minimizing the reliance on environmentally harmful solvents. More importantly, it is thought that the adoption of methods with stricter adherence to natural spinning conditions, called a biomimetic process, would yield fibers with mechanical properties similar to those of the native fibers. This is especially reasonable for spider silk, where attempts to spin silk using recombinant spidroin precursors via solidification in organic solvents usually produce inferior fibers that fail to replicate the impressive mechanical range of native fibers. The use of harsh solvents would lead to protein denaturation, leading to the loss of the self-assembly mechanisms that are orchestrated via the precise biochemical functions of the spidroin *N*-terminal domain (NTD) and *C*-terminal domain (CTD), which are needed to generate the hierarchical structure of spider silk fibers.

It is useful to briefly discuss some of the notable advances in the aqueous spinning of regenerated silk fibroin (RSF) obtained from *Bombyx mori* cocoon silk materials. Due to the ready availability and versatility of silkworm fibroin, diverse silk materials can be produced in aqueous solutions in the form of fibers [20], particles [21–23], foams [24], hydrogels [10–13], and films for a number of biomedical applications [2] (Fig. 8.2). To obtain artificial silk fibers, a variety of spinning techniques, such as wet spinning, dry spinning, electrospinning, hand-draw spinning, and microfluidic spinning, have been extensively developed [25]. The initial material can be easily obtained through membrane dialysis of the concentrated RSF solubilized by lithium bromide in aqueous solutions. As demonstrated by Shao and coworkers, a wet spinning system for RSF dissolved in water could be successfully established using aqueous ammonium sulfate as the coagulant at 60°C [26–29] with temperature being a critical factor [30]. By controlling the spinning parameters, regenerated silk fibers with improved mechanical properties relative to those of natural *B. mori* silks could be achieved [27]. Comparable results have also been achieved using an aqueous mixture of $CaCl_2$ and formic acid as the coagulation bath [31], although a $CaCl_2$-based system resulted in higher toughness at the cost of lower tensile strength [32]. Interestingly, in addition to various

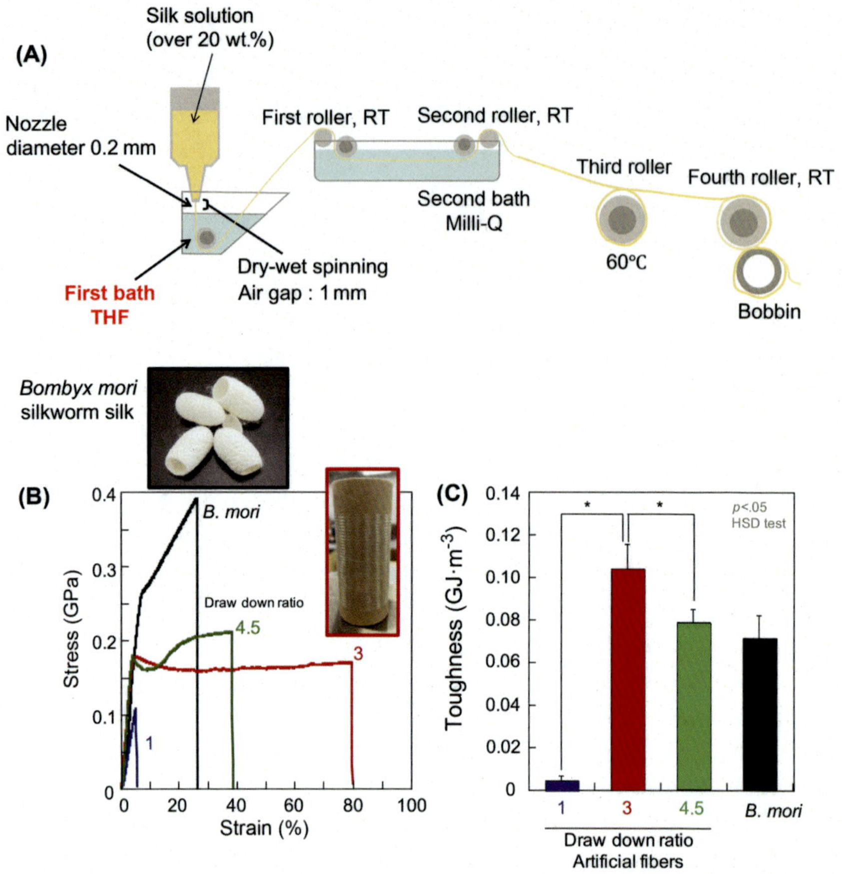

Figure 8.2 Spider-mimicking semiaqueous spinning system and also artificial silk threads. (A) Schematic illustration of the spinning system. The stress–strain curves (B) and toughness of the artificial silk threads and *Bombyx mori* silk [20]. HSD means Tukey's honestly significant difference test. RT denotes room temperature (25°C).

alcohols, glycerol, and polyethylene glycol (PEG) have also been presented as potential coagulants [33] (Fig. 8.3).

Unlike silkworm silk, spiders cannot be farmed because of their cannibalistic behaviors. The production of recombinant spidroin is a feasible solution to produce spider dragline silks on a larger scale for use in various applications. *Escherichia coli* is the most established host for producing recombinant spider silk protein, even up to the native size [35,36]. However, producing artificial spider silks that possess mechanical properties similar to those of native spider silks remains challenging. This is possibly because the molecular network and hierarchical structure of artificial spider silks do not

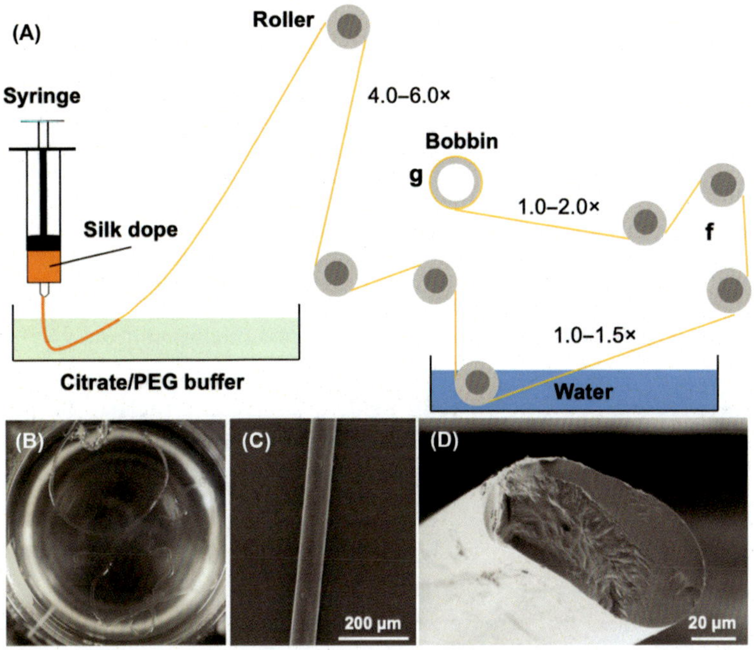

Figure 8.3 Spider-mimicking aqueous spinning system with a citrate buffer and also artificial silk threads. (A) Schematic illustration of the aqueous spinning system with a citrate buffer. (B) Coagulation of silk solution in citrate buffer. (C and D) Scanning electron microscopic images of the synthetic silk fibers prepared by the system. (D) Section of the fiber [34].

mimic those of the native material via NTD and CTD functions. The typical approach has been to convert silk protein in the form of a liquid feedstock (silk dope) into insoluble fibers by passing it through a coagulation bath consisting of a harsh organic solvent that acts as a dehydration agent and induces the formation of beta-sheet structures responsible for the high tensile strength of silk. One particular attraction of water-based methods is the potential to exploit the sensitivity of the spidroin terminal domain structures to subtle chemical stimuli to direct the self-assembly of hierarchical silk materials. A few examples of the wet spinning of fibers from recombinant spidroin solutions under aqueous conditions have been reported. Heidebrecht et al. [37] explored different fiber spinning strategies using engineered ADF3 constructs bearing variable spidroin domain combinations. Their method using "biomimetic spinning dope (BSD)" involved the dialysis of denatured and purified ADF3 into aqueous potassium phosphate buffer, followed by wet spinning in a water/isopropanol coagulation bath and poststretching, resulting in ductile fibers with high toughness values. Andersson et al. [38] produced a chimeric minispidroin based on the *Euprosthenops australis* MaSp1 sequence (33 kDa), with NTD and CTD and two repeat domains, which could be purified under native conditions and solubilized to a high concentration in Tris buffer at pH 8.0. The spidroin spinning dope was extruded into an aqueous bath of 500 mM sodium acetate and 200 mM NaCl at pH 5.0, resulting in extended solid fibers with a high beta-sheet content.

It is also possible in some cases to produce solid fibers directly from a highly concentrated aqueous solution of recombinant spidroin by simply drawing in air with forceps. This has been reported using a MaSp1/Flag hybrid construct, and the air-pulled fibers displayed a mechanical performance superior to that of wet-spun fibers; [39] similarly, fibers could be air-pulled from aqueous droplets of recombinant AcSp (consisting of repeat modules), and the results suggested an accompanying transition from alpha-helical to beta-sheet conformations [40]. In addition, DeSimone et al. [41] recently reported a mild process for preparing electrospun fiber mats from engineered *Araneus diadematus* fibroin 4 (ADF4) spidroin consisting of 16 repeat modules (48 kDa).

To induce self-assembly of spider silk under mild aqueous conditions as precursor spidroins migrate through the spinning ducts toward the spinnerets, several extrinsic parameters are thought to orchestrate the native silk spinning process, including acidification [42,43], ion gradients (primarily an exchange of sodium and chloride for potassium and phosphate ions) [44,45], dehydration [44,46,47], and shear forces [46,48]. Invaluable insights have been gained through investigations of the biochemical responses of isolated spidroin domains or sequence fragments toward specific external stimuli [42,49–56]. However, how multidomain spidroin proteins achieve coordinated structural transitions under simultaneous gradients has to date not been rigorously explored. It is also becoming increasingly clear that to replicate the physical properties of natural silk, it is essential to understand and control the protein self-assembly process using biomimetic methods [38,57,58]. To address these questions, Malay et al. generated an array of genetic constructs based on the conserved dragline silk protein major ampullate spidroin 2 (MaSp2) [59,60]. They used a rational design approach to generate chimeric high-yield constructs that incorporate the complete set of functional domains, that is, the NTD or *N*, tandem repeat domain (R*x*, where *x* denotes the number of repeat elements), and CTD or *C*, and truncated domain constructs. Full-domain recombinant MaSp2 was highly soluble and exhibited native-like secondary structure and biochemical behavior under dilute conditions, including constitutive complex formation via the CTD [52] and inducible dimerization via the NTD in response to acidification [49]. The array of MaSp2 domain variants was used to systematically explore structural transitions in response to different external stimuli associated with the in vivo silk spinning process. Notably, the addition of potassium phosphate at neutral pH to purified MaSp2 above a critical component concentration produced instantaneous but reversible turbidity. At the microscopic level, these cloudy samples exhibited myriad spherical droplets of heterogeneous size ($\sim$0.1–10 μm) undergoing dynamic fusion, hallmarks of LLPS. Analysis of the ion specificity for LLPS induction revealed a strict dependence on anionic species. Aside from phosphate ions—found in an increasing concentration gradient in the spider silk spinning ducts—multivalent anions such as citrate and sulfate also induced efficient MaSp2 condensation; in contrast, the cation species were not seen to exert any particular influence on the phase separation. Sequence determinants of the ion- and pH-induced nanofibril self-assembly from MaSp2 condensates were assessed via domain truncation analysis. Constructs bearing the full complement of N–R–C domains exhibited similar behaviors, transitioning from spherical liquid droplets at neutral pH toward self-assembled fibril networks with acidification. NTD dimerization via pH decrease was crucial for fibril network assembly [54], with both persisting in the LLPS state despite acidification. The CTD, which plays a central role in LLPS, also affects the formation of the fibrillar networks: constructs lacking CTD produced only loose, fragmented aggregates at pH 5, unlike the tight networks seen with the full-domain constructs. Thus our results indicate that the NTD and CTD play complementary roles in synergistic nanofibril self-assembly, with the complete set of spidroin functional domains required to mimic the natural spinning system. Based on these in vitro results, a biomimetic protocol was developed for generating silk-like fibers from recombinant MaSp2 that incorporates phase separation, acidification, and mechanical stress. Typically, concentrated recombinant MaSp2 is added onto a drop of 1.0 M potassium phosphate, pH 5, on a glass slide, yielding a condensed mass of protein fibrils. By manual drawing in air, the protein mass was transformed into extended fibers $\sim$10–20 μm in diameter. These fibers were shown to feature an internal structure of unidirectionally aligned nanofibrils, thus representing true hierarchical architecture (Fig. 8.4B,C). Notably, fibers made using the inactive mutant N^{A72R}-R12-C lacked any visible organization of internal structure but instead resembled solidified LLPS droplets (Fig. 8.4D).

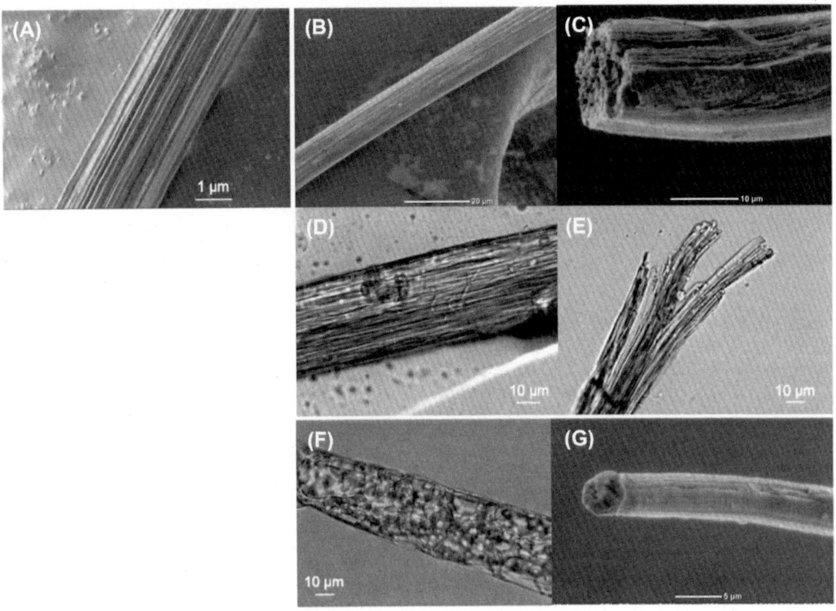

Figure 8.4 Microscopic images of silk fibers. (A) Field emission scanning electron microscopic (SEM) image of *Trichonephila clavata* dragline silk after proteolytic treatment [61]. (B,C) SEM images of the surface (B) and section (C) of artificial spider silk threads via LLPS process. (D,E) Optical microscope images of the artificial spider silk threads via LLPS process. (F,G) Optical microscope (F) and SEM images (G) of the artificial spider silk threads without LLPS process using the silk protein containing inactivated NTD [15].

The findings regarding the central role of LLPS in spider silk assembly indicate an intriguing connection with general mechanisms of biomolecular condensates. In recent years, there has been an explosion of work detailing the role of LLPS proteins in the formation of membraneless compartments (such as nucleoli or stress granules) and other intracellular condensates that play crucial roles in development and stress response [62,63]. With the emergence of phase separation playing a key role in spider silk self-assembly, expanded theoretical frameworks [64] can be used to elucidate these complex biophysical phenomena in future research.

8.1.2 Organic solvent system

Instead of aqueous spinning systems, organic solvent-mediated spinning is also possible. Since spidroin contains large amounts of hydrophobic amino acids such as alanine and glycine, organic solvents are therefore commonly used to induce strong solvent–protein interactions, which allow spidroin to dissolve at high concentrations [65]. According to previous reports, HFIP is often used to dissolve spidroin at concentrations of 10%–30% (w/v) [36,66]. However, there are several issues related to using organic solvents; one is the toxicity of the organic solvents, which can pose health risks, especially when the artificial silk fibers are used for biomedical applications. Importantly, strong solvent–protein interactions prevent the natural self-assembly of spidroin, influencing the mechanical properties of the artificial spider silk. Using HFIP and 88% formic acid (4:1 ratio) to dissolve synthetic MaSp1 and MaSp2 (molecular weight: 65 kDa) and isopropyl alcohol for the coagulation bath resulted in silk fibers with mechanical properties inferior to those of native fibers [67]. The use of the native-sized recombinant RD (96-mer) of spider silk protein can greatly improve the strength (~0.5 GPa) of artificial spider silk, even though it is still approximately half of the strength of the native silk [36]. The lower strength of this artificial silk fiber is possibly due to the absence of terminal domains and the formation of excess beta-sheets during coagulation with methanol. In the presence of phosphate buffer, spidroin undergoes LLPS to the low- and high-density phases [51]. The high-density phase of spidroin in phosphate buffer is the BSD. When 600% stretching was added directly after spinning, the silk fiber resulting from the BSD exhibited a toughness similar to that of native fibers [37]. In comparison, when spidroin was concentrated by dialyzing this protein against 30% (w/v) PEG, the toughness of the silk fiber was much lower than that of native fibers, even after applying 600% stretching after spinning. The great toughness of these artificial silks can be explained by the fact that the alanine sequences are located in the crystalline region and the glycine-rich regions are found in the amorphous region [68]. However, since the spinning process is different from the native process, no orientation order ($S = 0.01$) was found in the artificial silk fiber, which is probably responsible for the lower strength of this artificial silk fiber [68].

8.2 Implants, tubes, and sponges as scaffolds

Structural protein has high potential as an implant biomaterial, due to its mechanical and biological properties. Scaffolds for tissue engineering need the structure and biological function of the extracellular matrix. The natural extracellular matrix is a composite material with fibrous collagens embedded in proteoglycans. The collagen fibers are organized in a three-dimensional (3D) porous network that forms hierarchical structures from nanometer length-scale multifibrils to macroscopic tissue architectures [69]. The nanofibrous scaffold generated by electrospinning is composed of nanoscale fibers with microscale interconnected pores, resembling the topographic features of the extracellular matrix.

Silk fiber scaffolds formed by electrospinning have potential as scaffolds. SF fiber scaffolds containing bone morphogenetic protein-2 (BMP-2) and/or nanoparticles of hydroxyapatite (HAP) prepared via electrospinning have been studied for in vitro bone formation from human mesenchymal stem cells (hMSCs) [70]. The bioactivity of BMP-2 was retained after the aqueous-based electrospinning process, and the nanofibrous electrospun scaffolds with coprocessed BMP-2 supported high calcium deposition and enhanced transcript levels of bone-specific markers, indicating that the electrospun scaffolds were an efficient delivery system for HAP nanoparticles and BMP-2.

Similar to the fiber scaffold, porous materials are used as 3D scaffold. Silk solution can be processed to porous material, which can be used as 3D scaffolds for bone tissue regeneration due to their biocompatibility and mechanical properties [71,72]. The 3D SF scaffolds loaded with BMP-2 were successfully developed for sustained release of BMP-2 to induce human bone marrow stromal cells to undergo osteogenic differentiation when the seeded scaffolds were cultured in vitro and in vivo with osteogenic stimulants for 4 weeks [73].

Silk biomaterials show the potential use of slow structural protein-based 3D scaffolds and tubes loaded with bioactive molecules such as BMP, horseradish peroxidase (HRP), and adenosine for drug-releasing biomaterials. For the controlled release of bioactive molecules, 3D scaffold is loaded and/or immobilized. For example, HRP was immobilized on silk 3D scaffolds to prepare new functional scaffolds, including regional patterning of the gradients to control cell and tissue outcomes [74]. Adenosine release via silk-based implants to the brain for refractory epilepsy treatments was also reported [75,76]. Silk-based implants designed for the controlled release of adenosine demonstrated therapeutic ability, including the sustained release of adenosine for two weeks via slow degradation of silk, biocompatibility, and the controlled releases of adenosine [75]. Nerve growth factor (NGF)-loaded silk-based nerve conduits have been studied to guide the sprouting of axons and physically protect the axonal cone for peripheral nerve repair [77]. NGF release from the differently prepared SF-nerve conduits was prolonged over 3 weeks. Silk-based

3D scaffolds containing insulin-like growth factor I (IGF-I) were prepared for the controlled release of IGF-I in the context of cartilage repair [78]. As a result, chondrogenic differentiation of hMSCs was observed, starting after 2 weeks and more strongly after 3 weeks [78].

In addition to *B. mori* silk, other silks are studied widely. Tropical tasar silkworm *Antheraea mylitta* silk-based 3D matrices were also evaluated for in vitro drug release and for the study of cell-surface interactions [79]. *A. mylitta* silk contains Arg-Gly-Asp (RGD) sequences, which improve the cell adhesion property of the resultant materials. The silk-based matrices contained two different model compounds, bovine serum albumin (66 kDa) and fluorescein isothiocyanate (FITC)-insulin (3.9 kDa), to characterize release profiles. SF protein blended calcium alginate beads resulted in prolonged drug release without initial burst for 35 days as compared to calcium alginate beads without SF. Furthermore, silk-based matrices contributed to a significant enhancement in cell attachment, spreading, mitochondrial activity, and proliferation with feline fibroblasts.

8.3 Films and coatings

Films and coatings of structural proteins are physically stable and provide a biological surface to materials. Silk solution can be processed into film, thin film, and coating easily. Silk protein is expected to improve the cell adhesion and biocompatibility. Silk films have been used with a covalent decoration of functional peptides as implants for bone formation and drug delivery. For bone regeneration, BMP-2, RGD, and parathyroid hormone can be directly immobilized on silk films using carbodiimide chemistry [4,73]. Differentiation of human bone marrow-derived stem cells cultured with the decorated silk films was induced by immobilized BMP-2. Also, the utility of silk films to promote long-term adenosine release from adenosine kinase deficient embryonic stem cells has been investigated [80,81]. These studies demonstrated that SF constitutes a suitable material for the directed differentiation of embryonic stem cells and cell-mediated therapeutic release of bioactive molecules. Silk films decorated with bioactive molecules could therefore be used for local drug delivery via direct implantation.

Silk films have also been used to promote the stabilization of entrained molecules such as enzymes or therapeutic proteins [82]. Glucose oxidase, lipase, and HRP were entrapped in silk films over 10 months and significant activity if the enzymes were retained, even when stored at 37°C.

Silk films stabilize enzymes without the need for cryoprotectants, emulsifiers, covalent immobilization, or other techniques. In addition, the stabilization of enzymes in silk films is amenable to environmental distribution without refrigeration, and offers potential use in vivo, such as the controlled release of bioactive molecules.

There is a critical need in medicine and biomedical fields to develop versatile and straightforward methods to assemble robust, biocompatible, and functional biomaterial coatings. Silk-based coatings have been reported with metal, bioelectronics, titanium, and bone implants [83–86]. Recently, silk was reported as a potential food coating [87]. Coatings of *B. mori* silkworm SF provides biocompatible interfaces for various biomaterials [3,9,88,89], and also spider silk coating was reported with antibacterial activities [90]. The driving force of self-assembly to form coatings is mainly hydrophobic interactions. Silk-based coatings have been investigated using an aqueous stepwise deposition process with *B. mori* silk solution, which can control the structure and stability of the SF in layer-by-layer films [9]. The flexibility of silk coating can be tuned by controlling its crystallinity. The thickness of silk layer is directly related to the concentration of silk protein in aqueous solution. The secondary structure of SF in the coatings is regulated to control the biodegradation rate, which indicates that release of drugs from these coatings can be controlled via layer thickness, numbers of layers, and secondary structure of the layers.

Nanolayer coatings of silk protein containing pharmaceutical small molecular drugs and therapeutically relevant proteins have been prepared using the stepwise deposition method [88]. Multilayered silk-based coatings have also been developed and used as drug carriers and delivery systems to evaluate vascular cell responses to heparin, paclitaxel, and clopidogrel [88]. The coating thickness can be controlled by manipulating the silk coating solution concentration, and the number of coatings applied. Cell viability and adhesion with human coronary artery smooth muscle cells as well as human aortic endothelial cells on the drug-loaded silk-based coatings showed that paclitaxel and clopidogrel inhibited the cell proliferation. The silk multilayer coats containing heparin promoted human aortic endothelial cell proliferation while inhibiting human coronary artery smooth muscle cell proliferation. This result was the desired outcome for the prevention of restenosis. The coating of SF and adenosine powder was prepared for local and controlled delivery of the anticonvulsant adenosine from encapsulated reservoirs [91]. An increase in either coating thickness or crystallinity delayed and/or prevented adenosine burst, decreased average daily release rate, and increased the release duration. Thus the silk coating is an effective system for the controlled release of bioactive molecules and the cell attachments, based on its useful micromechanical properties and biological outcomes.

8.4 Hydrogel

Another basic application of regenerated protein materials is the creation of hydrogels. Hydrogels are desirable polymeric materials when performing drug delivery and tissue engineering due to their excellent biocompatibility due to their high water content. The role of water molecules in hydrogels has been studied by many investigators, and as a result, it has been shown that bound water, bulk water, and intermediate water are present in hydrogels. It is known that these water molecules significantly affect the physical properties of the gel. For example, bound water content is an essential factor in controlling the enzyme activity in the gel and the drug release rate.

Silk hydrogel prepared from an aqueous silk solution has three network structures: a physical gel in which a beta-sheet structure contributes to network formation, a chemical gel in which covalent bonds between tyrosine contribute a physical–chemical gel formed from both chemical and physical networks. Silk hydrogels are prepared from RSF solutions by physical and/or chemical cross-linking. In physically cross-linked silk hydrogels, the formation of stable beta-sheet nanocrystals, mimicking molecular events in native silk fibers, is induced by changes in the pH of the solvent or by introducing shear forces by ultrasonication or vortex mixing [6,12,13,92,93]. In contrast, in chemically cross-linked silk hydrogels, interchain cross-links are formed through enzymatic reactions, for example, tyrosinase inducing dityrosine oxidation [94] or 3,4-dihydroxy-L-phenylalanine (DOPA)-based covalent bond network formation via introduced catechol groups [95–98].

The gelation mechanism of silk solution was studied based on the beta-sheet structure content and pH of the silk solution, and the results indicated that the gelation is dependent on the beta-sheet content [99,100]. Silk-based hydrogels have also been investigated by Kaplan and coworkers, who found that the gelation of silk solution was induced by pH change, ultrasonication, or vortex [10–12,101,102]. Upon the gelation, a random-coil structure of silk molecules was shown to transform into a beta-sheet structure. Subsequently, the beta-sheet structures aggregated to form molecular network structures, namely, hydrogels [11,102]. hMSCs grew and proliferated over 21 days in the sonication-induced hydrogel prepared with 4% silk solution, indicating that the silk hydrogel is biocompatible and low-cytotoxic enough to be used in cell encapsulation [11,102]. The silk hydrogel has therefore been considered as a candidate for a noncytotoxic biomaterial. However, it is still necessary to clarify the effects of the state of water and the average molecular weights between cross-links of silk hydrogels on various biological activities, such as cell viability and adhesiveness.

A rapid and straightforward method for preparing silk hydrogels using ethanol, a poor solvent for silk protein, has been developed. The water state and cytotoxicity of the obtained silk hydrogels are significantly related [13]. Gelation of an aqueous silk solution can be described as a ternary phase diagram of silk polymer, water (solvent), and ethanol (poor solvent), as schematically shown in Fig. 8.5. As the solution/ethanol ratio increases, the phase is expected to change from solution phase (I) through gel phase (II) to aggregates of silk polymer phase (III). When the loss modulus G during gelation is measured, a slight decrease can be confirmed just before gelation.

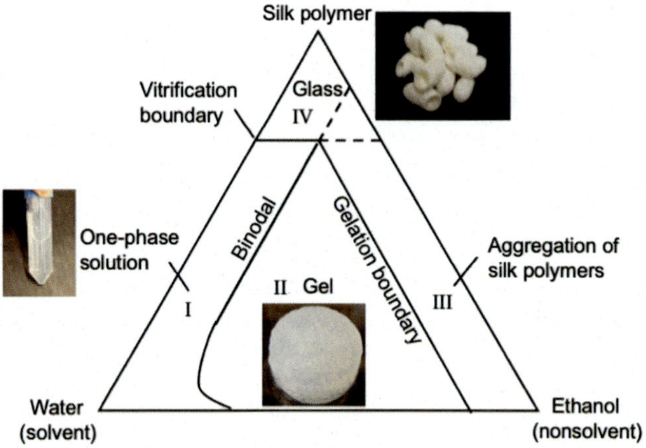

Figure 8.5 Schematic ternary phase diagram of the system: silk polymer–water (solvent)–ethanol (nonsolvent). Solutions in region I are a homogeneous liquid state. Silk polymer solutions in region II separated into two phases and their gelation occurs. In region III, silk polymer aggregates. Silk solutions in region IV are in a glass state. *Source: K. Numata, T. Katashima, T. Sakai, State of water, molecular structure, and cytotoxicity of silk hydrogels, Biomacromolecules 12 (6) (2011), 2137–2144; Copyright 2011 American Chemical Society [13].*

This phenomenon is considered to include the fibrillation of silk polymer due to the formation of beta-sheet and the subsequent aggregation.

The viability of hMSCs cultured on silk gel as a scaffold material has been studied widely. The viability of hMSCs cultured on silk gel increased significantly with increasing silk concentration of the gel. This is because of an increase in beta-sheet content, elastic modulus, mesh size, and bound water content. Considering the hydrophobicity of silk, the bound water content in addition to silk concentration of silk gel contributed to cell viability and played a more critical role than beta-sheet content and elastic modulus.

Furthermore, a new class of silk hydrogels with both chemical and physical cross-linking, namely, silk chemical/physical hydrogel (ChemPhysG) was recently reported [103]. In the chemical/physical gel, phase separation is mildly suppressed by the presence of chemical cross-linking, resulting in intermediate mechanical properties (see Fig. 3.14). Notably, the complete removal of water content from these silk hydrogels leads to a transformation into a resin with a high elastic modulus ($> 10^8$ Pa) and short elastic region without soft elasticity and wide deformability. In the resin, the amount of the beta-sheet structures increases with the decreasing water content. These structures are the origin of the glass-like energetic elasticity of silk resins.

8.5 Resin

Water molecules in protein materials can primarily be categorized into two types: free water and bound water. Free water is unbound water that is capable of freezing and that behaves similarly to bulk water. In contrast, bound water strongly interacts with protein molecules, cannot be frozen and differs from bulk water in terms of physical properties [13,104–106]. After the water content of structural protein is removed, its hydrogel and water-swollen material become a resin material.

Here is an example of the structural protein resin. In the case of silk protein, the bound water effects on the mechanical properties of silks have been widely studied by many groups [13,107–119]. The storage modulus and loss tangent of *B. mori* silk and *Trichonephila edulis* dragline [109,116] and the elastic modulus of *Antheraea pernyi* silkworm silk [108] and *Argiope trifasciata* spider silk [115] have been reported as being influenced by bound water. This is because bound water is considered to play a role in disrupting the hydrogen bonds between amorphous silk molecules, which enhances the mobility of the silk molecules and contributes

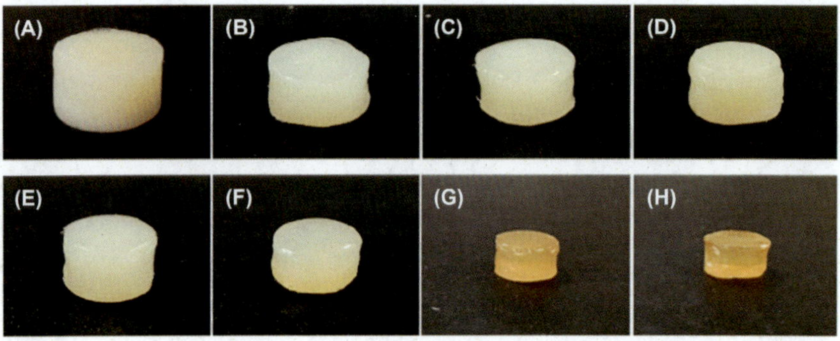

Figure 8.6 Silk-based chemical/physical hydrogel (ChemPhysG) (A) and silk resin prepared at RH 97% (B), RH 75% (C), RH 58% (D), RH 43% (E), RH 23% (F), RH 11% (G), and RH 6% (H) [103].

to glass transitions [107,109,110,114,120]. The existence of a critical humidity influences the mechanical properties such as Young's modulus. In the dry state [relative humidity (RH) less than 20%], the silk resin prepared from ChemPhysG is brittle. Its mechanical properties are a relatively high modulus, more than 150 MPa, and lower strength and toughness (Fig. 8.6). In the wet state (RH higher than 60%), the mechanical properties of silk resin are similar to hydrated ChemPhysG (hydrogel) in that it is a soft and brittle material. This is because of the plasticization effect of water molecules on the silk molecules. In the mild humidity (RH 20%–60%), the silk resin is elastic, similar to ChemPhysG, as well as the strongest and toughest among those silk resin materials (see Fig. 3.15). This mild condition was appropriate for realizing the original toughness of the silk molecules, that is, the toughness of the natural silk fibers.

On the other hand, silk-based physical hydrogel (PhysG) cannot become a tough resin, even when it is incubated at the mild humidity. This indicates that both the network structure and hydration state are necessary to enhance the toughness of the silk material. Toughness cannot be attained by physical cross-links alone. Chemical cross-links between dityrosine do not contribute significantly to the strength and toughness of chemical hydrogel (ChemG), while chemical cross-links are necessary with crystallization to yield the toughness of silk-based materials. This concept agrees with the literature on the importance of dityrosine interactions in native and regenerated silks [121–124]. Previously, silk hydrogels and sponges with high strength have been reported; [37,125,126] however, relatively high strength and elasticity were not achieved simultaneously. It is necessary to control both the network structure and hydration state to design tough, strong, and robust silk materials. The toughness of silk resin can be achieved by beta-sheet structures and covalent bonding between dityrosine as cross-linking points.

The cross-linking points in silk materials are considered to consist of beta-sheet crystals, which involve noncovalent and hydrophobic interactions [127,128]. ChemG did not contain enough beta-sheet structures to form a physical network, whereas PhysG, ChemPhysG, and the resins had beta-sheet structures. In theory, higher *Xc* and beta-sheet content lead to a higher number of cross-linking points and lower molecular weights between cross-links. In other words, lower molecular weights between cross-linking points generally lead to a higher density of cross-links, in particular, the formation of beta-sheet structures via hydrophobic interactions, inducing a higher crystallinity (degree of crystallinity) *Xc*. The wide-angle X-ray diffraction study of PhysG and ChemPhysG demonstrated that the sharpness of the peak (full width at half maximum, FWHM) was not significantly different, indicating the beta-sheet crystal sizes were maintained according to Vonk's theory [129]. The FWHM of the peaks from the beta-sheet structures in the hydrated resin decreased when the RH decreased, and hence the dehydration at lower RHs induces the beta-sheet formation. The silk protein concentrations of the network materials like hydrogel and elastomer do not affect *Xc* significantly, but do influence on Young's modulus. The cross-linking points of silk material consist mainly of beta-sheet crystals (physical cross-link) and covalent bonds between dityrosine (chemical cross-link). However, the number of dityrosine-mediated chemical cross-links is insufficient to increase Young's modulus, which can be observed from the number of tyrosines in the silk protein's amino acid sequence. In addition to beta-sheet crystals and chemical cross-links, entanglements and assembly via the hydrophobic interactions of silk molecules during dehydration could play a role as cross-linking points at high silk protein concentrations greater than 150 g/L (The molecular weight of *B. mori* silkworm silk heavy chain is approximately 350 kDa). Thus to design tough and robust silk materials, multiple cross-links, including beta-sheet crystals-mediated network, chemical cross-links, and entanglements and assembly via the hydrophobic interactions of silk molecules, must be combined and regulated. Realistically, network design, concentration, and hydration state of silk molecules are the keys to accomplish an artificial silk material with high strength and toughness.

8.6 Nanoparticles

Nanoparticles are mainly developed for drug delivery systems to achieve the controlled release of pharmaceutical molecules. Here we describe the example of silk-based nanoparticles. Silkworm silk-based nanoparticles have been studied and developed as one of the biocompatible drug delivery systems [130–132]. SF-based nanoparticles with an average diameter of less than 100 nm for local and controlled therapeutic curcumin delivery to cancer cells were fabricated by blending with noncovalent interactions to encapsulate curcumin various proportions with pure SF or SF with chitosan [130]. According to the literature, Silk nanoparticles from SF solutions of *B. mori* and tropical tasar silkworm *A. mylitta* were stable, spherical, negatively charged, 150–170 nm in average diameter, and showed no toxicity [131]. From the silk nanoparticles, the growth factor was released, resulting in significantly sustained release over 3 weeks. The silk-based nanoparticles containing curcumin showed a higher efficiency against breast cancer cells and demonstrated the potential to treat in vivo breast tumors by local, sustained, and long-term therapeutic delivery [132]. These reports indicate potential application as a growth factor and also other bioactive protein delivery systems.

The silk nanoparticles complexed with other biomacromolecules have been developed due to their excellent processability and biocompatibility [132–136]. Silk-based nanoparticles used for gene delivery have recently been reported to provide biodegradability, biocompatibility, high transfection efficiency, and DNase resistance [21,137,138]. The secondary structure of silk sequences can be used to regulate the enzymatic degradation rates, namely, higher beta-sheet structure content enhances the stability of the silk nanoparticles. The secondary structure is an important factor in controlling the release of bioactive molecules from silk nanoparticles [138]. Silk-based and also protein-based nanoparticles are therefore potentially useful as candidates for gene and drug carriers.

Dual-drug delivery systems, which are capable of controlling the release behaviors of multiple drugs, have recently become attractive for the combined administration of different drugs and the optimization of therapeutic effects [139–141]. One such dual-drug delivery system is based on hydrogel cross-linking with nanoparticles via polymeric networks, a method that has undergone impressive progress in terms of both syntheses and applications (Fig. 8.7) [142,143]. Hydrogels are 3D macromolecular networks used to functionally demonstrate the controlled

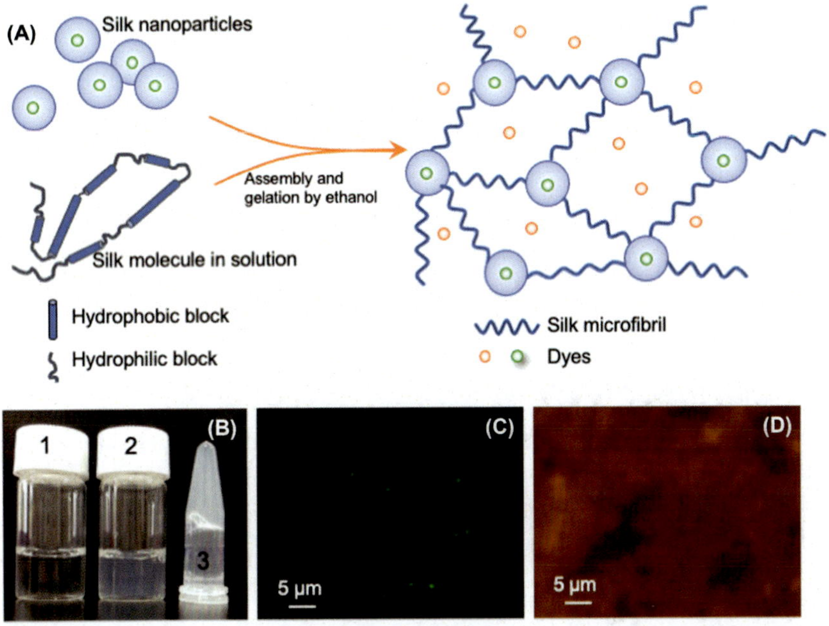

Figure 8.7 (A) Schematic representation of the formation of silk hydrogel containing silk nanoparticles for dual-drug release. Smaller particles denote drug molecules and larger particles indicate silk nanoparticles. Silk molecules in the solution formed microfibrils and molecular networks by the addition of ethanol. (B) Gelation behavior of the silk solution. Ethanol containing silk nanoparticles with incorporated FITC (B1) and silk solution of 60 g/L (B2) were mixed, resulting in gelation of the silk solution (B3). (C) Fluorescence microscopic image of silk nanoparticles containing FITC in the silk hydrogel and (D) of the silk hydrogel containing RhB in its molecular networks. *Source: Reprinted with permission from K. Numata, S. Yamazaki, N. Naga, Biocompatible and biodegradable dual-drug release system based on silk hydrogel containing silk nanoparticles, Biomacromolecules 13 (5) (2012) 1383–1389, Copyright 2012 American Chemical Society [22].*

release of cells and bioactive molecules such as drugs, antibodies, proteins, peptides, and genes [144–147]. A fully silk-based biocompatible and slowly biodegradable dual-drug delivery system consisted of silk hydrogel containing silk nanoparticles in silk-microfibril networks (Fig. 8.7) [22]. To confirm the dual-drug release profiles from silk hydrogel and nanoparticles, two different model drugs, FITC and rhodamine B (RhB), were loaded into the silk nanoparticles and the silk hydrogels, respectively [22]. The FITC signals are distributed as particles in the hydrogel (Fig. 8.7C), indicating that the silk nanoparticles are successfully entrapped in the silk hydrogel. Additionally, the RhB signals are observed across the entire area

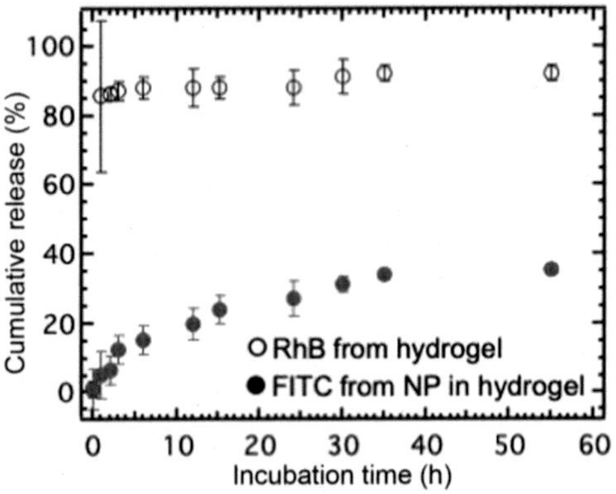

Figure 8.8 Dual-release behaviors of FITC and rhodamine B (RhB) from the silk nanoparticles (NP) and from the silk hydrogel in the presence of protease XIV at 37°C. *Source: Reprinted with permission from K. Numata, S. Yamazaki, N. Naga, Biocompatible and biodegradable dual-drug release system based on silk hydrogel containing silk nanoparticles, Biomacromolecules 13 (5) (2012) 1383–1389, Copyright 2012 American Chemical Society [22].*

of the hydrogel (Fig. 8.7D), demonstrating that the molecular networks or the silk hydrogel are filled with RhB. These observations confirmed that a silk hydrogel containing silk nanoparticles was successfully constructed for dual-drug release.

The dual release of model drugs from a silk hydrogel containing silk nanoparticles was confirmed in the presence of protease XIV (300 μg/mL) at 37°C (Fig. 8.8) [22]. Protease XIV is a protease degrading silk materials, including beta-sheet structures. The burst release of RhB from the silk hydrogel (approximately 90%) was recognized within 1 hour of incubation time. In contrast, slow and constant release of FITC from the silk nanoparticles, which were trapped in the silk network, was observed. After 5 days, the FITC was completely released due to the enzymatic degradation of the silk materials. These two different release behaviors successfully demonstrate the dual-drug release from silk hydrogels containing silk nanoparticles. The release of RhB from the silk hydrogel is relatively fast because of the fast degradation of noncrystalline silk molecules in the silk hydrogel by protease XIV, as reported previously [148]. The size of the silk hydrogel network has been reported to range from 5 to 50 μm, which is large enough for proteases to degrade the silk molecules of nanoparticles and hydrogels. According to previous literature [13,138], enzymatic degradation rate of silk-based materials can be regulated by physical properties, such as crystallinity, similar to general polymeric materials. Therefore, the release behavior of drugs from the silk-based dual-drug release system has the potential to be regulated by the physical properties of the silk nanoparticles and the network sizes of the silk hydrogel.

8.7 Microspheres

SF microspheres were processed using spray-drying, however, the sizes of the microspheres were above 100 μm, which is suboptimal for drug delivery [149]. Other methods to prepare silk microspheres include lipid vesicles as templates to efficiently load bioactive molecules for local controlled releases [150]. The lipid is subsequently removed by methanol or NaCl, resulting in silk microspheres consisting of the beta-sheet structure and approximately 2 μm in diameter. The silk microspheres loaded with HRP, used as a model drug, demonstrated controlled and sustained release of active enzyme over 10–15 days. Growth factor delivery via the silk microspheres in alginate gels was also reported to be more efficient in delivering BMP-2 than IGFs, probably due to the sustained release of the growth factor [151]. Additionally, growth factors successfully formed linear concentration gradients in scaffolds to control osteogenic and chondrogenic differentiation of hMSCs during culture. This silk microsphere/polymeric scaffold system is an option for the delivery of multiple growth factors with spatial control for in vitro and in vivo 3D cultures. In more recent studies, a new mode to generate micro- and nanoparticles from silk, based on blending with polyvinyl alcohol was reported [152]. This method simplifies the overall process compared with lipid templating and provides high yield and good control over the feature sizes, from 300 nm to 20 μm, depending on the ratio of polyvinyl alcohol/silk used.

8.8 Adhesive

8.8.1 Adhesive motif and peptide

Adhesive substances are essential materials in various commercial sectors such as the automobile and construction industries. In particular, biologically based and biodegradable adhesives, which can be used in different materials, are required for the recycling of various component parts of cars and houses. One of the candidate adhesive materials is a protein-mimic adhesive, which has recently been highlighted as an attractive biologically based and highly functional adhesive. The blue mussel (*Mytilus edulis*) foot protein 5 (Mefp-5) is an adhesive protein that can be found on the surface of the mussel adhesive plaque, and is mainly composed of L-glycine, L-lysine, and DOPA [153,154]. According to previous studies, which have analyzed thousands of adhesive pads, adhesion is mainly achieved by the redox-chemistry of DOPA [155,156]. Artificially designed materials containing DOPA have also been developed and have demonstrated that adhesive function was directly correlated to DOPA content [98]. Without lysine, linear and branched DOPA and PEG copolymers were shown to have adhesive functions [157–159]. The effects of the sequence of mussel proteins have also been investigated using random copolypeptides containing DOPA [160,161], suggesting that DOPA is the main factor underlying adhesive function. In addition to DOPA, Mefp-5 is composed of L-lysine (approximately 20 mol%), which is necessary to form a network of structures between catechol and amine groups [98,155]. From these reports it is suggested that DOPA and lysine are essential for biomimetic adhesive design; however, the application of these adhesive copolymers is limited, in terms of yield, because their synthesis requires multiple reaction steps. To clear these issues, chemoenzymatic synthesis of adhesive peptides containing DOPA and L-lysine has been reported via two enzymatic reactions, namely, chemoenzymatic polymerization of L-tyrosine and L-lysine by *Papaya* peptidase I (papain) as well as the enzymatic conversion of tyrosine to DOPA by tyrosinase (Fig. 8.9). The synthesis was characterized by the yield, degree of polymerization, and the composition of the polypeptide. Furthermore, the conversion of tyrosine to DOPA by tyrosinase was evaluated quantitatively by nuclear magnetic resonance (^{1}H-NMR) spectroscopy and amino acid analysis. The peptides produced, consisting

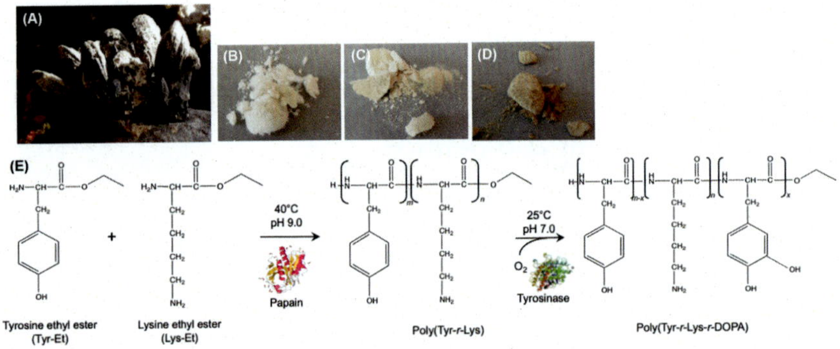

Figure 8.9 Biochemical synthesis of adhesive polypeptides. Pictures of (A) marine mussel, (B) P(75%Tyr-25%Lys), (C) P(50%Tyr-25%Lys-25%DOPA), and (D) P(45%DOPA-30%Tyr-25%Lys). (E) Reaction schemes for the enzymatic synthesis of poly(L-tyrosine-*r*-3,4-dihydroxy-L-phenylalanine-*r*-L-lysine) [P(Tyr-DOPA-Lys)], via poly(L-tyrosine-*r*-L-lysine) [P(Tyr-Lys)], from L-tyrosine ethyl ester (Tyr-Et) and L-lysine ethyl ester (Lys-Et) [97].

DOPA, L-lysine, and L-tyrosine, were evaluated for adhesive function in terms of adhesion strength and the Young's modulus.

To evaluate the adhesive functions of the DOPA-containing synthetic polypeptide, P(Tyr-DOPA-Lys), made up with different DOPA content, its adhesion strength was characterized at different pH levels of 6, 10, and 12. The results show that the adhesive function of P(50%Tyr-25%DOPA-25% Lys) increased significantly with increasing pH, with no adhesiveness detected under acidic conditions, but notably higher adhesive function at pH 12 (Fig. 8.10). The polypeptide P(50%Tyr-25%DOPA-25%Lys) was not

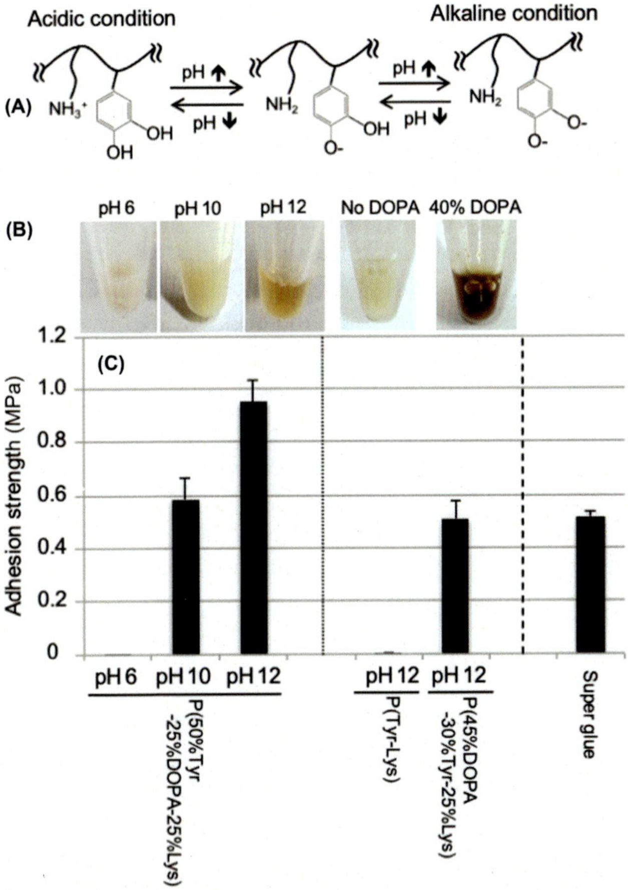

Figure 8.10 The (A) chemical interactions, (B) physical state and color, and (C) adhesive strengths, measured at different pH levels and using super glue as a control, of poly(L-tyrosine-*r*-3,4-dihydroxy-L-phenylalanine-*r*-L-lysine) [P(Tyr-DOPA-Lys)] and poly (L-tyrosine-*r*-L-lysine) [P(Tyr-Lys)]. Data represent means ± standard deviations from three independent experiments [97].

dissolved in water at pH 6, but was dissolved at pH 10 as a white solution. At pH 12, the solution of P(50%Tyr-25%DOPA-25%Lys) was brown in color and had an adhesive strength of approximately 0.95 MPa, which is higher than that of Super Glue. In addition, to clarify the effects of the content of DOPA, the adhesion strengths of P(Tyr) and P(45%DOPA-30%Tyr-25%Lys) were characterized at pH 12. The resultant adhesive strengths show that DOPA was required for adhesive function, however, 45 mol% DOPA appeared to be too high a content to yield the best adhesive strength against mica substrates. This may be because the molar balance between DOPA and lysine is one of the most important factors in the adhesive properties of mussel-like adhesive peptides.

In nature, mussel-derived adhesion protein, Mefp-5, is composed mainly of 20.7 mol% L-glycine, 19.5 mol% L-lysine, and 25.5 mol% DOPA, that is, the molar ratio between DOPA and lysine is approximately 5:4 [98]. Furthermore, the effect of the molar ratio of DOPA to lysine on adhesion strength has been studied; the adhesion strength against steel has been reported to be highest when the molar ratio of DOPA: lysine was 1:4 [162]. Based on these studies, excess DOPA does not appear to contribute significantly to adhesive strength, while the presence of deprotonated lysine (deprotonated at the primary amine group) positively affected the adhesion resulting from DOPA interactions. Conversely, it has also been reported that peptides containing DOPA, without lysine also have adhesive properties, whereas lysine functions to form a network of structures between the amine group of lysine and the catechol group of DOPA [98,155].

To clarify the influence of lysine on adhesive function, the Young's modulus (the elastic modulus) of the adhesive peptides was evaluated. The network structure, in particular the molecular weight between crosslinking points, affects the elastic modulus [13,163]. The adhesive peptides containing 25% and 45% DOPA at pH 12 demonstrated similar elastic moduli, which were around 6.2 MPa (Fig. 8.11A). Without DOPA, the peptides did not have adhesive function or elastic moduli. On the other hand, the peptide with 25% DOPA and 25% lysine at pH 10 showed lower Young's modulus, suggesting that the alkaline (deprotonated) condition induced more molecular network formation via the deprotonation of the primary amine group of lysine (Fig. 8.11B). The elastic modulus increased with an increase in the lysine content, indicating that increased lysine formation of more molecular networks, thus resulting in a higher modulus. Based on these results, DOPA is an essential component

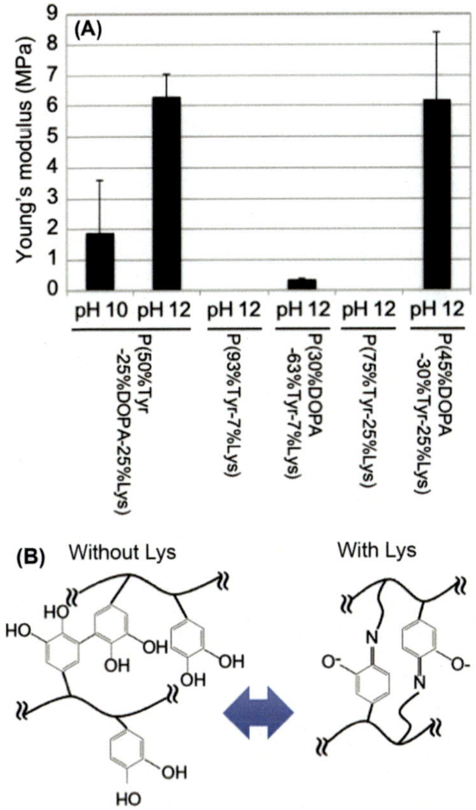

Figure 8.11 (A) Determination of Young's moduli (MPa) of poly(L-tyrosine-*r*-3,4-dihydroxy-L-phenylalanine-*r*-L-lysine) [P(Tyr-DOPA-Lys)] and poly(L-tyrosine-*r*-L-lysine) [P(Tyr-Lys)] at pH 10, and/or 12. (B) The suggested chemical interactions between side-chain groups with and without lysine [97].

adhesive function, that is, molecular interactions between peptides and surfaces, whereas lysine is the component to form the molecular networks of the peptide.

8.8.2 Silk as adhesive

To add adhesive functions into silk proteins, the tyrosine residues of silkworm silk were successfully converted to DOPA unit using tyrosinase as a biocatalyst (Fig. 8.12) [164]. Adhesive functions of DOPA-modified silk fibroin (DOPA-SF) between several material surfaces including mica, paper, polypropylene (PP), wood, and silk film were elucidated by lap shear tests. Fourier transform infrared (FTIR) measurements demonstrated

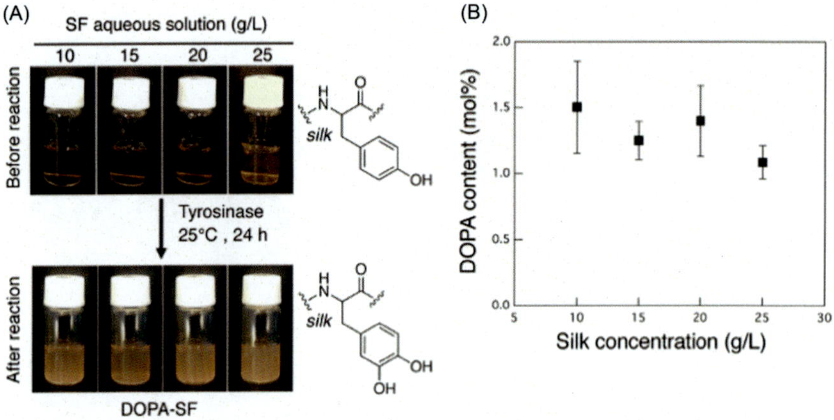

Figure 8.12 (A) Tyrosinase-catalyzed modification of SF to DOPA-SF at different concentrations, together with the photographs of solutions. The caps of grass sample bottles were opened during the reaction. (B) The DOPA content in DOPA-SF after modification [164].

that the adhesion strength of DOPA-SF was not directly related to the beta-sheet formation of silk molecules. This eco-friendly and facile method offers a new perspective to fabricate natural adhesive materials for various application areas.

The adhesion function was evaluated by lap shear tests. Based on their stress—strain curves, the adhesion properties, including breaking strain, adhesive strength, adhesive fracture energy, and Young's modulus at different silk concentrations and pH levels, are summarized in Fig. 8.13. It was obvious that both adhesive strength and fracture energy were improved by the introduction of DOPA at any silk concentration and pH level, whereas maximum values (0.97 MPa and 19.1 $kJ \cdot m^{-3}$, respectively) were observed for 15 g/L of DOPA-SF at pH 10. The increase in the Young's modulus of DOPA-SF compared to that of nonmodified SF was also observed, probably because the original outstanding physical properties of silk have started to appear for DOPA-SF since the interface of mica and silk more strongly interact. These results suggest that the adhesive function of SF was improved by DOPA modification, and also DOPA modification is important from the viewpoint of not only improving its adhesive function but also widening the variety of applicable surfaces.

The adhesive properties of DOPA-SF were evaluated for different surfaces including paper, PP, wood, and silk film [164]. As mentioned above, SF solution itself exhibited good adhesion properties for several surfaces such

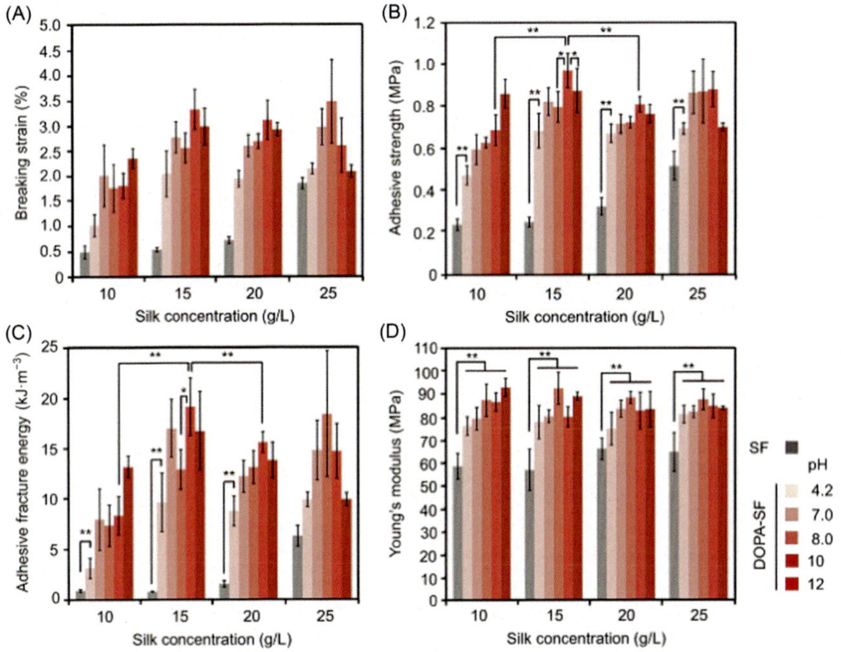

Figure 8.13 Adhesion properties of DOPA-SF. (A) Breaking strain, (B) adhesive strength, (C) adhesive fracture energy, and (D) Young's modulus of DOPA-SF for mica substrate at different silk concentrations and pH levels. *Significant differences between groups at $P < .06$. **Significant differences between groups at $P < .01$ [164].

as skin. Thus several surfaces, whose homogeneity to SF seem to be low, were chosen to emphasize the significance of DOPA modification. DOPA-SF solution (15 g/L) at different pH levels was coated on substrates. Fig. 8.14 shows stress–strain curves of pasted samples and a summary of adhesive function for these substrates, respectively. Note that the overlay area of silk film (0.5×10^2 mm^2) was changed to three times smaller than that of others (1.5×10^2 mm^2) because the rupture of samples took place at other adhesive points by lap shear tests when the samples were prepared under original conditions. DOPA-SF exhibited higher adhesive function than SF for not only mica but also for all tested substrates, especially at higher pH values. The highest adhesive strength was observed for silk film even though the overlay area was decreased. Meanwhile, the Young's modulus determined with paper, PP, and wood was almost constant for both SF and DOPA-SF in contrast to mica. This outcome occurred because DOPA-SF adhered to these surfaces relatively weakly, and cleavage on the surface took place before the modulus of silk itself had appeared in contrast to mica. Improvement in

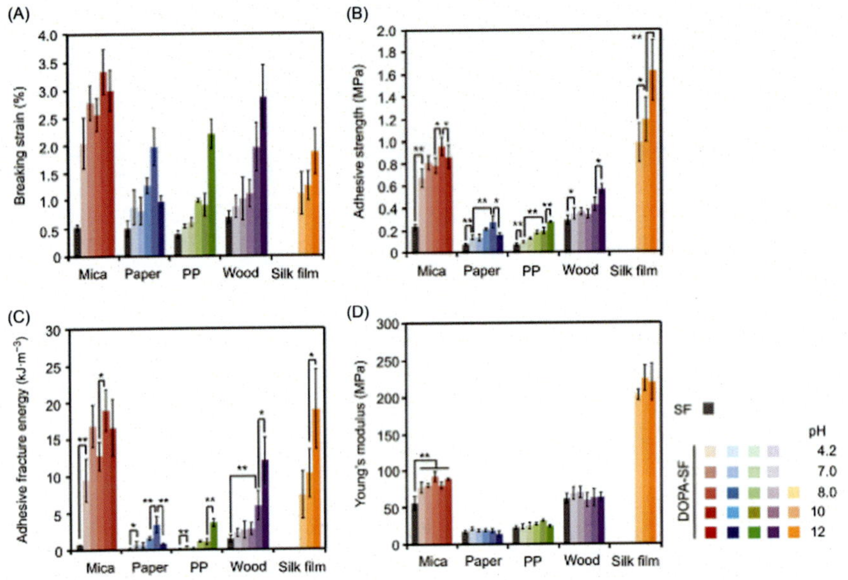

Figure 8.14 (A) Breaking strain, (B) adhesive strength, (C) adhesive fracture energy, and (D) Young's modulus of DOPA-SF for mica, paper, PP, wood, and silk film at different pH levels. Silk concentration: 15 g/L. *Significant differences between groups at $P < .06$. **Significant differences between groups at $P < .01$ [164].

adhesive strengths was observed for these surfaces; however, it was not significant. Conversely, the Young's modulus determined with silk film was approximately 200 MPa, which was the highest for all the materials studied. Together with the results of adhesive strengths, it was concluded that DOPA-SF functioned against various surfaces, most strongly interacted with silk film, and exhibited outstanding adhesiveness.

Tyrosinase-catalyzed modification of SF from the *B. mori* silkworm was successfully achieved to produce DOPA-modified SF. Amino acid composition analysis revealed 25% of tyrosine residues in SF were converted to DOPA resulting in a DOPA-SF solution with c. 1.3 mol% of DOPA moiety. Although the content of DOPA was not high, significant improvement of the adhesive function of DOPA-SF for mica surface was confirmed by lap shear tests. A higher adhesion function was observed for higher pH levels. DOPA-SF exhibited excellent adhesion not only for mica but also for several types of substrates including paper, PP, wood, and silk film. It was also found that the adhesion strength was not assisted by the formation of a beta-sheet structure of DOPA-SF based on ATR (attenuated total reflection)-FTIR measurements. This simple but powerful approach offers a new perspective to fabricate natural adhesive materials with an environmentally benign process.

In addition to the enzymatic reaction, the chemical modification of silk is also an excellent strategy to induce the adhesive functions of silk molecules. The serine side chains of silk were chemically functionalized with substituents (−COOH, −OH, and $-CH_3$) with different abilities to hydrogen bond (−COOH > −OH > $-CH_3$). Serine is the most abundant amino acid in the protein's primary sequence capable of hydrogen bond formation. The hydrophilic amino acids of silk molecule consist of serine (~12%), tyrosine (~5.5%), and aspartate and glutamate (cumulated to ~1.5%); all other constituent amino acids (glycine, alanine, valine, etc.) are hydrophobic and chemically inert. SF was successfully modified with iodoacetic acid, 3-chloro-1-propanol and 1-chlorobutane to yield carboxy-SF (SF−COOH), hydroxy-SF (SF−OH), and methyl-SF ($SF-CH_3$), respectively.

Modified silks were then assessed for their adhesive capabilities to sheep skin/leather via single lap sheer tests, in parallel with structural assessments. The results support the aforementioned adhesion mechanism hypothesis, highlight the role of hydrogen bonding in adhesion and inform on subsequent design approaches necessary to impart reversibility into silk adhesives via temporary and controlled hydrogen bond disruption. The SF−OH, even though it bears the same terminal −OH as the original serine, might be unfavorably positioned for inter- or intramolecular hydrogen bonding due to the length of the substituent. The $SF-CH_3$ is not expected to be able to form beta-sheets because of the aforementioned considerations. The SF−COOH, has a shorter substituent and increased ability to form hydrogen bonds with adjacent atoms. These observations are important, especially in the context of our previous findings, where we showed that adhesive interactions with a substrate could be further enhanced via postadhesion induction of beta-sheets. In the context of the FTIR data presented herein, the chemically modified silks, with the exception of carboxy-SF, could not be further enhanced for adhesion via secondary structure induction.

8.9 Composite

8.9.1 Coating for textile

To enhance the water resistance of silk materials, the fluoropolymer coating of the silk materials. *B. mori* silkworm silk textile and silk film were coated with the fluoropolymer. The coating of the fluoropolymer on the silk films was approximately 50-μm thick based on the cross section scanning electron microscopic observation of the fluoropolymer-coated silk films. The fluoropolymer-coated silk films show a contact angle of approximately 90°. The water vapor permeability of the fluoropolymer-coated silk films was approximately 110 g/m^2/day. These properties of the fluoropolymer-coated silk films indicate the potential advantages of a fluoropolymer coating on silk materials [165].

To show the excellent water resistance of the fluoropolymer-coated silk-based materials, shrinkage tests, which were similar to the washing and drying process used for clothing, were performed with the silk textiles. Although the silk textile was examined by shrinkage test, the silk film was not done because shrinkage test is designed for clothes. We coated *B. mori* silkworm silk textile with fluoropolymer and subjected it to the shrinkage test (Fig. 8.15). The coated textile showed great size stability during the washing/drying cycles (Fig. 8.15A,C). On the other hand, the silk textile without the fluoropolymer coating did not show size stability (Fig. 8.15B). After the third drying stage, the sizes had decreased by approximately 5% (Fig. 8.15D). Thus the fluoropolymer coating can enhance the water resistance of silk materials in practical applications.

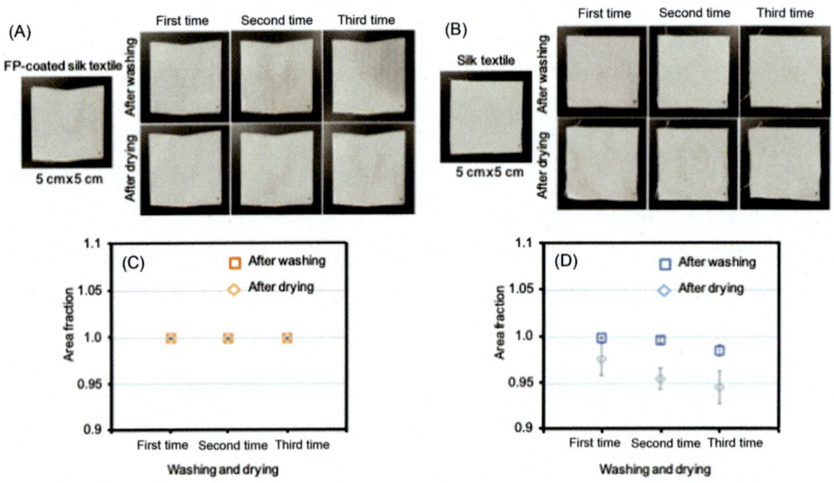

Figure 8.15 Shrinkage tests of silk textiles with and without a fluoropolymer (FP) coating. Pictures of fluoropolymer-coated silk textile (A) and mock silk textile (B) before and during the washing/drying cycles. The area changes of the FP-coated silk textiles (C) and mock silk (D) during the washing and drying processes. The original area was 50 × 50 mm. Each data point is expressed as the mean and standard deviation [165].

8.9.2 Fiber-reinforced plastic

In recent years, natural plant fiber–reinforced composites have become increasingly popular due to their "green" credentials and low-density characteristics. Conversely, natural fibers from animals, such as silk (from silkworms and spiders), have been examined extensively but have found limited structural engineering material applications because of the insufficient strength and stiffness [166]. Comparing the three most popular reinforcement fibers (glass fiber, carbon fiber, and flax fiber) with *B. mori*, *A. pernyi*, and *Nephila edulis* spider dragline silks, it can be concluded that silk fibers have the lowest density and the highest breaking energy among these fibers. In particular, the breaking energy of wild *A. pernyi* silk reached 150–200 $MJ \cdot m^{-3}$, which demonstrated an energy absorption ability among all silks [167–169]. A few studies demonstrated that *B. mori* silk fiber–reinforced plastic composites (SFRPs) could outperform traditionally reinforced composites as structural materials [170,171]. A previous work showed that *B. mori* silk fabric could be used at a high volume fraction of 70% in epoxy resin composites owing to its superior compressibility; furthermore, these composites also displayed an enhanced impact strength that was four times higher than the unreinforced epoxy matrix [172].

Recently, hybrid fiber–reinforced plastic composite (HFRP) laminates using *A. pernyi* silk and carbon fiber fabrics were fabricated [173]. Using hybrid fibers with complementary strength and toughness is an effective approach to new structural composites with desirable mechanical performances. Wild *A. pernyi* silk fibers exhibit high toughness which originate from alpha-helical/random-coil conformation structures and abundant microfiber morphology (Figs. 8.16 and 8.17). However, the insufficient strength and stiffness hinder its application in structural composites. High-stiffness and strength carbon fibers are commonly used to reinforce epoxy resin composites. In this work, interply hybridization of silk and carbon fibers is applied to reinforce epoxy–matrix composites. With increased carbon fiber content, the quasistatic tensile and flexural stiffness and strength increased following the rule of mixture, and the maximum load in the unnotched Charpy impact test increased significantly. On the other hand, more silk fiber could increase the ductility, the overall breaking energy in flexural test, and the impact strength of hybrid fiber composites. Remarkably, the impact strength of the composite that comprised equal volumes of carbon and silk fibers achieved 98 $kJ \cdot m^{-2}$ impact strength, which was twice greater of that from purely carbon fiber–reinforced

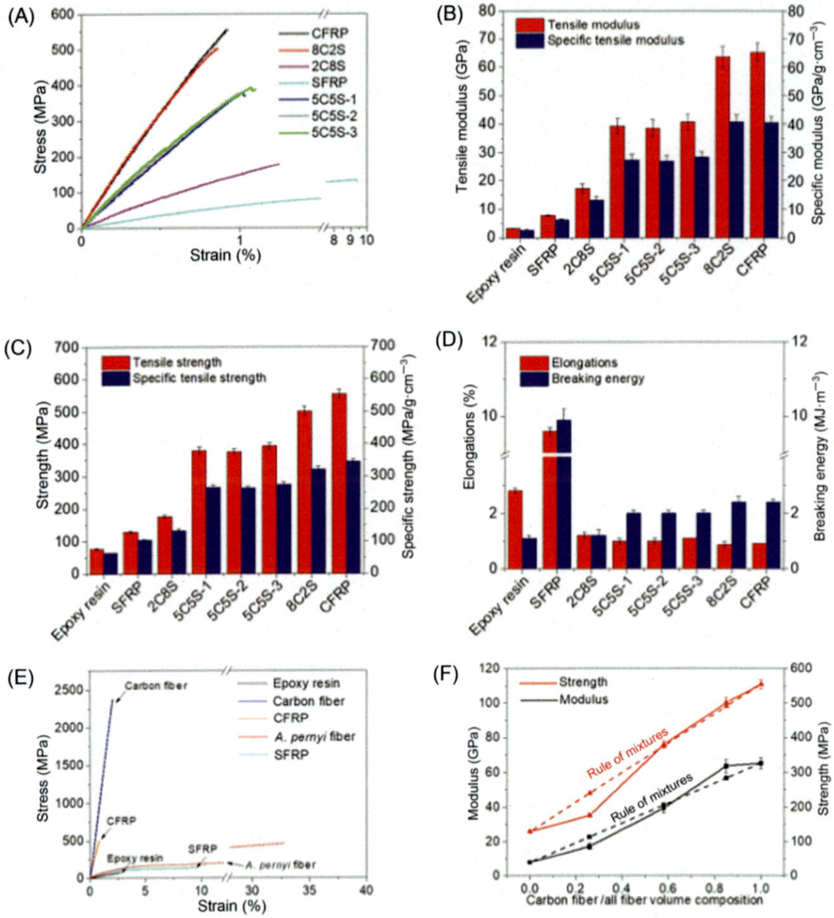

Figure 8.16 Tensile properties of carbon fiber–reinforced composite (CFRP), silk fiber–reinforced plastic composite (SFRP), and hybrid fiber from carbon fiber and silk-reinforced composite. (A) Typical tensile engineering stress–strain curves; (B) modulus and specific tensile modulus; (C) strength and specific tensile strength; and (D) elongation and breaking energy. Note: Breaking energy is calculated as the area under the stress–strain curve and the unit is converted from MPa to $MJ \cdot m^{-3}$ for easy comparison with the literature. (E) Direct comparison of tensile stress–strain behavior of pure epoxy resin, two reinforcement fibers, 69 vol.% CFRP and 51 vol.% SFRP. (F) Comparison of the tensile modulus and strength of the HFRP composites with increasing silk volume composition. The silk volume compositions of 8C2S, 5C5S, and 2C8S were calculated as 15%, 42%, and 74%, respectively [173].

composites ($44\ kJ \cdot m^{-2}$). This work demonstrates that tough natural silk fibers and strong synthetic fibers could integrate in composites for tailored mechanical properties, especially improved impact performance.

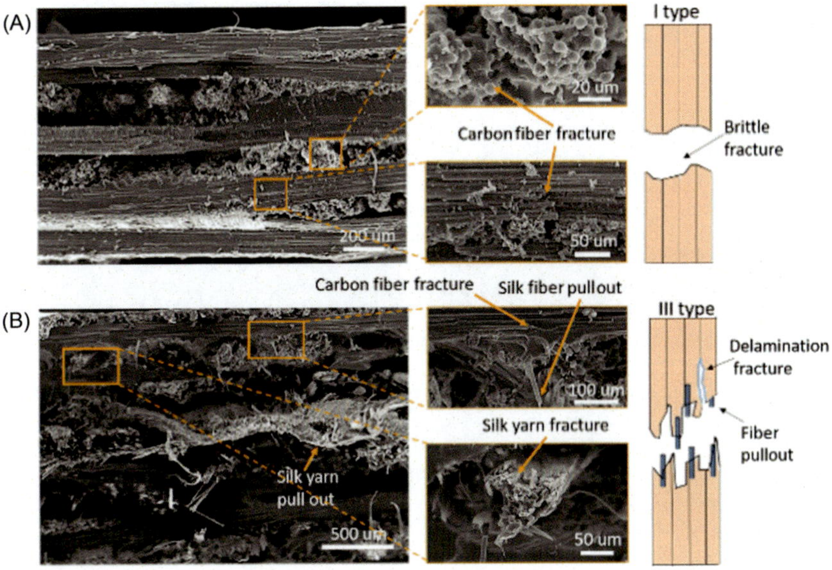

Figure 8.17 Scanning electron microscopic (SEM) images and schematic diagram of fracture surfaces after a uniaxial tensile testing of the (A) SEM image of carbon fiber–reinforced composite (CFRP), (B) SEM image of hybrid fiber from carbon fiber and silk-reinforced composite (HFRP 5C5S-1 composites), and (C) schematic diagram of fracture surfaces of HFRP [173].

References

[1] D.N. Rockwood, R.C. Preda, T. Yucel, X.Q. Wang, M.L. Lovett, D.L. Kaplan, Materials fabrication from Bombyx mori silk fibroin, Nat. Protoc. 6 (10) (2011) 1612–1631.

[2] K. Numata, D.L. Kaplan, Silk-based delivery systems of bioactive molecules, Adv. Drug Delivery Rev. 62 (15) (2010) 1497–1508.

[3] X. Wang, E. Wenk, X. Hu, G.R. Castro, L. Meinel, X. Wang, et al., Silk coatings on PLGA and alginate microspheres for protein delivery, Biomaterials 28 (28) (2007) 4161–4169.

[4] S. Sofia, M.B. McCarthy, G. Gronowicz, D.L. Kaplan, Functionalized silk-based biomaterials for bone formation, J. Biomed. Mater. Res. 54 (1) (2001) 139–148.

[5] M. Li, W. Tao, S. Lu, C. Zhao, Porous 3-D scaffolds from regenerated Antheraea pernyi silk fibroin, Polym. Adv. Technol. 19 (3) (2008) 207–212.

[6] U.J. Kim, J. Park, H.J. Kim, M. Wada, D.L. Kaplan, Three-dimensional aqueous-derived biomaterial scaffolds from silk fibroin, Biomaterials 26 (15) (2005) 2775–2785.

[7] K. Makaya, S. Terada, K. Ohgo, T. Asakura, Comparative study of silk fibroin porous scaffolds derived from salt/water and sucrose/hexafluoroisopropanol in cartilage formation, J. Biosci. Bioeng. 108 (1) (2009) 68–75.

[8] H.J. Jin, J. Park, V. Karageorgiou, U.J. Kim, R. Valluzzi, D.L. Kaplan, Water-stable silk films with reduced beta-sheet content, Adv. Funct. Mater. 15 (8) (2005) 1241–1247.

[9] X.Y. Wang, H.J. Kim, P. Xu, A. Matsumoto, D.L. Kaplan, Biomaterial coatings by stepwise deposition of silk fibroin, Langmuir 21 (24) (2005) 11335–11341.

[10] A. Matsumoto, J. Chen, A.L. Collette, U.J. Kim, G.H. Altman, P. Cebe, et al., Mechanisms of silk fibroin sol-gel transitions, J. Phys. Chem. B 110 (43) (2006) 21630–21638.
[11] X.Q. Wang, J.A. Kluge, G.G. Leisk, D.L. Kaplan, Sonication-induced gelation of silk fibroin for cell encapsulation, Biomaterials 29 (8) (2008) 1054–1064.
[12] T. Yucel, P. Cebe, D.L. Kaplan, Vortex-induced injectable silk fibroin hydrogels, Biophys. J. 97 (7) (2009) 2044–2050.
[13] K. Numata, T. Katashima, T. Sakai, State of water, molecular structure, and cytotoxicity of silk hydrogels, Biomacromolecules 12 (6) (2011) 2137–2144.
[14] H.J. Jin, S.V. Fridrikh, G.C. Rutledge, D.L. Kaplan, Electrospinning Bombyx mori silk with poly(ethylene oxide), Biomacromolecules 3 (6) (2002) 1233–1239.
[15] A.D. Malay, T. Suzuki, T. Katashima, N. Kono, K. Arakawa, K. Numata, Spider silk self-assembly via modular liquid-liquid phase separation and nanofibrillation, Sci. Adv. 6 (45) (2020) eabb6030.
[16] Y.W. Liu, J. Ren, S.J. Ling, Bioinspired and biomimetic silk spinning, Compos. Commun. 13 (2019) 85–96.
[17] J. Cheng, S.H. Lee, Development of new smart materials and spinning systems inspired by natural silks and their applications, Front. Mater. 2 (2016).
[18] C.C. Guo, C.M. Li, X. Mu, D.L. Kaplan, Engineering silk materials: from natural spinning to artificial processing, Appl. Phys. Rev. 7 (1) (2020) 011313.
[19] T. Lefevre, M. Auger, Spider silk as a blueprint for greener materials: a review, Int. Mater. Rev. 61 (2) (2016) 127–153.
[20] K. Yazawa, A.D. Malay, N. Ifuku, T. Ishii, H. Masunaga, T. Hikima, et al., Combination of amorphous silk fiber spinning and postspinning crystallization for tough regenerated silk fibers, Biomacromolecules 19 (6) (2018) 2227–2237.
[21] K. Numata, B. Subramanian, H.A. Currie, D.L. Kaplan, Bioengineered silk protein-based gene delivery systems, Biomaterials 30 (29) (2009) 5775–5784.
[22] K. Numata, S. Yamazaki, N. Naga, Biocompatible and biodegradable dual-drug release system based on silk hydrogel containing silk nanoparticles, Biomacromolecules 13 (5) (2012) 1383–1389.
[23] R. Rajkhowa, E.S. Gil, J. Kluge, K. Numata, L. Wang, X. Wang, et al., Reinforcing silk scaffolds with silk particles, Macromol. Biosci. 10 (6) (2010) 599–611.
[24] Y. Feng, X. Li, Q. Zhang, D. Ye, M. Li, R. You, et al., Fabrication of porous silk fibroin/cellulose nanofibril sponges with hierarchical structure using a lithium bromide solvent system, Cellulose 26 (2) (2019) 1013–1023.
[25] A. Koeppel, C. Holland, Progress and trends in artificial silk spinning: a systematic review, ACS Biomater. Sci. Eng. 3 (3) (2017) 226–237.
[26] G. Fang, Y. Huang, Y. Tang, Z. Qi, J. Yao, Z. Shao, et al., Insights into silk formation process: correlation of mechanical properties and structural evolution during artificial spinning of silk fibers, ACS Biomater. Sci. Eng. 2 (11) (2016) 1992–2000.
[27] G. Zhou, Z. Shao, D.P. Knight, J. Yan, X. Chen, Silk fibers extruded artificially from aqueous solutions of regenerated Bombyx mori silk fibroin are tougher than their natural counterparts, Adv. Mater. 21 (3) (2009) 366–370.
[28] H. Zhou, Z.-z Shao, X. Chen, Wet-spinning of regenerated silk fiber from aqueous silk fibroin solutions: influence of calcium ion addition in spinning dope on the performance of regenerated silk fiber, Chin. J. Polym. Sci. 32 (1) (2014) 29–34.
[29] G. Fukuhara, T. Nakamura, C. Yang, T. Mori, Y. Inoue, Diastereodifferentiating photocyclodimerization of 2-anthracenecarboxylate tethered to cellulose scaffold, J. Org. Chem. 75 (12) (2010) 4307–4310.
[30] Z. Chen, H. Zhang, Z. Lin, Y. Lin, J.H. van Esch, X.Y. Liu, Programing performance of silk fibroin materials by controlled nucleation, Adv. Funct. Mater. 26 (48) (2016) 8978–8990.

[31] F. Zhang, Q. Lu, X. Yue, B. Zuo, M. Qin, F. Li, et al., Regeneration of high-quality silk fibroin fiber by wet spinning from CaCl2–formic acid solvent, Acta Biomater. 12 (2015) 139–145.
[32] R. Madurga, A.M. Gañán-Calvo, G.R. Plaza, J.M. Atienza, G.V. Guinea, M. Elices, et al., Comparison of the effects of post-spinning drawing and wet stretching on regenerated silk fibers produced through straining flow spinning, Polymer 150 (2018) 311–317.
[33] S. Ling, L. Zhou, W. Zhou, Z. Shao, X. Chen, Conformation transition kinetics and spinnability of regenerated silk fibroin with glycol, glycerol and polyethylene glycol, Mater. Lett. 81 (2012) 13–15.
[34] J.M. Chen, Y. Ohta, H. Nakamura, H. Masunaga, K. Numata, Aqueous spinning system with a citrate buffer for highly extensible silk fibers, Polym. J. 53 (2020) 179–189.
[35] C.H. Bowen, B. Dai, C.J. Sargent, W. Bai, P. Ladiwala, H. Feng, et al., Recombinant spidroins fully replicate primary mechanical properties of natural spider silk, Biomacromolecules 19 (9) (2018) 3853–3860.
[36] X.-X. Xia, Z.-G. Qian, C.S. Ki, Y.H. Park, D.L. Kaplan, S.Y. Lee, Native-sized recombinant spider silk protein produced in metabolically engineered *Escherichia coli* results in a strong fiber, Proc. Natl. Acad. Sci. 107 (32) (2010) 14059–14063.
[37] A. Heidebrecht, L. Eisoldt, J. Diehl, A. Schmidt, M. Geffers, G. Lang, et al., Biomimetic fibers made of recombinant spidroins with the same toughness as natural spider silk, Adv. Mater. 27 (13) (2015) 2189–2194.
[38] M. Andersson, Q. Jia, A. Abella, X.-Y. Lee, M. Landreh, P. Purhonen, et al., Biomimetic spinning of artificial spider silk from a chimeric minispidroin, Nat. Chem. Biol. 13 (3) (2017) 262.
[39] F. Teulé, W.A. Furin, A.R. Cooper, J.R. Duncan, R.V. Lewis, Modifications of spider silk sequences in an attempt to control the mechanical properties of the synthetic fibers, J. Mater. Sci. 42 (21) (2007) 8974–8985.
[40] L. Xu, J.K. Rainey, Q. Meng, X.-Q. Liu, Recombinant minimalist spider wrapping silk proteins capable of native-like fiber formation, PLoS One 7 (11) (2012).
[41] E. DeSimone, T.B. Aigner, M. Humenik, G. Lang, T. Scheibel, Aqueous electrospinning of recombinant spider silk proteins, Mater. Sci. Eng. C 106 (2020) 110145.
[42] M. Andersson, G. Chen, M. Otikovs, M. Landreh, K. Nordling, N. Kronqvist, et al., Carbonic anhydrase generates CO_2 and H^+ that drive spider silk formation via opposite effects on the terminal domains, PLoS Biol. 12 (8) (2014) e1001921.
[43] F. Vollrath, D.P. Knight, X.W. Hu, Silk production in a spider involves acid bath treatment, Proc. R. Soc. London. Ser. B Biol. Sci. 265 (1398) (1998) 817–820.
[44] E.K. Tillinghast, S.F. Chase, M.A. Townley, Water extraction by the major ampullate duct during silk formation in the spider, *Argiope aurantia* Lucas, J. Insect Physiol. 30 (7) (1984) 591–596.
[45] D.P. Knight, F. Vollrath, Changes in element composition along the spinning duct in a *Nephila* spider, Die Naturwissenschaften 88 (4) (2001) 179–182.
[46] F. Vollrath, D.P. Knight, Liquid crystalline spinning of spider silk, Nature 410 (6828) (2001) 541–548.
[47] N. Kojic, M. Kojic, S. Gudlavalleti, G. McKinley, Solvent removal during synthetic and *Nephila* fiber spinning, Biomacromolecules 5 (5) (2004) 1698–1707.
[48] J. Sparkes, C. Holland, Analysis of the pressure requirements for silk spinning reveals a pultrusion dominated process, Nat. Commun. 8 (1) (2017) 594.
[49] G. Askarieh, M. Hedhammar, K. Nordling, A. Saenz, C. Casals, A. Rising, et al., Self-assembly of spider silk proteins is controlled by a pH-sensitive relay, Nature 465 (7295) (2010) 236–238.
[50] L. Eisoldt, J.G. Hardy, M. Heim, T.R. Scheibel, The role of salt and shear on the storage and assembly of spider silk proteins, J. Struct. Biol. 170 (2) (2010) 413–419.

[51] J.H. Exler, D. Hümmerich, T. Scheibel, The amphiphilic properties of spider silks are important for spinning, Angew. Chem. (Int. Ed. Engl.) 46 (19) (2007) 3559–3562.
[52] F. Hagn, L. Eisoldt, J.G. Hardy, C. Vendrely, M. Coles, T. Scheibel, et al., A conserved spider silk domain acts as a molecular switch that controls fibre assembly, Nature 465 (7295) (2010) 239–242.
[53] D. Huemmerich, C.W. Helsen, S. Quedzuweit, J. Oschmann, R. Rudolph, T. Scheibel, Primary structure elements of spider dragline silks and their contribution to protein solubility, Biochemistry 43 (42) (2004) 13604–13612.
[54] K. Jaudzems, G. Askarieh, M. Landreh, K. Nordling, M. Hedhammar, H. Jornvall, et al., pH-dependent dimerization of spider silk N-terminal domain requires relocation of a wedged tryptophan side chain, J. Mol. Biol. 422 (4) (2012) 477–487.
[55] N.A. Oktaviani, A. Matsugami, A.D. Malay, F. Hayashi, D.L. Kaplan, K. Numata, Conformation and dynamics of soluble repetitive domain elucidates the initial beta-sheet formation of spider silk, Nat. Commun. 9 (1) (2018) 2121.
[56] S. Rammensee, U. Slotta, T. Scheibel, A.R. Bausch, Assembly mechanism of recombinant spider silk proteins, Proc. Natl. Acad. Sci. U S A 105 (18) (2008) 6590–6595.
[57] A. Rising, J. Johansson, Toward spinning artificial spider silk, Nat. Chem. Biol. 11 (5) (2015) 309–315.
[58] F. Vollrath, D. Porter, C. Holland, There are many more lessons still to be learned from spider silks, Soft Matter 7 (20) (2011) 9595–9600.
[59] M.B. Hinman, R.V. Lewis, Isolation of a clone encoding a second dragline silk fibroin. *Nephila clavipes* dragline silk is a two-protein fiber, J. Biol. Chem. 267 (27) (1992) 19320–19324.
[60] A.D. Malay, K. Arakawa, K. Numata, Analysis of repetitive amino acid motifs reveals the essential features of spider dragline silk proteins, PLoS One 12 (8) (2017) e0183397.
[61] H. Sogawa, K. Nakano, A. Tateishi, K. Tajima, K. Numata, Surface analysis of native spider draglines by FE-SEM and XPS, Front. Bioeng. Biotech. 8 (2020) 231.
[62] S. Alberti, A. Gladfelter, T. Mittag, Considerations and challenges in studying liquid-liquid phase separation and biomolecular condensates, Cell 176 (3) (2019) 419–434.
[63] Y. Shin, C.P. Brangwynne, Liquid phase condensation in cell physiology and disease, Science 357 (6357) (2017) eaaf4382.
[64] C.P. Brangwynne, P. Tompa, R.V. Pappu, Polymer physics of intracellular phase transitions, Nat. Phys. 11 (11) (2015) 899.
[65] I.C. Um, C.S. Ki, H. Kweon, K.G. Lee, D.W. Ihm, Y.H. Park, Wet spinning of silk polymer. II. Effect of drawing on the structural characteristics and properties of filament, Int. J. Biol. Macromol. 34 (1-2) (2004) 107–119.
[66] A.E. Brooks, S.M. Stricker, S.B. Joshi, T.J. Kamerzell, C.R. Middaugh, R.V. Lewis, Properties of synthetic spider silk fibers based on Argiope aurantia MaSp2, Biomacromolecules 9 (6) (2008) 1506–1510.
[67] C.G. Copeland, B.E. Bell, C.D. Christensen, R.V. Lewis, Development of a process for the spinning of synthetic spider silk, ACS Biomater. Sci. Eng. 1 (7) (2015) 577–584.
[68] A.M. Anton, A. Heidebrecht, N. Mahmood, M. Beiner, T. Scheibel, F. Kremer, Foundation of the outstanding toughness in biomimetic and natural spider silk, Biomacromolecules 18 (12) (2017) 3954–3962.
[69] K.E. Kadler, D.F. Holmes, J.A. Trotter, J.A. Chapman, Collagen fibril formation, Biochem. J. 316 (1996) 1–11.
[70] C.M. Li, C. Vepari, H.J. Jin, H.J. Kim, D.L. Kaplan, Electrospun silk-BMP-2 scaffolds for bone tissue engineering, Biomaterials 27 (16) (2006) 3115–3124.
[71] G.H. Altman, F. Diaz, C. Jakuba, T. Calabro, R.L. Horan, J.S. Chen, et al., Silk-based biomaterials, Biomaterials 24 (3) (2003) 401–416.

[72] M. Santin, A. Motta, G. Freddi, M. Cannas, In vitro evaluation of the inflammatory potential of the silk fibroin, J. Biomed. Mater. Res. 46 (3) (1999) 382–389.
[73] V. Karageorgiou, M. Tomkins, R. Fajardo, L. Meinel, B. Snyder, K. Wade, et al., Porous silk fibroin 3-D scaffolds for delivery of bone morphogenetic protein-2 in vitro and in vivo, J. Biomed. Mater. Res. A 78 (2) (2006) 324–334.
[74] C.P. Vepari, D.L. Kaplan, Covalently immobilized enzyme gradients within three-dimensional porous scaffolds, Biotechnol. Bioeng. 93 (6) (2006) 1130–1137.
[75] A. Wilz, E.M. Pritchard, T. Li, J.Q. Lan, D.L. Kaplan, D. Boison, Silk polymer-based adenosine release: therapeutic potential for epilepsy, Biomaterials 29 (26) (2008) 3609–3616.
[76] T.F. Li, G.Y. Ren, D.L. Kaplan, D. Boison, Human mesenchymal stem cell grafts engineered to release adenosine reduce chronic seizures in a mouse model of CA3-selective epileptogenesis, Epilepsy Res. 84 (2-3) (2009) 238–241.
[77] L. Uebersax, M. Mattotti, M. Papaloizos, H.P. Merkle, B. Gander, L. Meinel, Silk fibroin matrices for the controlled release of nerve growth factor (NGF), Biomaterials 28 (30) (2007) 4449–4460.
[78] L. Uebersax, H.P. Merkle, L. Meinel, Insulin-like growth factor I releasing silk fibroin scaffolds induce chondrogenic differentiation of human mesenchymal stem cells, J. Control. Release 127 (1) (2008) 12–21.
[79] B.B. Mandal, S.C. Kundu, Calcium alginate beads embedded in silk fibroin as 3D dual drug releasing scaffolds, Biomaterials 30 (28) (2009) 5170–5177.
[80] L. Uebersax, D.E. Fedele, C. Schumacher, D.L. Kaplan, H.P. Merkle, D. Boison, et al., The support of adenosine release from adenosine kinase deficient ES cells by silk substrates, Biomaterials 27 (26) (2006) 4599–4607.
[81] C. Szybala, E.M. Pritchard, T.A. Lusardi, T.F. Li, A. Wilz, D.L. Kaplan, et al., Antiepileptic effects of silk-polymer based adenosine release in kindled rats, Exp. Neurol. 219 (1) (2009) 126–135.
[82] S.Z. Lu, X.Q. Wang, Q. Lu, X. Hu, N. Uppal, F.G. Omenetto, et al., Stabilization of enzymes in silk films, Biomacromolecules 10 (5) (2009) 1032–1042.
[83] M. Hawker, C.C. Guo, D.L. Barreiro, F. Martin-Martinez, F. Omenetto, M. Buehler, et al., Transient probe-type bioelectronics implants: solvent-free strategy to coat transient metals with silk, Abstract of the Papers of the American Chemical Society, Vol. 256, 2018.
[84] J.J. Li, S.I. Roohani-Esfahani, K. Kim, D.L. Kaplan, H. Zreiqat, Silk coating on a bioactive ceramic scaffold for bone regeneration: effective enhancement of mechanical and in vitro osteogenic properties towards load-bearing applications, J. Tissue Eng. Regen. Med. 11 (6) (2017) 1741–1753.
[85] R. Elia, C.D. Michelson, A.L. Perera, M. Harsono, G.G. Leisk, G. Kugel, et al., Silk electrogel coatings for titanium dental implants, J. Biomater. Appl. 29 (9) (2015) 1247–1255.
[86] R. Elia, C.D. Michelson, A.L. Perera, T.F. Brunner, M. Harsono, G.G. Leisk, et al., Electrodeposited silk coatings for bone implants, J. Biomed. Mater. Res. B 103 (8) (2015) 1602–1609.
[87] B. Marelli, M.A. Brenckle, D.L. Kaplan, F.G. Omenetto, Silk fibroin as edible coating for perishable food preservation, Sci. Rep. 6 (2016) 25263.
[88] X.Y. Wang, X. Hu, A. Daley, O. Rabotyagova, P. Cebe, D.L. Kaplan, Nanolayer biomaterial coatings of silk fibroin for controlled release, J. Control. Release 121 (3) (2007) 190–199.
[89] X. Wang, X. Zhang, J. Castellot, I. Herman, M. Iafrati, D.L. Kaplan, Controlled release from multilayer silk biomaterial coatings to modulate vascular cell responses, Biomaterials 29 (7) (2008) 894–903.
[90] A.R. Franco, E.M. Fernandes, M.T. Rodrigues, F.J. Rodrigues, M.E. Gomes, I.B. Leonor, et al., Antimicrobial coating of spider silk to prevent bacterial attachment on silk surgical sutures, Acta Biomater. 99 (2019) 236–246.

[91] E.M. Pritchard, C. Szybala, D. Boison, D.L. Kaplan, Silk fibroin encapsulated powder reservoirs for sustained release of adenosine, J. Control. Release 144 (2) (2010) 159–167.
[92] A. Matsumoto, J. Chen, A.L. Collette, U.-J. Kim, G.H. Altman, P. Cebe, et al., Mechanisms of silk fibroin sol – gel transitions, J. Phys. Chem. B 110 (43) (2006) 21630–21638.
[93] X. Wang, J.A. Kluge, G.G. Leisk, D.L. Kaplan, Sonication-induced gelation of silk fibroin for cell encapsulation, Biomaterials 29 (8) (2008) 1054–1064.
[94] B.P. Partlow, C.W. Hanna, J.R. Kovacina, J.E. Moreau, M.B. Applegate, K.A. Burke, et al., Highly tunable elastomeric silk biomaterials, Adv. Funct. Mater. 24 (29) (2014) 4615–4624.
[95] K.A. Burke, D.C. Roberts, D.L. Kaplan, Silk Fibroin Aqueous-Based Adhesives Inspired by Mussel Adhesive Proteins, ACS Publications, 2015.
[96] L.A. Burzio, J.H. Waite, Cross-linking in adhesive quinoproteins: studies with model decapeptides, Biochemistry 39 (36) (2000) 11147–11153.
[97] K. Numata, P.J. Baker, Synthesis of adhesive peptides similar to those found in blue mussel (Mytilus edulis) using papain and tyrosinase, Biomacromolecules 15 (8) (2014) 3206–3212.
[98] J.H. Waite, X. Qin, Polyphosphoprotein from the adhesive pads of Mytilus edulis, Biochemistry 40 (9) (2001) 2887–2893.
[99] Z.H. Ayub, M. Arai, K. Hirabayashi, Quantitative structural-analysis and physical-properties of silk fibroin hydrogels, Polymer 35 (10) (1994) 2197–2200.
[100] Z.H. Ayub, M. Arai, K. Hirabayashi, Mechanism of the gelation of fibroin solution, Biosci. Biotechnol. Biochem. 57 (11) (1993) 1910–1912.
[101] X.A. Hu, Q.A. Lu, L. Sun, P. Cebe, X.Q. Wang, X.H. Zhang, et al., Biomaterials from ultrasonication-induced silk fibroin-hyaluronic acid hydrogels, Biomacromolecules 11 (11) (2010) 3178–3188.
[102] U.J. Kim, J.Y. Park, C.M. Li, H.J. Jin, R. Valluzzi, D.L. Kaplan, Structure and properties of silk hydrogels, Biomacromolecules 5 (3) (2004) 786–792.
[103] K. Numata, N. Ifuku, H. Masunaga, T. Hikima, T. Sakai, Silk resin with hydrated dual chemical-physical cross-links achieves high strength and toughness, Biomacromolecules 18 (6) (2017) 1937–1946.
[104] Y.S. Kim, L. Dong, M.A. Hickner, T.E. Glass, V. Webb, J.E. McGrath, State of water in disulfonated poly(arylene ether sulfone) copolymers and a perfluorosulfonic acid copolymer (Nafion) and its effect on physical and electrochemical properties, Macromolecules 36 (17) (2003) 6281–6285.
[105] I.D. Kuntz, Hydration of macromolecules. III. Hydration of polypeptides, J. Am. Chem. Soc. 93 (2) (1971) 514–516.
[106] K.Y. Lee, W.S. Ha, DSC studies on bound water in silk fibroin/S-carboxymethyl kerateine blend films, Polymer 40 (14) (1999) 4131–4134.
[107] N. Agarwal, D.A. Hoagland, R.J. Farris, Effect of moisture absorption on the thermal properties of Bombyx mori silk fibroin films, J. Appl. Polym. Sci. 63 (3) (1997) 401–410.
[108] C. Fu, D. Porter, Z. Shao, Moisture effects on Antheraea pernyi silk's mechanical property, Macromolecules 42 (20) (2009) 7877–7880.
[109] J. Guan, D. Porter, F. Vollrath, Thermally induced changes in dynamic mechanical properties of native silks, Biomacromolecules 14 (3) (2013) 930–937.
[110] X. Hu, D. Kaplan, P. Cebe, Effect of water on the thermal properties of silk fibroin, Thermochim. Acta 461 (1) (2007) 137–144.
[111] X. Hu, K. Shmelev, L. Sun, E.S. Gil, S.H. Park, P. Cebe, et al., Regulation of silk material structure by temperature-controlled water vapor annealing, Biomacromolecules 12 (5) (2011) 1686–1696.

[112] W. Huang, S. Krishnaji, O.R. Tokareva, D. Kaplan, P. Cebe, Influence of water on protein transitions: morphology and secondary structure, Macromolecules 47 (22) (2014) 8107–8114.
[113] W. Huang, S. Krishnaji, O.R. Tokareva, D. Kaplan, P. Cebe, Influence of water on protein transitions: thermal analysis, Macromolecules 47 (22) (2014) 8098–8106.
[114] C. Mo, P. Wu, X. Chen, Z. Shao, The effect of water on the conformation transition of Bombyx mori silk fibroin, Vib. Spectrosc. 51 (1) (2009) 105–109.
[115] G.R. Plaza, G.V. Guinea, J. Pérez-Rigueiro, M. Elices, Thermo-hygro-mechanical behavior of spider dragline silk: glassy and rubbery states, J. Polym. Sci. Part B Polym. Phys. 44 (6) (2006) 994–999.
[116] Q. Yuan, J. Yao, L. Huang, X. Chen, Z. Shao, Correlation between structural and dynamic mechanical transitions of regenerated silk fibroin, Polymer 51 (26) (2010) 6278–6283.
[117] T. Asakura, M. Demura, Y. Watanabe, K. Sato, H-1 pulsed NMR-study of Bombyx-mori silk fibroin—dynamics of fibroin and of absorbed water, J. Polym. Sci. Pol. Phys. 30 (7) (1992) 693–699.
[118] M. Ishida, T. Asakura, M. Yokoi, H. Saito, Solvent-induced and mechanical-treatment-induced conformational transition of silk fibroins studied by high-resolution solid-state C-13 NMR-spectroscopy, Macromolecules 23 (1) (1990) 88–94.
[119] H. Yoshimizu, T. Asakura, The structure of Bombyx-mori silk fibroin membrane swollen by water studied with ESR, C-13-NMR, and FT-IR spectroscopies, J. Appl. Polym. Sci. 40 (9-10) (1990) 1745–1756.
[120] X. Hu, D. Kaplan, P. Cebe, Dynamic protein – water relationships during β-sheet formation, Macromolecules 41 (11) (2008) 3939–3948.
[121] D.J. Raven, C. Earland, M. Little, Occurrence of dityrosine in tussah silk fibroin and keratin, Biochim. Biophys. Acta 251 (1) (1971) 96.
[122] C.S. Wang, N.N. Ashton, R.B. Weiss, R.J. Stewart, Peroxinectin catalyzed dityrosine cross-linking in the adhesive underwater silk of a casemaker caddisfly larvae, Hysperophylax occidentalis, Insect Biochem. Mol. Biol. 54 (2014) 69–79.
[123] D. Balasubramanian, R. Kanwar, Molecular pathology of dityrosine cross-links in proteins: Structural and functional analysis of four proteins, Mol. Cell Biochem. 234 (1) (2002) 27–38.
[124] B.P. Partlow, M. Bagheri, J.L. Harden, D.L. Kaplan, Tyrosine templating in the self-assembly and crystallization of silk fibroin, Biomacromolecules 17 (11) (2016) 3570–3579.
[125] B.B. Mandal, A. Grinberg, E.S. Gil, B. Panilaitis, D.L. Kaplan, High-strength silk protein scaffolds for bone repair, Proc. Natl. Acad. Sci. U S A 109 (20) (2012) 7699–7704.
[126] K. Numata, S. Yamazaki, T. Katashima, J.-A. Chuah, N. Naga, T. Sakai, Silk-pectin hydrogel with superior mechanical properties, biodegradability, and biocompatibility, Macromol. Biosci. 14 (6) (2014) 799–806.
[127] Y. Termonia, Molecular modeling of spider silk elasticity, Macromolecules 27 (25) (1994) 7378–7381.
[128] A.T. Nguyen, Q.L. Huang, Z. Yang, N.B. Lin, G.Q. Xu, X.Y. Liu, Crystal networks in silk fibrous materials: from hierarchical structure to ultra performance, Small 11 (9-10) (2015) 1039–1054.
[129] C.G. Vonk, Computerization of Rulands X-ray method for determination of crystallinity in polymers, J. Appl. Crystallogr. 6 (1973) 148–152.
[130] V. Gupta, A. Aseh, C.N. Rios, B.B. Aggarwal, A.B. Mathur, Fabrication and characterization of silk fibroin-derived curcumin nanoparticles for cancer therapy, Int. J. Nanomed. 4 (1) (2009) 115–122.

[131] J. Kundu, Y.I. Chung, Y.H. Kim, G. Taeb, S.C. Kundu, Silk fibroin nanoparticles for cellular uptake and control release, Int. J. Pharm. 388 (1-2) (2010) 242–250.

[132] B.B. Mandal, S.C. Kundu, Self-assembled silk sericin/poloxamer nanoparticles as nanocarriers of hydrophobic and hydrophilic drugs for targeted delivery, Nanotechnology 20 (35) (2009).

[133] R. Anumolu, J.A. Gustafson, J.J. Magda, J. Cappello, H. Ghandehari, L.F. Pease, Fabrication of highly uniform nanoparticles from recombinant silk-elastin-like protein polymers for therapeutic agent delivery, ACS Nano 5 (7) (2011) 5374–5382.

[134] Y.Q. Zhang, W.D. Shen, R.L. Xiang, L.J. Zhuge, W.J. Gao, W.B. Wang, Formation of silk fibroin nanoparticles in water-miscible organic solvent and their characterization, J. Nanopart. Res. 9 (5) (2007) 885–900.

[135] Y.Q. Zhang, Y.J. Wang, H.Y. Wang, L. Zhu, Z.Z. Zhou, Highly efficient processing of silk fibroin nanoparticle-L-asparaginase bioconjugates and their characterization as a drug delivery system, Soft Matter 7 (20) (2011) 9728–9736.

[136] L. Zhu, R.P. Hu, H.Y. Wang, Y.J. Wang, Y.Q. Zhang, Bioconjugation of neutral protease on silk fibroin nanoparticles and application in the controllable hydrolysis of sericin, J. Agr. Food Chem. 59 (18) (2011) 10298–10302.

[137] K. Numata, J. Hamasaki, B. Subramanian, D.L. Kaplan, Gene delivery mediated by recombinant silk proteins containing cationic and cell binding motifs, J. Control. Release 146 (1) (2010) 136–143.

[138] K. Numata, D.L. Kaplan, Silk-based gene carriers with cell membrane destabilizing peptides, Biomacromolecules 11 (11) (2010) 3189–3195.

[139] J. Wei, F. Chen, J.W. Shin, H. Hong, C. Dai, J. Su, et al., Preparation and characterization of bioactive mesoporous wollastonite—polycaprolactone composite scaffold, Biomaterials 30 (6) (2009) 1080–1088.

[140] T.P. Richardson, M.C. Peters, A.B. Ennett, D.J. Mooney, Polymeric system for dual growth factor delivery, Nat. Biotechnol. 19 (11) (2001) 1029–1034.

[141] J. Lehar, A.S. Krueger, W. Avery, A.M. Heilbut, L.M. Johansen, E.R. Price, et al., Synergistic drug combinations tend to improve therapeutically relevant selectivity, Nat. Biotechnol. 27 (7) (2009) 659–666.

[142] W.E. Hennink, C.F. van Nostrum, Novel cross-linking methods to design hydrogels, Adv. Drug. Deliv. Rev. 54 (1) (2002) 13–36.

[143] S.C. Glotzer, M.J. Solomon, Anisotropy of building blocks and their assembly into complex structures, Nat. Mater. 6 (8) (2007) 557–562.

[144] G. Huang, J. Gao, Z. Hu, J.V. St John, B.C. Ponder, D. Moro, Controlled drug release from hydrogel nanoparticle networks, J. Control. Release 94 (2-3) (2004) 303–311.

[145] M.S. Muthu, M.K. Rawat, A. Mishra, S. Singh, PLGA nanoparticle formulations of risperidone: preparation and neuropharmacological evaluation, Nanomedicine 5 (3) (2009) 323–333.

[146] C. Wang, N.T. Flynn, R. Langer, Controlled structure and properties of thermoresponsive nanoparticle-hydrogel composites, Adv. Mater. 16 (13) (2004) 1074.

[147] J.T. Zhang, S.W. Huang, Y.N. Xue, R.X. Zhuo, Poly(N-isopropylacrylamide) nanoparticle-incorporated PNIPAAm hydrogels with fast shrinking kinetics, Macromol. Rapid Comm. 26 (16) (2005) 1346–1350.

[148] K. Numata, P. Cebe, D.L. Kaplan, Mechanism of enzymatic degradation of beta-sheet crystals, Biomaterials 31 (10) (2010) 2926–2933.

[149] T. Hino, M. Tanimoto, S. Shimabayashi, Change in secondary structure of silk fibroin during preparation of its microspheres by spray-drying and exposure to humid atmosphere, J. Colloid Interf. Sci. 266 (1) (2003) 68–73.

[150] X.Q. Wang, E. Wenk, A. Matsumoto, L. Meinel, C.M. Li, D.L. Kaplan, Silk microspheres for encapsulation and controlled release, J. Control. Release 117 (3) (2007) 360–370.

[151] X.Q. Wang, E. Wenk, X.H. Zhang, L. Meinel, G. Vunjak-Novakovic, D.L. Kaplan, Growth factor gradients via microsphere delivery in biopolymer scaffolds for osteochondral tissue engineering, J. Control. Release 134 (2) (2009) 81–90.

[152] X.Q. Wang, T. Yucel, Q. Lu, X. Hu, D.L. Kaplan, Silk nanospheres and microspheres from silk/pva blend films for drug delivery, Biomaterials 31 (6) (2010) 1025–1035.

[153] M.J. Harrington, A. Masic, N. Holten-Andersen, J.H. Waite, P. Fratzl, Iron-clad fibers: a metal-based biological strategy for hard flexible coatings, Science 328 (5975) (2010) 216–220.

[154] H.G. Silverman, F.F. Roberto, Understanding marine mussel adhesion, Mar. Biotechnol. 9 (6) (2007) 661–681.

[155] L.A. Burzio, J.H. Waite, Cross-linking in adhesive quinoproteins: studies with model decapeptides, Biochemistry 39 (36) (2000) 11147–11153.

[156] L.O. Burzio, V.A. Burzio, T. Silva, L.A. Burzio, J. Pardo, Environmental bioadhesion: themes and applications, Curr. Opin. Biotechnol. 8 (3) (1997) 309–312.

[157] J.L. Dalsin, B.H. Hu, B.P. Lee, P.B. Messersmith, Mussel adhesive protein mimetic polymers for the preparation of nonfouling surfaces, J. Am. Chem. Soc. 125 (14) (2003) 4253–4258.

[158] N. Holten-Andersen, M.J. Harrington, H. Birkedal, B.P. Lee, P.B. Messersmith, K. Y.C. Lee, et al., pH-induced metal-ligand cross-links inspired by mussel yield self-healing polymer networks with near-covalent elastic moduli, Proc. Natl. Acad. Sci. U S A 108 (7) (2011) 2651–2655.

[159] K. Huang, B.P. Lee, D.R. Ingram, P.B. Messersmith, Synthesis and characterization of self-assembling block copolymers containing bioadhesive end groups, Biomacromolecules 3 (2) (2002) 397–406.

[160] M. Yu, T.J. Deming, Synthetic polypeptide mimics of marine adhesives, Macromolecules 31 (15) (1998) 4739–4745.

[161] M.E. Yu, J.Y. Hwang, T.J. Deming, Role of L-3,4-dihydroxyphenylalanine in mussel adhesive proteins, J. Am. Chem. Soc. 121 (24) (1999) 5825–5826.

[162] J. Wang, C. Liu, X. Lu, M. Yin, Co-polypeptides of 3,4-dihydroxyphenylalanine and L-lysine to mimic marine adhesive protein, Biomaterials 28 (23) (2007) 3456–3468.

[163] P.J. Flory, Principles of Polymer Chemistry, Cornell University Press, Ithaca, NY, 1953.

[164] H. Sogawa, N. Ifuku, K. Numata, 3,4-Dihydroxyphenylalanine (DOPA)-containing silk fibroin: its enzymatic synthesis and adhesion properties, ACS Biomater. Sci. Eng. 5 (11) (2019) 5644–5651.

[165] K. Numata, N. Ifuku, A. Isogai, Silk composite with a fluoropolymer as a water-resistant protein-based material, Polymers (Basel) 10 (4) (2018) 459.

[166] D.U. Shah, D. Porter, F. Vollrath, Can silk become an effective reinforcing fibre? A property comparison with flax and glass reinforced composites, Compos. Sci. Technol. 101 (2014) 173–183.

[167] C.J. Fu, D. Porter, X. Chen, F. Vollrath, Z.Z. Shao, Understanding the mechanical properties of Antheraea pernyi silk—from primary structure to condensed structure of the protein, Adv. Funct. Mater. 21 (4) (2011) 729–737.

[168] J.M. Gosline, P.A. Guerette, C.S. Ortlepp, K.N. Savage, The mechanical design of spider silks: from fibroin sequence to mechanical function, J. Exp. Biol. 202 (23) (1999) 3295–3303.

[169] F. Vollrath, B. Madsen, Z.Z. Shao, The effect of spinning conditions on the mechanics of a spider's dragline silk, Proc. R. Soc. B Biol. Sci. 268 (1483) (2001) 2339–2346.

[170] H.Y. Cheung, K.T. Lau, X.M. Tao, D. Hui, A potential material for tissue engineering: silkworm silk/PLA biocomposite, Compos. Part B Eng. 39 (6) (2008) 1026–1033.

[171] M.P. Ho, K.T. Lau, H. Wang, D. Bhattacharyya, Characteristics of a silk fibre reinforced biodegradable plastic, Compos. Part B Eng. 42 (2) (2011) 117–122.
[172] K. Yang, R.O. Ritchie, Y.Z. Gu, S.J. Wu, J. Guan, High volume-fraction silk fabric reinforcements can improve the key mechanical properties of epoxy resin composites, Mater. Des. 108 (2016) 470–478.
[173] K. Yang, J. Guan, K. Numata, C.G. Wu, S.J. Wu, Z.Z. Shao, et al., Integrating tough Antheraea pernyi silk and strong carbon fibres for impact-critical structural composites, Nat. Commun. 10 (2019).

Questions for this chapter

1. Protein materials are generally sensitive to water and humidity. How can we stabilize protein materials against water?
2. How can we process protein into specific material forms, even though the natural proteins do not show any melting behaviors like thermoplastics?

CHAPTER 9

Applications as bulk material and future perspective

9.1 Industrial applications

In the academic as well as industrial fields, protein-based materials have been studied widely by many researchers and also well reviewed by several groups [1–5]. Generally speaking, the natural structural proteins and their regenerated materials are considered to show specific physical and mechanical properties, as listed in Table 9.1. However, it is still challenging to reproduce the native physical properties and biological functions of a structural protein using regenerated or recombinant proteins, even though structural proteins can be synthesized by various methods, as introduced in Chapter 2, Synthesis. One of the major reasons is that the hierarchical structure and natural biological arrangement are difficult to reproduce by artificial processes based on the current scientific understanding. The polymeric material process needs to be developed and optimized to realize the desired physical and biological properties of the structural protein material like tough silk-based materials. In this chapter, several examples of the structural proteins that have already been commercialized or produced in bulk scale as a polymeric material as well as materials containing a structural protein are introduced. The industrial use of structural proteins in consumer products, such as cosmetics and haircare date back to the 1990s by major companies such as BASF, Danisco/Genencor, DuPont, and L'Oreal, based on silk-elastin-like polymers and resilins [6]. However, this chapter focuses on more structural material rather than the cosmetic and haircare products.

9.1.1 Spider silk

AMSilk GmbH is a German spinoff company based on scientific studies by Prof. Thomas Scheibel and colleagues and is the first industrial supplier of recombinant silk biopolymers across a range of applications including structural materials and also consumable products. AMSilk exploits *Araneus diadematus* (European garden spider) fibroin 4 (ADF4)-inspired

Biopolymer Science for Proteins and Peptides
DOI: https://doi.org/10.1016/B978-0-12-820555-6.00010-0

Table 9.1 Typical mechanical/physical properties and potential materials for structural proteins reviewed herein [4].

Protein	Mechanical/physical aspect	Potential material products
Silkworm silk	Strength	Fiber, textile (nanofiber), resin, gel, nano/microparticle, sponge, tube
Spider silk	Strength and toughness	Fiber, textile (nanofiber), resin, form, gel, nano/microparticle, sponge, tube
Collagen	Strength and softness	Gel, film, surface modification, scaffold, emulsion
Elastin	Elasticity	Gel, nanomaterials including nanofiber, scaffold
Resilin	Elasticity and resilience	Rubber, gel, film
Keratin	Hardness and strength	Sponge, resin
Reflectin	Optical property for reflection	Film

silks to generate high quality materials for apparel, cosmetic, industrial, and structural applications. Biosteel fiber is their leading product, and it is at various developmental stages for application in the footwear, automotive, and aircraft industries. AMSilk proceeded with a biomedical trial using their silk products including silk coating with ADF4-based silk (BioShield-S1) (SILKline) in several human subjects. According to the report, ADF4-coated silicone implants showed no acute toxicity or immunogenicity and they reduced postoperative inflammation and minimized implant-induced capsule thickness and contraction when compared to uncoated control implants [7].

Another important industrial pioneer is Spiber Technologies AB (Stockholm, Sweden). They produce Spiber silk biomaterials, which can be processed into a range of material formats, such as fibers, scaffolds, films, coatings, and foams. Their work is also applied to the preclinical Research and Development pipeline exploiting spider silk for biomedical applications including cell—matrix interactions, culture matrices, surface modifications, and electrochemical meshes [8,9]. Academic and industrial scientists affiliated with Spiber Technologies AB have been working on the molecular mechanisms of spider silk spinning to develop better biomimetic fibers [10,11]. The pH-sensitive *N*-terminal domain of spider silk (major ampullate spidroin) enables the silk fibroin to stay in solution even in the concentrated conditions. Based on this phenomenon, they developed SolvNT, which is a recombinant fusion technology of the *N*-terminal domain. SolvNT can convert an aggregation-prone therapeutic protein to its hypersoluble counterpart [12].

A rising star company in the spider silk field is Bolt Threads Inc. (Emeryville, CA, United States), which produces recombinant silk proteins by yeast expression systems and also developed silk fibers by wet spinning (Engineered Silk). Their recombinant silk is produced on a scale that makes it a viable contender for broad applications. Bolt Threads has already introduced and partially commercialized several Engineered Silk fiber and Bolt Microsilk products (Fig. 9.1). The Bolt Threads Microsilk tie is composed of 100% Microsilk. In 2017 50 ties were sold at US$314 each, and they sold out immediately. Best Made Co. x Bolt Threads Cap of Courage was released in 2017 with 100 colors and 10 patterns at US$198 each, and they sold out immediately as well. These hats are a composite of Microsilk and American Rambouillet wool. The Stella McCartney x Bolt Threads Gold Microsilk Dress is made of 100% Microsilk and was designed by Stella McCartney. It was presented and exhibited at "Items: Is Fashion Modern?" at the Museum of Contemporary Art in New York.

Spiber Inc. is a Japanese venture company and produces artificial silk fibers, QMONOS and Brewed Protein. They developed artificial silk fibers with various molecular designs and amino acid sequences for the practical use of apparel and materials. For instance, Spiber Inc. Moon Parka and T-shirts made of structural protein fibers and textiles were commercialized in 2019 (Fig. 9.2). In addition, Spiber Inc. exhibited artificial silk protein-based resin, film, and other materials, although these materials have not yet been commercialized as of January 2021.

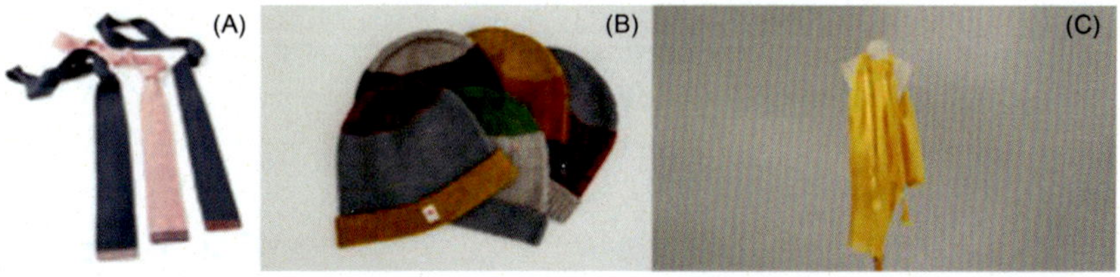

Figure 9.1 Products from Bolt Threads: Bolt Threads Microsilk tie (A), Best Made Co. x Bolt Threads Cap of Courage (B), and Stella McCartney x Bolt Threads Gold Microsilk Dress (C). *Source: The photographs were kindly provided by Dr. Lindsay Wray, Bolt Threads [4].*

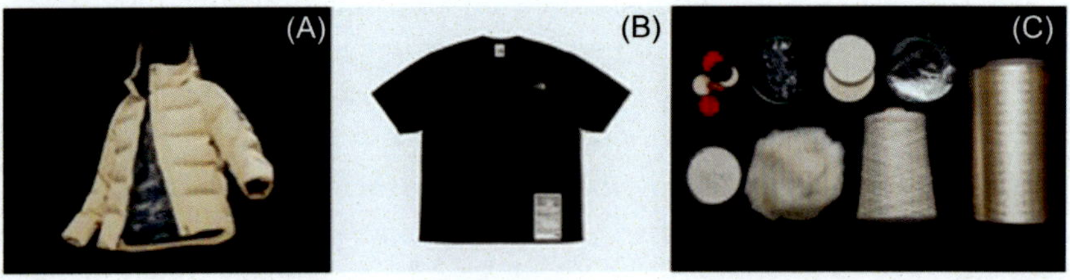

Figure 9.2 Appearance of Moon Parka (A), T-shirt (B), and structural protein materials such as resins and films (C). *Source: Provided by Hiroyuki Nakamura, Spiber Inc. [4]*

9.1.2 Silkworm silk

Silkworm silk fibers have been used as sutures, which has encouraged the use of silkworm silk materials for tissue engineering fields. One of the silkworm silk-based commercial materials is SERI Surgical Scaffold (Food and Drug Administration approved), the silk surgical mesh produced by Serica Technologies Inc. (Massachusetts, United States). Clinical products are manufactured by unwinding *Bombyx mori* silk cocoons fibers. The clinical performance of silkworm silk sutures has been reviewed in the literature [13,14]. The SERI Surgical Scaffold technology is derived from the various studies by David Kaplan (Tufts University). The current SERI Surgical Scaffold indications are for abdominal wall reconstruction and investigational plastic surgery applications, including total body contouring, brachioplasty, abdominoplasty, mastopexy, and breast reconstruction [14]. SERI Surgical Scaffold was examined in a sheep model of two-stage breast reconstruction, resulting in progressive degradation and vascularization of the silk mesh. The details are described in the report [15]. After 1 year, the SERI Surgical Scaffold had stimulated extensive type I collagen deposition, and the resulting tissue was mechanically robust. Clinical hernia repair in a horse showed incomplete SERI Surgical Scaffold degradation at 2 years postimplantation, but no hernia relapse [16].

Another silkworm silk product for clinical use is silk garments (DermaSilk and DreamSkin) for the treatment of dermatological conditions, especially atopic dermatitis [17,18] and acne vulgaris [19]. Mechanical skin irritation by harsh, rough (e.g., wool), and short (e.g., cotton) textile fibers are considered to cause atopic dermatitis [14]. Besides, the skin of atopic dermatitis patients is often colonized with *Staphylococcus aureus*, and the extent of colonization correlates with the severity of the disease. In contrast to wool and cotton, *B. mori* silkworm silk fibers are very long (up to 1500 m) and smooth, so they minimize mechanical irritation when knitted into clothing. Another commercial product for the treatment of atopic dermatitis is sericin-free silk fibers that are chemically functionalized with 3-trimethylsilylpropyl-dimethyloctadecyl ammonium chloride (AEM 5700/5772; AEGIS), known as DermaSilk. The silk of DermaSilk is highly purified to minimize the risk of contact dermatitis [20].

Silkworm silk-based materials are suitable for skin wound healing because of their hemostatic properties, low inflammatory potential, and permeability to oxygen and water. Also, silkworm silk functions as a barrier to bacterial colonization. In China, a silk fibroin sponge attached to a silicone membrane named Sidaiyi is a first-generation wound dressing approved by the China Food and Drug Administration [21]. Thin silk films have also been used in prospective human clinical trials to repair acute and chronic tympanic membrane perforations [22]. Daewoong-Bio (South Korea) produces the silk patches, Tympasil, using reverse-engineered *B. mori* silk fibroin. They performed the first clinical trial with acute traumatic tympanic membrane perforation with a silk film [23]. The silk patch significantly shortened the healing time in comparison to control samples [24]. A similar improved treatment for chronic traumatic tympanic membrane perforation with a silk patch was previously reported in animal studies [22].

9.1.3 Casein

Casein is not a structural protein but is applied to bulk materials. Casein plastic was invented in Germany in 1898 [4]. Acid is added to milk to solidify casein into a material. The precipitated casein becomes a thermoplastic with an appearance similar to ivory. This is called casein plastic or lactic casein and is used industrially as a material for impressions, clothing buttons, etc. (Fig. 9.3). Precisely, it is not a thermoplastic plastic, but it is

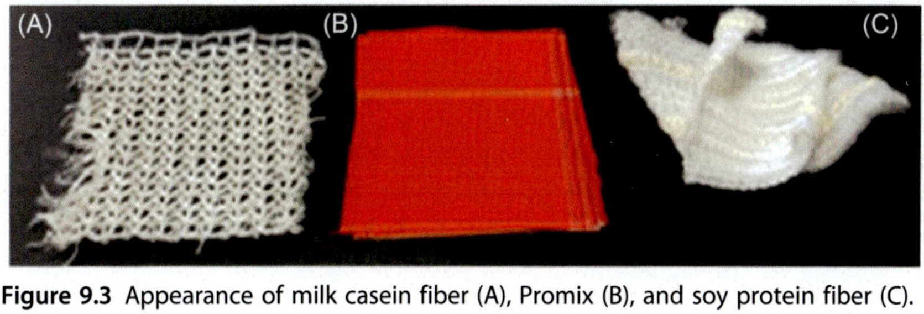

Figure 9.3 Appearance of milk casein fiber (A), Promix (B), and soy protein fiber (C).

recognized as a biodegradable, biologically derived material. A fiber made by graft polymerization of acrylonitrile on casein is called as Promix (Fig. 9.3), which was developed by Toyobo Co., Ltd. in the 1970s. The texture is similar to silk. Materials that combine Promix and bioplastics such as polylactic acid are also commercially available.

9.1.4 Soy protein fiber

Fibers have been produced using protein extracted from plants that contain a lot of protein such as soybeans. Textile materials have been developed from peanuts, corn, and soybeans. These proteins are difficult to be fibrillated as a single substance, and are commercialized by compounding with polyester or nylon, for example, a composite fiber of 20% soybean, 20% nylon, and 60% polyester.

9.2 Future perspective

In this textbook, various structural proteins have been introduced and explained in terms of chemistry, material science, biology, and also polymer science. A structural protein can be a globular or fibrillar protein and also is defined as "a protein that possesses a characteristic amino acid sequence or motif that repeats and forms a skeleton or contributes to the mechanical properties of a living organism, cell, or material."

Even though structural proteins such as silk and elastin are used as structural materials in nature, they have not been put into practical use as bulk-scale structural materials for human use. The challenges of using structural proteins as structural materials are distributed through their synthesis, purification, processing, and so on. In the future, many researchers and scientists are expected to enter this field and establish innovative synthetic methods and molecular designs.

The remarkable mechanical properties, versatile processing in an aqueous environment, biocompatibility, and controlled degradation of structural proteins suggest they are attractive biomedical materials for controlled delivery of bioactive molecules as well as tissue engineering. Hybrid or composite protein-based materials have not been extensively studied yet, but should provide applicable mechanical, thermal, and biological properties for drug/gene delivery and tissue engineering, medical imaging, and regenerative medicine.

As one of the structural proteins, spider silk has excellent mechanical properties due to its hierarchical structures. There have been many attempts to understand and utilize the excellent physicochemical properties of silk-based materials. To control and utilize the hierarchical structures responsible for silk characteristics, it is crucial to precisely synthesize silk-like polymers, analyze their high-order structures, and understand their dynamics. During this process, we need to extract information about the natural proteins to design novel materials.

References

[1] X. Hu, P. Cebe, A.S. Weiss, F. Omenetto, D.L. Kaplan, Protein-based composite materials, Mater. Today 15 (5) (2012) 208–215.

[2] N.C. Abascal, L. Regan, The past, present and future of protein-based materials, Open. Biol. 8 (10) (2018).

[3] J.C. van Hest, D.A. Tirrell, Protein-based materials, toward a new level of structural control, Chem. Commun. 19 (2001) 1897–1904.

[4] K. Numata, How to define and study structural proteins as biopolymer materials, Polym. J. 52 (2020) 1043–1056.

[5] K. Numata, Poly(amino acid)s/polypeptides as potential functional and structural materials, Polym. J. 47 (8) (2015) 537–545.

[6] D.N. Breslauer, Recombinant protein polymers: a coming wave of personal care ingredients, ACS Biomater. Sci. Eng. 6 (11) (2020) 5980–5986.

[7] P.H. Zeplin, N.C. Maksimovikj, M.C. Jordan, J. Nickel, G. Lang, A.H. Leimer, et al., Spider silk coatings as a bioshield to reduce periprosthetic fibrous capsule formation, Adv. Funct. Mater. 24 (18) (2014) 2658–2666.

[8] C. Muller, M. Hamedi, R. Karlsson, R. Jansson, R. Marcilla, M. Hedhammar, et al., Woven electrochemical transistors on silk fibers, Adv. Mater. 23 (7) (2011) 898.

[9] L. Nileback, J. Hedin, M. Widhe, L.S. Floderus, A. Krona, H. Bysell, et al., Self-assembly of recombinant silk as a strategy for chemical-free formation of bioactive coatings: a real-time study, Biomacromolecules 18 (3) (2017) 846–854.

[10] G. Askarieh, M. Hedhammar, K. Nordling, A. Saenz, C. Casals, A. Rising, et al., Self-assembly of spider silk proteins is controlled by a pH-sensitive relay, Nature 465 (7295) (2010) 236–U125.

[11] M. Andersson, Q.P. Jia, A. Abella, X.Y. Lee, M. Landreh, P. Purhonen, et al., Biomimetic spinning of artificial spider silk from a chimeric minispidroin, Nat. Chem. Biol. 13 (3) (2017) 262.

[12] N. Kronqvist, M. Sarr, A. Lindqvist, K. Nordling, M. Otikovs, L. Venturi, et al., Efficient protein production inspired by how spiders make silk, Nat. Commun. 8 (2017). Article number 15504.

[13] G.H. Altman, F. Diaz, C. Jakuba, T. Calabro, R.L. Horan, J.S. Chen, et al., Silk-based biomaterials, Biomaterials 24 (3) (2003) 401–416.

[14] C. Holland, K. Numata, J. Rnjak-Kovacina, F.P. Seib, The biomedical use of silk: past, present, future, Adv. Healthc. Mater. 8 (1) (2019) e1800465.

[15] J.E. Gross, R.L. Horan, M. Gaylord, R.E. Olsen, L.D. McGill, J.M. Garcia-Lopez, et al., An evaluation of SERI surgical scaffold for soft-tissue support and repair in an ovine model of two-stage breast reconstruction, Plast. Reconstr. Surg. 134 (5) (2014) 700e–704e.

[16] J. Haupt, J.M. Garcia-Lopez, K. Chope, Use of a novel silk mesh for ventral midline hernioplasty in a mare, BMC Vet. Res. 11 (2015) 379.

[17] D.Y. Koller, G. Halmerbauer, A. Bock, G. Engstler, Action of a silk fabric treated with AEGIS in children with atopic dermatitis: a 3-month trial, Pediatr. Allergy Immunol. 18 (4) (2007) 335–338.
[18] K.S. Thomas, L.E. Bradshaw, T.H. Sach, J.M. Batchelor, S. Lawton, E.F. Harrison, et al., Silk garments plus standard care compared with standard care for treating eczema in children: a randomised, controlled, observer-blind, pragmatic trial (CLOTHES Trial), PLoS Med. 14 (4) (2017) e1002280.
[19] C. Schaunig, D. Kopera, Silk textile with antimicrobial AEM5772/5 (Dermasilk): a pilot study with positive influence on acne vulgaris on the back, Int. J. Dermatol. 56 (5) (2017) 589–591.
[20] A. Inoue, I. Ishido, A. Shoji, H. Yamada, Textile dermatitis from silk, Contact Dermat. 37 (4) (1997) 185.
[21] W. Zhang, L.K. Chen, J.L. Chen, L.S. Wang, X.X. Gui, J.S. Ran, et al., Silk fibroin biomaterial shows safe and effective wound healing in animal models and a randomized controlled clinical trial, Adv. Healthc. Mater. 6 (10) (2017).
[22] J.H. Lee, D.K. Kim, H.S. Park, J.Y. Jeong, Y.K. Yeon, V. Kumar, et al., A prospective cohort study of the silk fibroin patch in chronic tympanic membrane perforation, Laryngoscope 126 (12) (2016) 2798–2803.
[23] J.H. Lee, J.S. Lee, D.K. Kim, C.H. Park, H.R. Lee, Clinical outcomes of silk patch in acute tympanic membrane perforation, Clin. Exp. Otorhinolaryngol. 8 (2) (2015) 117–122.
[24] J. Kim, C.H. Kim, C.H. Park, J.N. Seo, H. Kweon, S.W. Kang, et al., Comparison of methods for the repair of acute tympanic membrane perforations: silk patch vs. paper patch, Wound Repair Regen. 18 (1) (2010) 132–138.

CHAPTER 10

Experimental details

This chapter describes the practical examples of the experimental procedures that are introduced in this textbook. If readers need more details, the author recommends checking out the original papers cited here.

10.1 Chemical synthesis

10.1.1 Solid-phase peptide synthesis

The solid-phase method is one of the most common techniques used to synthesize relatively short polypeptides and peptides of up to about 50 amino acids [1,2]. In solid-phase peptide synthesis reaction, the *N*-terminal amino acid is attached to a solid matrix with the carboxyl group and the amine group is protected. The amine group undergoes a deprotection step revealing an *N*-terminal amine. This is followed by a coupling reaction between the activated carboxyl group of the next amino acid and the amine group of the immobilized residue. Side chains of several amino acids contain functional groups, which may interfere with the formation of amide bonds and must be protected. This process may continue through iterative cycles until the polypeptide has reached its desired chain length. The polypeptide is then cleaved off the resin and purified. One example of a solid-phase peptide synthesis is shown here: ZnO-binding β-annulus peptide [3]. The peptide-Ser (tBu)-Alko-PEG resin was synthesized on fluorenylmethyloxycarbonyl protecting group (Fmoc)-Ser(tBu)-Alko-PEG resin (Watanabe Chemical Industries, Ltd., Hiroshima, Japan; 0.21 mmol/g) using Fmoc-based coupling reactions (4 equivalent Fmoc amino acids). Solutions of (1-cyano-2-ethoxy-2-oxoethylidenaminooxy)dimethylamino-morpholino-carbenium hexafluorophosphate and diisopropylamine in *N*-methyl-2-pyrrolidone (NMP) were used as coupling reagents. A solution of 20% piperidine in *N*,*N*-dimethylformamide (DMF) was used for Fmoc deprotection. The progress of the coupling reaction and Fmoc deprotection was confirmed using TNBS (2,4,6-trinitrobenzene sulfonic acid) and the Chloranil Test Kit (Tokyo Chemical Industry Co., Ltd., Tokyo, Japan). The peptidyl resin was washed with NMP. The peptide was deprotected and cleaved from the resin by

Biopolymer Science for Proteins and Peptides
DOI: https://doi.org/10.1016/B978-0-12-820555-6.00007-0

treatment with a mixture of trifluoroacetic acid (TFA)/thioanisole/water/1,2-ethanedithiol/triisopropylsilane = 8.15/0.5/0.5/0.25/0.1 at room temperature for 3 h. The reaction mixture was filtered to remove the resin, and the filtrate was concentrated under vacuum. The peptide was precipitated by adding methyl *tert*-butyl ether (MTBE) to the residue, and the supernatant was decanted. After the washing with MTBE was repeated three times, the precipitated peptide was dried under vacuum. The crude product was purified by reverse-phase high-performance liquid chromatography (Inertsil WP300 C18; GL Science Inc., Tokyo, Japan), eluting with a linear gradient of acetonitrile (CH_3CN)/water containing 0.1% TFA (23.5/76.5 to 25.5/74.5 over 100 min). The fraction containing the desired peptide was lyophilized to give 6.2 mg of a flocculent solid (6.2% yield). Matrix-assisted laser desorption/ionization time-of-flight mass spectrometry (MALDI-TOF-MS) (matrix: α-cyano-4-hydroxycinnamic acid (α-CHCA)): $m/z = 3181$ $[M]^+$.

10.1.2 Liquid-phase peptide synthesis

10.1.2.1 Synthesis of dimeric peptide Nyl3Ala-OEt

N-Boc-Nylon3 was synthesized according to a previously reported procedure [4]. To a flask equipped with an addition funnel and a stir bar were added *N*-Boc-Nylon3 (25.3 g, 134 mmol), L-alanine ethyl ester hydrochloride (20.5 g, 134 mmol), 1-hydroxybenzotriazole (HOBt) monohydrate (18.0 g, 134 mmol), triethylamine (18.6 mL, 134 mmol), and chloroform (125 mL) at −10°C under nitrogen. A solution of 1-ethyl-3-(3-dimethylaminopropyl)carbodiimide (water-soluble carbodiimide, WSCI) hydrochloride (25.6 g, 134 mmol) in chloroform (125 mL) was added dropwise over 30 min, and the resulting mixture was stirred at −10°C for 30 min and then at 25°C for 24 h. The mixture was washed sequentially with water, 5% $NaHCO_3$ aqueous, and brine. The organic layer was dried with Na_2SO_4 and concentrated using a rotary evaporator. The product was dried in vacuo to give Boc-Nyl3Ala-OEt with a yield of 36.2 g (94.0%). The obtained Boc-Nyl3Ala-OEt was then subjected to deprotection of the Boc group. To the solution of Boc-Nyl3Ala-OEt (34.0 g, 118 mmol) in dichloromethane (84.0 mL) was slowly added TFA (45.1 mL, 590 mmol) at 0°C under nitrogen. The mixture was stirred at 0°C for 10 min and then at 25°C for 24 h. After the solvent was removed under reduced pressure, the crude product was dissolved in dioxane/HCl (4.0 M, 68.0 mL). The solution was poured into diethyl ether. The precipitate was filtered, washed with diethyl ether, and dried in vacuo to afford Nyl3Ala-OEt as a white hygroscopic solid. The yield was 21.5 g (97.0%).

10.1.3 α-Amino acid-*N*-carboxylic anhydride (NCA) ring-opening polymerization

As described in Chapter 2, Synthesis, NCA ring-opening polymerization is a powerful technique to synthesize polypeptides [5,6]. It is possible to synthesize various polypeptides by devising an initiator amine [5], while this technique can be used to synthesize polypeptides composed of one or a few amino acid residue(s) rather than sequentially controlled polypeptides. Here shows an example of synthesis of PEG-poly(β-benzyl-L-aspartate) (PEG-PBLA) [7]. PEG-PBLA was synthesized from the ring-opening polymerization of β-benzyl-L-aspartate *N*-carboxy-anhydride (BLA-NCA). NCA preparation was according to a previous report by Daly and Poche [8]. Polymerization of BLA-NCA was initiated by the terminal primary amino group of α-methoxy-ω-amino poly(ethylene glycol) under argon atmosphere in distilled DMF.

10.2 Biochemical synthesis

10.2.1 Proteinase K-catalyzed chemoenzymatic synthesis of oligo(L-cysteine) (OligoCys) without side-chain protection

The proteinase K-catalyzed oligomerization of L-Cys-OEt was carried out according to previous studies [9,10]. In these reactions, the side-chain protection is not needed in contrast to the general chemical peptide synthesis (Fig. 10.1).

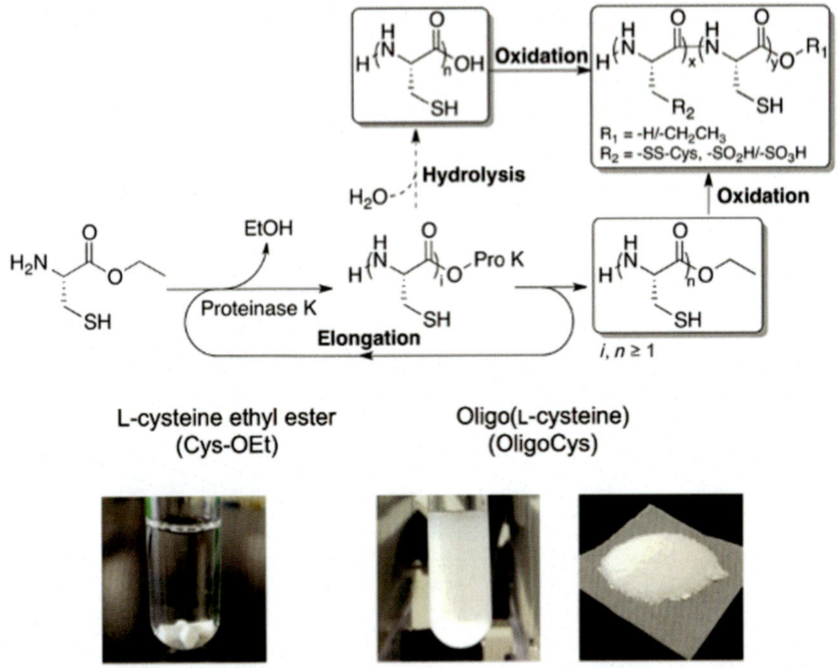

Figure 10.1 Mechanism of proteinase K-catalyzed chemoenzymatic synthesis of OligoCys and oxidation by H_2O_2 [10].

Typically, L-Cys-OEt · HCl (1114.0 mg, 6.0 mmol) was dissolved in 5 mL of 1.0 M potassium phosphate buffer (pH 8), and potassium hydroxide was added to adjust the initial pH to 8.0. The solutions were then transferred into a 25-mL glass reaction tube and proteinase K (23 units/mL) was added into monomer solutions prewarmed at 40°C. Reactions were performed with gentle stirring in EYELA ChemiStation (Tokyo, Japan) at 40°C for 1.5 h. The blank reaction without adding protease K was also performed. Next, 1 mL of 6 M HCl solution was added to inactivate the enzyme, and the reaction mixture was cooled to room temperature. Precipitates were collected by centrifugation (10,000 × *g*, 25°C for 5 min), washed once with 0.1 M HCl and twice with Milli-Q water, and then lyophilized to obtain OligoCys as white powders.

10.2.2 Chemoenzymatic polymerization-mediated synthesis of multiblock copolypeptides

10.2.2.1 Synthesis of the polyAla by chemoenzymatic polymerization

A solution of alanine ethyl ester hydrochloride (5.53 g, 36 mmol) in phosphate buffer (12 mL, 1.0 M, pH = 8.0) and methanol (3.0 mL) was placed in a glass tube equipped with a stirring bar. To this was added a solution of papain (1.80 g) in phosphate buffer (12.8 mL) in one portion. The final concentrations of alanine and papain were 1.0 M and 50 mg mL^{-1}, respectively, in a total volume of 36 mL. The mixture was stirred at 800 rpm and 40°C for 2 h using an EYELA ChemiStation PPS-5511 (Tokyo Rikakikai Co. Ltd., Tokyo, Japan). After cooling to room temperature, the precipitate was collected by centrifuging at 7000 rpm and 4°C for 10 min. The crude product was washed twice with deionized water, centrifuged, and lyophilized to give a polyalanine with an ethyl ester terminal as a white powder. The yield was 0.678 g (27%) (Scheme 10.1).

10.2.2.2 Synthesis of poly(Gly-r-Leu) by chemoenzymatic polymerization

A solution of glycine ethyl ester hydrochloride and leucine ethyl ester hydrochloride in phosphate buffer (3.0 mL, 1.0 M, pH = 8.0) was placed

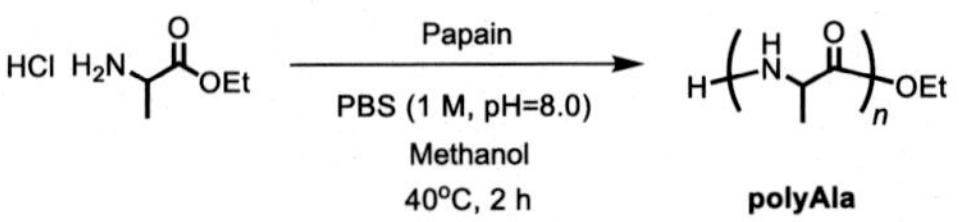

Scheme 10.1 Chemoenzymatic polymerization of alanine ethyl ester using papain.

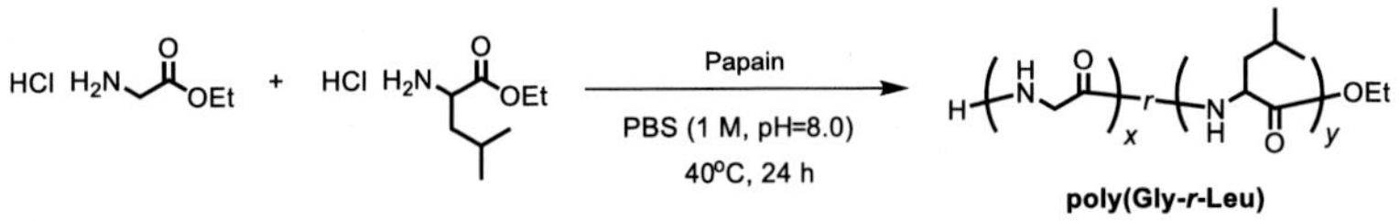

Scheme 10.2 Chemoenzymatic copolymerization of glycine and leucine ethyl esters using papain.

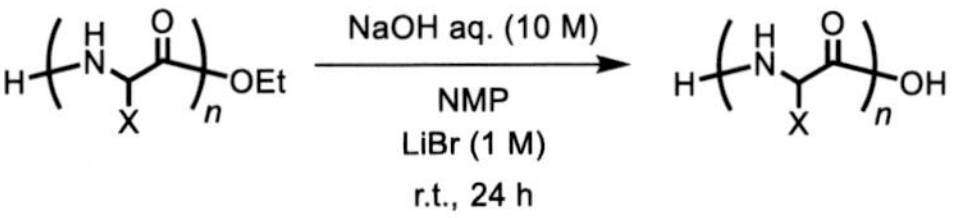

Scheme 10.3 Terminal hydrolysis of polypeptide fragments using NaOH.

in a glass tube equipped with a stirring bar. To this was added a solution of papain (0.30 g) in phosphate buffer in one portion. The concentration of papain was 50 mg mL in a total volume of 6.0 mL. The mixture was stirred at 800 rpm and 4°C for 24 h. After cooling to room temperature, the precipitate was collected by centrifuging at 7000 rpm and 4°C for 15 min. The polymerization was performed on the various feed ratios of glycine and leucine, where the final concentration of the amino acids was kept constant at 1.0 M (Scheme 10.2).

10.2.2.3 Hydrolysis of the terminal ester in polypeptides

A representative procedure using polyAla follows. To a 20 mL flask equipped with a stirring bar and a stopcock were added polyAla (ethyl ester terminal, 0.290 g), lithium bromide (0.521 g, 6 mmol), and NMP (6 mL) under nitrogen. With vigorous stirring, an aqueous solution of sodium hydroxide (10 M, 0.25 mL) was added to the solution. After the mixture was stirred at room temperature for 24 h, water was added to precipitate the polyAla. The crude product was collected by centrifuging at 7000 rpm and 4°C for 15 min, and washed with water followed by centrifugation. A white powdery product was obtained after lyophilization. The yield was 0.258 g (94%). Poly(Gly-*r*-Leu) was also hydrolyzed by the same procedure above (Scheme 10.3).

10.2.2.4 Post-polycondensation of polypeptide fragments using water-soluble carbodiimide

PolyAla (50 mg) and poly(Gly-*r*-Leu) (50 mg) with a hydrolyzed *C*-terminal were dissolved in NMP (1 mL) containing lithium bromide (0.0868 g, 1 mmol) in a 10 mL flask equipped with a stirring bar and a three-way

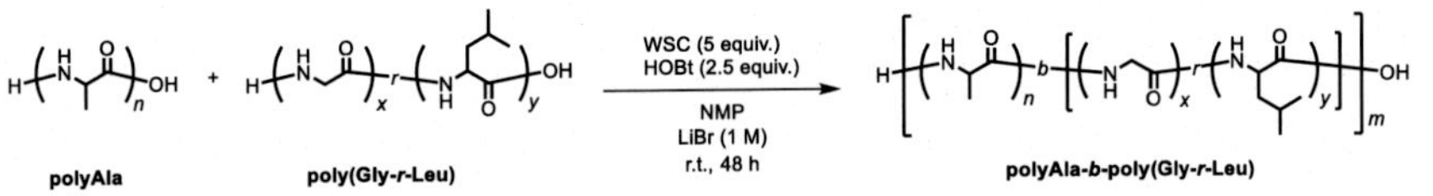

Scheme 10.4 Polycondensation of polypeptide fragments using WSCI as a condensing agent.

stopcock under nitrogen. To this were added HOBt monohydrate (0.068 g, 0.5 mmol) and 1-ethyl-3-(3-dimethylaminopropyl)carbodiimide hydrochloride (WSCI) (0.177 g, 1 mmol), and the mixture was stirred at room temperature for 48 h under nitrogen. The mixture was poured in water and the precipitate was collected by centrifugation at 9000 rpm and 4°C for 15 min. The crude product was washed with water and methanol followed by centrifugation twice, and lyophilized to afford white powder. The yield was 79.7 mg (80%). The number- (M_n) and weight average molecular weight (M_w) of the polymer was estimated by gel permeation chromatography (GPC) measurement eluted with NMP containing LiBr (10 mM) as 1600 and 2000, respectively (Scheme 10.4).

10.2.2.5 Post-polycondensation of polypeptide fragments using polyphosphoric acid (PPA)

A general synthetic procedure using PPA is described in the following. A solution of polyAla and poly(Gly-*r*-Leu) (100 mg in total) in NMP (1 mL) containing lithium bromide (0.0868 g, 1 mmol) was added to PPA (0.15 g) placed in a 10 mL flask equipped with a stirring bar and a stopcock under nitrogen. The solution was stirred at 120°C for 24 h under nitrogen. After cooling to room temperature, the mixture was poured in water and stirred for 1 h. The precipitate was collected by centrifugation at 9000 rpm and 4°C for 15 min. The crude product was washed with water followed by centrifugation twice, and lyophilized to afford slightly brown powder.

10.2.3 Chemoenzymatic synthesis of DOPA-containing peptides

Chemoenzymatic syntheses of poly(L-tyrosine) [P(Tyr)] and poly(L-tyrosine-*r*-L-lysine) [P(Tyr-Lys)] were performed using the activated papain as a catalyst to monitor the degree of polymerization, number-average molecular weight (M_n), and composition for 24 h, based on our previous reaction conditions [11]. Briefly, a 0.6 M solution of an amino acid ethyl ester mixture (Tyr-Et:Lys-Et ratio of 1:0, 5:1, 1:1, and 1:5) was made up in a 1.0 M borate buffer solution (pH 9.5) and was prepared in an

EYELA ChemiStation (Tokyo Rikakikai Co., Tokyo, Japan) at 40°C. The activated papain (7.0 mg/mL) was added to the reaction. Following the reaction, the reaction mixture was centrifuged at 15,000 × *g* and the resultant pellet was washed twice with cold Milli-Q (4°C). For the characterization of P(Tyr) and P(Tyr-Lys), the washed pellet was lyophilized and directly used for the tyrosinase catalyzed reaction. To convert tyrosine to 3,4-dihydroxy-L-phenylalanine (DOPA), 10 mg of the synthesized polypeptides, P(Tyr) and P(Tyr-Lys), were dissolved in a modified phosphate buffer (20 mM boric acid, 150 mM NaCl, 0.1 M ascorbic acid, and 0.1 M phosphate buffer, pH 7). Tyrosinase from mushroom (final concentration: 1300 U/mL) was used as the enzymatic catalyst. The reaction was performed at 25°C for 0.5−12 h, as described previously [12,13]. The reaction mixture was washed twice with cold Milli-Q water (4°C), to remove the enzyme, and was then lyophilized to remove any traces of water, resulting in a powder of poly(L-tyrosine-*r*-3,4-dihydroxy-L-phenylalanine-*r*-L-lysine).

10.2.4 Chemoenzymatic synthesis of Aib-containing peptides

10.2.4.1 Synthesis of AibAla-OEt HCl salt

To a flask equipped with an addition funnel and a stir bar were added alpha-(Boc-amino)isobutyric acid (10.16 g, 50 mmol), L-alanine ethyl ester hydrochloride (7.68 g, 50 mmol), L-HOBt monohydrate (6.76 g, 50 mmol), triethylamine (7.0 mL, 50 mmol), and chloroform (50 mL) at −10°C under nitrogen. A solution of L-ethyl-3-(3-dimethylaminopropyl) carbodiimide (WSCI) hydrochloride (9.59 g, 50 mmol) in chloroform (50 mL) was added dropwise over 30 min, and the resulting mixture was stirred at −10°C for 30 min and then at 25°C for 24 h. The mixture was washed sequentially with water, 5% $NaHCO_3$ aqueous, and brine. The organic layer was dried with Na_2SO_4 and concentrated using a rotary evaporator. The product was dried in vacuo to give Boc-AibAla-OEt in a yield of 13.9 g (92%). The obtained Boc-AibAla-OEt was then subjected to deprotection of the Boc group. To the solution of Boc-AibAla-OEt (7.56 g, 25 mmol) in dichloromethane (15 mL) was slowly added TFA (9.6 mL, 0.125 mol) at 0°C under nitrogen. The mixture was stirred at 0° C for 10 min and then at 25°C for 24 h. After the solvent was removed under reduced pressure, the crude product was dissolved in dioxane/HCl (4 M, 10 mL). The solution was poured into diethyl ether. The precipitate was filtered, washed with diethyl ether, and dried in vacuo to afford AibAla-OEt as a white hygroscopic solid. The yield was 5.47 g (92%).

10.2.4.2 Synthesis of AlaAibAla-OEt HCl salt

To a flask equipped with an addition funnel and a stir bar were added *N*-Boc-L-alanine (1.59 g, 8.4 mmol), AibAla-OEt hydrochloride (2.0 g, 8.4 mmol), HOBt monohydrate (1.28 g, 8.4 mmol), triethylamine (1.2 mL, 8.4 mmol), and chloroform (10 mL) at −10°C under nitrogen. A solution of WSCI hydrochloride (1.61 g, 8.4 mmol) in chloroform (10 mL) was added dropwise over 30 min, and the resulting mixture was stirred at −10°C for 30 min and then at 25°C for 24 h. The mixture was washed sequentially with water, 5% $NaHCO_3$ aqueous, and brine. The organic layer was dried with Na_2SO_4 and concentrated using a rotary evaporator. The product was dried in vacuo to give Boc-AlaAibAla-OEt in a yield of 2.45 g (78%). The obtained Boc-AlaAibAla-OEt was subsequently subjected to deprotection of the Boc group. To the solution of Boc-AlaAibAla-OEt (2.45 g, 6.6 mmol) in dichloromethane (6.0 mL) was slowly added TFA (9.6 mL, 0.125 mol) at 0°C under nitrogen. The mixture was stirred at 0°C for 10 min and then at 25°C for 24 h. After the solvent was removed under reduced pressure, the crude product was dissolved in dioxane/HCl (4 M, 3 mL). The solution was poured into diethyl ether. The precipitate was filtered, washed with diethyl ether, and dried in vacuo to afford AlaAibAla-OEt as a white hygroscopic solid. The yield was 1.98 g (97%).

10.2.4.3 Chemoenzymatic polymerization of Aib-containing monomers

The general procedure for the chemoenzymatic polymerization of Aib-containing monomers was as follows (Scheme 10.5) [14]. To a glass tube

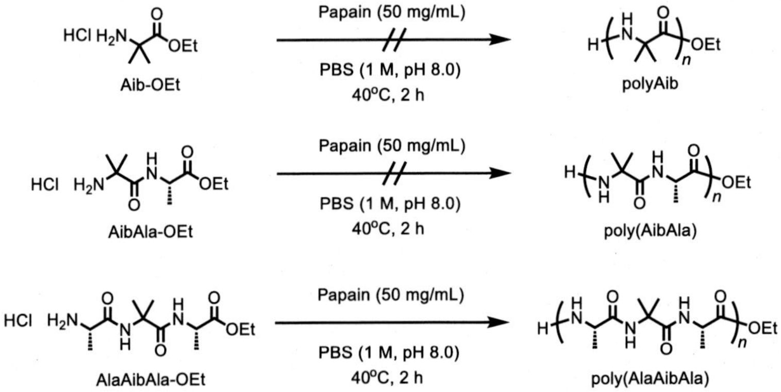

Scheme 10.5 Chemoenzymatic polymerization of Aib-containing monomers using papain [14].

equipped with a stir bar were added AlaAibAla-OEt HCl salt (0.155 g, 0.5 mmol) and phosphate buffer (1 mL, 1 M, pH = 8.0), and the mixture was stirred at 40°C until all of the substrates were completely dissolved. To this solution was added a solution of papain (0.100 g) in phosphate buffer (0.5 mL) in one portion. The final concentrations of AlaAibAla-OEt and papain were 0.25 M and 50 mg/mL, respectively. The mixture was stirred at 800 rpm and 40°C for 6 h. After the mixture cooled to room temperature, the precipitate was collected by centrifugation at 7000 rpm for 10 min at 4°C. The crude product was washed twice with deionized water and lyophilized to provide the oligopeptide as a white solid. The yield was 0.037 g (33%).

10.3 Biological synthesis

10.3.1 Protein expression by recombinant *Escherichia coli*

10.3.1.1 Design and cloning of the sequence of interest

The spider silk repeat unit was selected based on the consensus repeat (SGRGGLGGQGAGAAAAAGGAGQGGYGGLGSQGT) derived from the native sequence of the dragline protein major ampullate spidroin 1 (MaSp1) sequence from the spider *Nephila clavipes* (Accession P19837) [15]. The 6-mer containing six contiguous copies of this repeat was developed through the transfer of cloned inserts to pET-30a, which had been modified with a linker carrying the restriction sites *Nhe*I and *Spe*I. The double-stranded DNAs of the sequences were ligated into pET-30a to generate pET-30-6-mer by DNA ligase (New England Biolabs Inc, Ipswich, MA, United States).

10.3.1.2 Protein expression and purification

The construct pET30-6-mer was used to transform the *E. coli* strains BL21, and protein expression carried out by heat shock using chemical competent cells or electroporation. The transformed cells were cultivated in lysogeny broth (LB) broth containing kanamycin (50 μg/mL) at 37°C. Protein expression was induced by the addition of 0.5 mM IPTG (Sigma-Aldrich, St. Louis, MO, United States) when the optical density $(OD)_{600}$ nm reached 0.6. After approximately 4 h of protein expression, cells were harvested by centrifugation at 13,000 × *g*. The cell pellets were resuspended in denaturing buffer (100 mM NaH_2PO_4, 10 mM Tris HCl, 8 M urea, pH 8.0) and lysed by stirring for 12 h followed by centrifugation at 13,000 × *g* at 4°C for 30 min. His-tag purification of the proteins was performed by addition of Ni-NTA agarose resin (Qiagen, Valencia, CA, United States) and 20 mM imidazole to the supernatant (batch purification) under denaturing conditions. After washing the column with denaturing buffer at pH 6.3, the proteins were eluted with denaturing buffer at pH 4.5 (without imidazole). Purified samples were extensively dialyzed against Milli-Q water. For dialysis, Slide-A-Lyzer Cassettes (Pierce, Rockford, IL, United States) with molecular-weight cut-off (MWCO) of 3500 were used. The purified protein was confirmed by sodium dodecyl sulfate–polyacrylamide gel electrophoresis (SDS-PAGE), high-performance liquid chromatography (HPLC), and/or liquid chromatography with tandem mass spectrometry (LC/MS/MS).

10.3.2 Protein expression by photosynthetic bacteria

The marine photosynthetic purple nonsulfur bacterium *Rhodovulum sulfidophilum* DSM1374/ATCC35886/W4 can be obtained from the American Type Culture Collection. For general cultivation purposes, as described in a previous report [16], *R. sulfidophilum* was maintained under photoheterotrophic growth conditions on marine agar (MA) or in marine broth (MB) (BD Difco) at 30°C with continuous far-red LED light (730 nm, 20–30 W m^{-2}). Culture medium for recombinant strains of *R. sulfidophilum* was supplemented with 100 mg L^{-1} kanamycin for plasmid maintenance purposes. *E. coli* DH5α (TaKaRa Bio) was used for general cloning purposes and maintained on LB agar or in LB (BD Difco) at 37°C under aerobic conditions with shaking at 180 rpm. For the purpose of plasmid conjugation into *R. sulfidophilum*, *E. coli* S17-1 was used as a donor strain and maintained in the same way as *E. coli* DH5α [17].

10.3.2.1 Plasmid construction and conjugation into Rhodovulum sulfidophilum

Polymerase chain reaction (PCR) amplifications of the *trc* promoter and *MaSp1* gene were performed using KOD-plus DNA polymerase (TOYOBO) with primers that added suitable restriction sites. The RBS sequence "AGGAGA," derived from the region upstream of the *pufQ* gene or *puf* operon, was added downstream of the *trc* promoter. *E. coli*-codon-optimized *MaSp1* gene sequences from the spider *N. clavipes*, together with His-Tag, S-Tag, thrombin, and enterokinase sequences, were amplified from pET-30a-MaSp1 [17]. Both *trc* promoter and *MaSp1* gene sequences were digested with appropriate restriction enzymes, purified, and then ligated to the broad-host-range vector pBBR1MCS-2 with Ligation high Ver.2 (TOYOBO). Bacterial transformation of the newly constructed pBBR1-P_{trc}-MaSp1 into *E. coli* S17-1 was performed according to standard protocols [17]. Positive transformants were detected by colony PCR and double-confirmed with DNA sequencin. *E. coli* S17-1 harboring plasmid was inoculated into 5 mL of LB medium supplemented with 50 mg L^{-1} kanamycin and incubated at 37°C for 16 h at 180 rpm. *R. sulfidophilum* was inoculated into 15 mL of MB and incubated at 30°C with continuous far-red light (730 nm, 30 W m^{-2}) for 30 h. Both bacterial cultures were centrifuged at 10,000 × *g* for 3 min and resuspended in fresh culture medium (LB for *E. coli* S17-1 and MB for *R. sulfidophilum*). Then, a bacterial suspension mixture was prepared according to a 1:1 ratio of *E. coli* S17-1 and *R. sulfidophilum*. Approximately 200 μL of the cell mixtures was spotted on an MA plate and incubated at 30°C with continuous far-red light (730 nm, 30 W m^{-2}) for 24 h. Then, the regrown cells were scraped out and resuspended with 5 mL of fresh MB. Approximately 100 μL of cell suspension was spread on MA containing 100 mg L^{-1} kanamycin and 100 mg L^{-1} potassium tellurite. The plate was incubated at 30°C with continuous far-red light (730 nm, 30 W m^{-2}) for seven days. Positive conjugants were detected by colony PCR and double-confirmed by DNA sequencing.

10.3.2.2 Expression of MaSp1 under photoheterotrophic conditions

Recombinant *R. sulfidophilum* containing pBBR1-P_{trc}-MaSp1 was precultured in 15 mL of MB supplemented with 100 mg L^{-1} kanamycin for two days at 30°C with continuous far-red light (730 nm, 30 W m^{-2}) until the OD_{660} reached ~1.2 [17]. After that, approximately 5 mL (10% v/v) of inoculum was transferred into 45 mL of fresh MB supplemented with 100 mg L^{-1} kanamycin for 4 days of incubation at 30°C with continuous far-red light (730 nm, 30 W m^{-2}) until the OD_{660} indicated that the stationary growth phase had been reached (OD_{660} ~2.0). The bacterial cells were harvested by centrifugation at 10,000 × *g* for 10 min at 4°C, and the supernatant was discarded. For resuspension, 5 mL of lysis buffer (10 mM Tris, 8 M urea and 100 mM NaH_2PO_4, pH 7.4) was added to the cell pellet for every 1 g of wet cells. The cell suspensions were stirred vigorously for 12 h and then centrifuged at 10,000 × *g* for 30 min. Finally, total soluble proteins in the supernatant fraction were collected and then quantified by a Pierce BCA Protein Assay Kit (Thermo Fisher Scientific).

10.3.2.3 Confirmation of protein expressed in the photosynthetic bacteria

Soluble proteins were resolved via SDS-PAGE by using a 16.5% Mini-PROTEAN Tris-Tricine precast gel (Bio-Rad). The gel was first stained with fixation buffer (25% (v/v) ethanol and 15% (v/v) formaldehyde) for 30 min before proceeding with Coomassie Brilliant Blue (CBB)-R250 (FUJIFILM Wako) staining for 1 h.

Western blotting was performed by electrophoretically transferring proteins from an SDS-PAGE gel to an Immuno-Blot PVDF (polyvinylidene difluoride) membrane (0.2 μm pore size) (Bio-Rad) using a Trans-Blot SD Semi-Dry Transfer Cell (Bio-Rad). The blotting procedures were conducted according to the His•Tag western reagents protocol (Novagen Biosciences). In brief, membrane staining with Ponceau S (Beacle) was performed after electroblotting to confirm successful protein transfer before proceeding with blocking with 5% (w/v) milk/phosphate-buffered saline for 12 h. The membrane was first probed with 0.2 $\mu g\, mL^{-1}$ monoclonal His•Tag primary antibody (Merck Millipore) for 2 h and then probed with 0.1 $\mu g\, mL^{-1}$ goat antimouse IgG horseradish peroxidase (HRP) secondary antibody (Abcam) for 2 h. Finally, the membrane was treated with SuperSignal West Pico PLUS Chemiluminescent Substrate (Thermo Fisher Scientific), and chemiluminescence images were taken using a LAS-3000 imager (Fujifilm). Band intensities on the western blots were analyzed using Fiji/ImageJ version 1.52p.

10.4 Protein sample preparations

10.4.1 Preparation of silk solution

Silk solutions were prepared according to a previously reported method [18,19]. Briefly, silkworm cocoons of *B. mori* were cut and boiled for 30 min in a 0.02 M Na_2CO_3 solution and then washed with Milli-Q water to remove sericin proteins and wax. The extracted silk proteins were dried and dissolved in a 9.3 M LiBr solution at 60°C for 2 h at a concentration of 200 g/L. The silk solution was dialyzed with Milli-Q water for at least 4 days using a dialysis membrane (Pierce Snake Skin MWCO 3500; Thermo Fisher Scientific, Waltham, MA, United States). Dialysis was completed when the conductivity of the dialysis solution was identical to that of Milli-Q water. The silk solution was concentrated by dialysis against a 20 wt.% poly(ethylene glycol) solution (number-average molecular weight: 10,000 g/mol), [20] resulting in a 270 g/L silk solution.

Next, 1 mL of the silk solution was dried at 60°C for 24 h, and the resultant silk film was weighed to determine the silk concentration in the solution. The concentrated silk solution was diluted to prepare solutions of different silk concentrations (75, 100, 150, 200, and 250 g/L).

10.4.2 Preparation of silk hydrogels

To prepare physical gel, the silk solutions were poured into silicone cylinders that were 15 mm in diameter and 7.5 mm in height, and the cylinders were sealed with clips and dialysis membranes (Spectra/por 100–500 MWCO, Spectrum Laboratories, Inc., Rancho Dominguez, CA, United States). The MWCO of the dialysis membranes was sufficiently low to retain the silk proteins in the solution. The sealed silicon cylinders containing the silk solutions were immersed in ethanol for 12 h at 25°C, resulting in the formation of physical gel.

To obtain chemical gel, HRP and hydrogen peroxide were used with the silk solutions according to a previous report [21]. HRP type VI, lyophilized powder (Sigma-Aldrich, St. Louis, MO, United States) was dissolved into Milli-Q at 1000 U/mL (10 mg/mL). The silk solutions at various concentrations were mixed with the HRP solution at 30 μL/mL of the silk solution. The silk/HRP solution was mixed with 10 μL/mL of 165 mM hydrogen peroxide solution (Wako Pure Chemicals Industries Ltd., Osaka, Japan; the final concentration was 1.65 mM) to initiate oxidation reactions between the silk molecules. After 24 h, chemical gel was obtained.

To prepare chemical–physical gel, chemical gel was immersed in an ethanol solution (20%, 40%, 60%, 80%, and 100%) for 24 h to induce the formation of beta-sheet structures and physical molecular networks. The resultant chemical–physical gel from different silk concentrations was used for protein resin preparations.

10.4.3 Preparation of silk powder

To obtain silk powder samples, silk fibroin solution was prepared as described above. The silk solution was lyophilized to afford the silk powder.

10.4.4 Film preparation

Silk powder was dissolved into hexafluoroisopropanol (HFIP) to generate a silk HFIP solution (25 g/L). It took approximately 24 h to dissolve the silk powder in the HFIP at 25°C. To prepare the film samples, the silk

HFIP solution was cast on a Teflon petri dish. After drying for 16 h, films with a thickness of approximately 50 μm were obtained.

Silk films were prepared by mixing silk fibroin (SF) solution (8% w/v) with an equal volume (1:1 v/v ratio) of modified silk solution (0.5% w/v). Aliquots of 0.5 mL were than cast into wells of a 24-well plate. Samples were left to dry on the bench overnight at environmental temperature and humidity. For thickness measurements were conducted with a 0–1″ outside micrometer (Chicago Brand, Medford, OR, United States).

10.4.5 Preparation of silk samples with different water contents

Protein samples were dried in a vacuum oven at 40°C for 24 h to remove water without changing the solid-state structure. The samples were incubated at 25°C for 48 h under various relative humidities (RHs), which were controlled by the saturated water vapor of 10 types of salt, according to a previous method [22]. The 10 salts used in this study were lithium bromide (RH 6%), lithium chloride (RH 11%), potassium acetate (RH 23%), magnesium chloride (RH 33%), potassium carbonate (RH 43%), sodium bromide (RH 58%), potassium iodide (RH 69%), sodium chloride (RH 75%), potassium chloride (RH 84%), and potassium sulfate (RH 97%).

10.4.6 Preparation of silk nanoparticle

Silk solutions with different concentrations were treated at 120°C at 0.1 MPa for 20 min [23]. The treated silk solution was centrifuged at 1500 × *g* for 1 min to remove aggregations and its supernatant was mixed with ethanol for 2 h. The solvent was replaced with Milli-Q water, and then solutions containing silk particles were obtained. The absence of ethanol in the silk nanoparticle was confirmed by differential scanning calorimetry (DSC, Pyris 1; Perkin-Elmer, Waltham, MA, United States). DSC measurements were performed using a Perkin-Elmer Pyris 1 equipped with a cooling accessory. Silk nanoparticles containing fluorescent dyes, that is, Rhodamine B, Texas Red and fluorescein isothiocyanate, were prepared by adding each fluorescent dye (final concentration of 0.1 g/L) into the silk solution before applying the heating treatment. The other steps were identical to those described above. The encapsulation efficiency of each fluorescent dye in the silk nanoparticles was determined by gravimetric analysis of each dye solution.

10.5 Quantification of structural proteins

There are various natural and synthetic structural proteins and structural protein-like polymers, as introduced above. The content of structural proteins is always important when considering the biomass origin or biodegradability of a material. Furthermore, structural protein quantification is necessary for determining the content of structural proteins in blends and/or composite materials for commercialization. The methods for quantifying structural proteins are the same as those typically used for protein quantification; however, a suitable method must be identified based on the structure, shape, morphology, and size of the material to be measured.

10.5.1 Quantification

When the whole material is composed of proteins, there is no need to quantify the protein content. However, when in composite materials with other materials, such as polymer resin, the protein content in the material must be determined. If the structural protein is separable from other resins in a composite/blend material, the structural protein can be quantified by the weight ratio (Fig. 10.2); that is, the separated structural protein parts are quantified by weight [24]. If the structural protein parts are inseparable, the structural protein needs to be extracted for the content to be quantified.

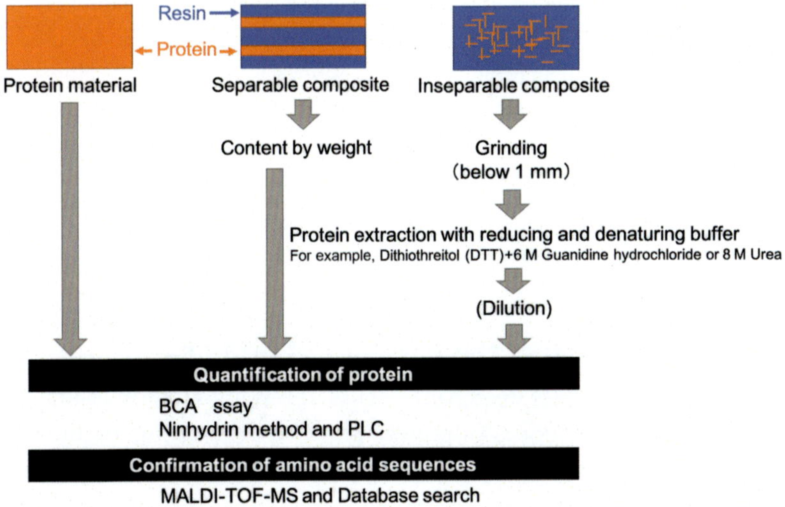

Figure 10.2 Overview of the procedures for quantifying structural proteins in protein-containing materials.

To extract the protein from a composite/blend material, a pretreatment is necessary prior to extraction. The material is first finely pulverized and subsequently treated with a reducing agent in a denaturing buffer. Typical examples of reducing agents are 10 mM dithiothreitol (DTT), mercaptoethanol, and Tris(2-carboxyethyl)phosphine, whereas examples of denaturing buffer are 6 M guanidine hydrochloride, 8 M urea, and formic acid. However, each reducing agent and denaturing buffer combination must be considered with caution because dangerous reactions can occur such as that between urea and chlorine bleach.

To quantify a protein, namely, to determine the concentration of protein in a solution, the UV method, Biuret method, Lowry method, bicinchoninic acid (BCA) method, Bradford method, etc., can be used. The UV method involves detecting the absorption at approximately 280 nm, which is the major absorption of tyrosine and tryptophan, but this strategy is not suitable for the quantification of structural proteins because it depends on the tyrosine and tryptophan content of the protein. The Biuret method is based on the detection of Cu^{+}, which is produced when Cu^{2+} reacts with a protein under alkaline conditions, is an appropriate technique, as are the Lowry method and BCA method, which are advanced variations of the Biuret method. However, the Lowry method is easily affected by a reducing agent such as DTT and is not suitable for this purpose considering the extraction conditions for structural proteins. On the other hand, the BCA method is inhibited in the presence of approximately 1 mM DTT, approximately 4 M guanidine hydrochloride, and approximately 3 M urea but can be used for quantification if the structural protein solution is sufficiently dilute. The Bradford method is conducted under acidic conditions, while the BCA method uses alkaline conditions. Many structural proteins have lower solubility under acidic conditions, making BCA quantification under alkaline conditions the more suitable method. When using the BCA method, both the denaturation buffer and the protein extract must be sufficiently diluted prior to use.

10.5.2 Confirmation of the presence of protein

To confirm that the test material contains some amount of protein, the amino acids and amide bonds of proteins must be detected by the ninhydrin method [25–27]. In the ninhydrin method, the sample is dissolved in a protein denaturing solution, and the resulting solution is used for amino acid analysis. For amino acid composition analysis, a protein sample is hydrolyzed with hydrochloric acid, and each amino acid is quantified by HPLC.

Using this strategy, 17 types of amino acids can be confirmed because tryptophan, asparagine, and glutamine cannot be quantified individually. The contents of asparagine and glutamine are combined with those of aspartic acid and glutamic acid, respectively, due to the hydrolysis of asparagine and glutamine during the sample preparation process. Tryptophan is also degraded during preparation, and hence, other protocols for determining tryptophan are needed. When posttranslationally modified peptides and amino acids, such as hydroxyproline, DOPA, and dityrosine, in structural proteins need to be quantified, standard substances corresponding to the target amino acids must be added to the mixture [11].

10.5.3 Confirmation of structural proteins among various proteins

Pure structural protein can be quantified by the general protein quantification method, as described above. However, if structural proteins and other proteins are mixed, they cannot be distinguished by the abovementioned method. To not only quantify a protein but also confirm that it is a structural protein, after treating the extracted protein with a degrading enzyme such as trypsin or separating the protein by SDS-PAGE and extracting the protein from the gel, MALDI-TOF-MS should be performed, and databases can be searched for possible sequences. This method requires the protein to have a known sequence, and it is difficult to quantify a novel structural protein prepared by chemical synthesis or genetic recombination technology using this strategy. In addition, shotgun proteomics as well as targeted proteomics using selected reaction monitoring and data-independent acquisition methods are alternative methods. However, modification with synthetic polymers will significantly prevent those characterizations. Thus, in protein quantification, it is generally difficult to specifically quantify proteins with unknown structures. Furthermore, impurity (contaminant) proteins contained in the protein materials synthesized by microbial fermentation must be considered in these analyses.

10.6 Physical characterization

10.6.1 Mechanical test

10.6.1.1 Tensile tests

The tensile tests of the film samples were performed by a mechanical testing apparatus (EZ-LX/TRAPEZIUM X, Shimadzu, Kyoto, Japan) [28]. The initial length of the film sample was approximately 5 mm.

The extension speed was 10 mm/min, and a 500 N load cell was used. The strength at break, Young's modulus, elongation at break, and toughness were obtained based on the resultant stress–strain curves.

10.6.1.2 Compression test of silk hydrogels and resins

The silk hydrogels and resins were also prepared in a Teflon cylinder that was 7.5 mm in diameter and 7.5 mm in height for the compression tests. The compression tests were conducted using a mechanical testing apparatus (EZ-Test, Shimadzu, Kyoto, Japan) at a velocity of 0.75 mm/min.

10.6.1.3 Adhesion test

Protein/peptide solutions (10 wt.%), consisting of variable DOPA content and made up with different pH levels of 6, 10, and 12, were applied in equal volumes between two freshly cleaved sheets of mica, which were subsequently pasted together. After 24 h, adhesive shear strength of the pasted sample was characterized by a mechanical testing apparatus (EZ-Test). The measurements were performed at approximately RH 40% and 25°C. The breaking strain, adhesive strength, adhesive fracture energy, and Young's modulus were analyzed based on the stress–strain curves of the prepared samples. The experiments were conducted five times for each condition. The adhesive function between paper, polypropylene (PP), and wood was also characterized in the same conditions. The sample preparation of silk film was varied because pasted samples were ruptured at overlapping points by lap shear tests when samples were prepared in the same conditions as others. A silk film of 10 mm width × 10 mm length was first pasted on a flat wood sheet (10 mm width × 40 mm length) with a super glue, and 20 μL of DOPA-SF solution was added to the surface of the silk film. Curing conditions were maintained in the same manner as others, although the overlay was changed to 10 mm width × 5 mm length (total overlay area: 0.5×10^2 mm^2). The initial distance between grips was also changed to 35 mm. The elongation rate was kept at 10 mm/min.

10.6.1.4 Rheological characterization

Silk hydrogel was set between the cone plates of a rheometer (MCR501; Anton Paar, Austria). The oscillatory shear rheological properties, that is, the storage elastic modulus (G') and the loss elastic modulus (G''), during beta-sheet formation by ethanol were measured at 25°C. The strain and the frequency were 1% and 1 Hz, respectively.

Silk solution was characterized with the different setting. The time development of strain $\gamma(t, \sigma)$ for phase-separated silk under constant stress, σ, was measured at 25°C with the stress-controlled rheometer MCR 301 (Anton Paar). All measurements were performed using a cone-and-plate fixture. The diameter of the cone was 10 mm, and the cone angle was 2 degrees. The applied stress ranged from 1.0 to 5.0 Pa, which was preliminarily checked to be within the range of linear viscoelasticity. From the obtained strain data, the creep compliance, $J(t)$, was evaluated using the equation:

$$J(t) = \frac{\gamma(t, \sigma)}{\sigma}$$

Purified silk solution (16 μL of 116.8 mg mL^{-1}) was pipetted onto the plate, and 4 μL of 1.0 M potassium phosphate buffer at pH values ranging from 8 to 5 was added. The creep measurements were initiated approximately 10 min after the cone fixture was set. For the pH 5 experiment, we conducted oscillatory measurements to investigate the time development of storage and loss moduli (G' and G'', respectively) with strain $\gamma = 1\%$ and angular frequency $\omega = 10 \text{ rad} \cdot s^{-1}$ (prior to creep test initiation).

10.6.1.5 Fluorescence recovery after photobleaching (FRAP)

Samples were prepared by pipetting 1.5 μL of N-R12-C (20 mg mL^{-1} in 0.15 M NaCl, with a 2:1 ratio of unlabeled:DyLight-488-labeled protein) into a 1.5 μL drop of 1.0 M potassium phosphate at the appropriate pH on a glass slide with a customized chamber. Intradroplet FRAP was performed using a Zeiss LSM 880 confocal microscope with a 63 × /1.4 numerical aperture oil immersion objective. For samples prepared at pH 6–8, approximately equal-sized spherical protein droplets (~5–10 μm in diameter) resting on the glass surface were selected for analysis. For samples prepared at pH 5 and 5.5, which formed fibrillar networks, discrete clumps of protein condensates of similar dimensions were analyzed. The sample fluorescence was bleached in a 0.55-μm-diameter spot in condensates with a size 5–10 times the bleach diameter, and time-lapse images were collected at approximately 773 ms intervals for >5 min. For each test, three regions of interest were analyzed, including the bleached spot, the overall droplet region, and a background region, using Fiji software [29]. Image drift along the *xy* plane, when present, was corrected using the ImageJ Template Matching plugin. The results represent replicates of several experiments (pH 8, $N = 6$; pH 7, $N = 6$; pH 6.5, $N = 7$; pH 6,

$N = 8$; pH 5.5, $N = 6$; pH 5, $N = 6$). The data were analyzed using easyFRAP [30], with full-scale normalized curves used to calculate the mobile fraction (Mf) and half-maximal recovery time ($t_{1/2}$) following biexponential curve fitting.

10.6.2 Advancing contact angle of water

The wettability of the cast film surface was estimated by the advancing contact angle (θ_{adv}) measurement with distilled water using a FACE Contact Angle Meter CA-X (Kyowa Interface Science, Saitama, Japan) according to a previous procedure [31]. Cast films of the samples were prepared on glass substrates by solution casting from the silk and fluoropolymer HFIP solutions as described above. The θ_{adv} value was calculated as the average of 10 data obtained at different points on the surface ($n = 5$).

10.6.3 Water vapor barrier test

The water vapor permeability rates of the film samples were determined at 37°C using a Mocon Permatran-W model 1/50 G (Modern Controls, Minneapolis, MN, United States) under standard conditions (ASTM 3985) [32]. Each measurement was continued until the water vapor permeability rate reached a stable value.

10.6.4 Shrinkage test

B. mori silk textiles were kindly provided by Spiber Inc. (Tsuruoka, Japan). The silk textiles were immersed in Lumiflon LF600X (50 wt.% fluoropolymer, 26 wt.% xylene, 24 wt.% ethylbenzene) for 1 min and then were dried in air at 25°C and an RH of approximately 40%. Square silk textile samples (approximately 50 × 50 mm) with and without a fluoropolymer coating were used for the shrinkage test. The samples were immersed in hot water (40°C) for 10 min and were dried at 25°C and an RH of 65% for 16 h. This washing and drying cycle was performed three times. To evaluate the shrinkage of the samples, the squares of the samples were measured, and the changes in the area of the samples were determined. The test was performed three times, and the results are expressed as the mean values and standard deviations.

10.6.5 Viscosity measurements

To study the change in the molecular weight of silk, the kinematic viscosities of silk solutions were measured by using an Ubbelohde viscometer (viscometer constant of 0.003 $mm^2 \cdot s^{-2}$; Sibata, Japan) at 40°C.

Temperature control was performed using a thermostatic water bath (Sibata, Japan). The cocoon raw fiber, degummed fiber, and silk film were dissolved in 9.3 M LiBr at a concentration of 10 g/L, which was followed by two cycles of centrifugation at 12,700 × g at 4°C for 20 min. The supernatant (13 mL) was then used for the viscosity measurement.

10.7 Thermal analyses

Thermogravimetric analysis (TGA) and DSC of silk materials were measured simultaneously by using TGA/DSC2 (METTLER TOLEDO, Greifensee, Switzerland). The material (3−5 mg) was weighed in an aluminum pan and equilibrated at the desired RH. The material was heated at 20°C/min from 30°C to 500°C under a nitrogen atmosphere in triplicate. The lid of the aluminum pan had a pinhole to allow water removal and to prevent the pan from bursting due to an increase in the internal pressure during the heating process. The device was calibrated with an empty cell to form a baseline and with indium to characterize the heat flow and temperature of the system. The silk materials were incubated with the saturated salt solutions until just before each measurement was made. The water contents of each silk material was calculated as the total amount of water obtained by the TGA data divided by the total weight of the silk material. The curves calculated from the least-squares method were applied to the TGA curves. DSC measurements of silk materials at the subzero temperatures were performed by DSC8500 (Perkin-Elmer) at the scan rate of 20°C/min from −150°C to 30°C.

Fast-scanning DSC of silk materials was conducted at the scan rate of 6000°C/min using Flash DSC1 (METTLER TOLEDO) device. The emply sensor was conditioned five times, according to the procedure in the manufacturer's manual. Silicone oil was placed on the sensor to improve the adhesion between the sensor and the silk materials [33,34]. Under an optical microscope, a small piece of silk material was placed on the 100 × 100 μm^2 active area of the chip sensor. The measurement was performed under an inert atmosphere of nitrogen gas at a flow rate of 50 mL/min.

10.8 Structural characterization

10.8.1 Nuclear magnetic resonance analysis

To confirm the chemical modifications of silk proteins, proton nuclear magnetic resonance spectral data were obtained using a Varian VNMRS

500 MHz (Agilent Technologies, Palo Alto, CA, United States) at 20°C. Samples were dissolved in D_2O and all spectra were referenced to the residual solvent peak (H_2O) at $\delta = 4.65$ ppm. Peak assignments corresponding to the silk backbone: δ 0.72 (br, Val γ), 1.21 (br, Ala β), 1.90 (br, Val β), 2.74–2.85 (br, Asp/Tyr β), 3.77 (m, Ser β/Gly α), 4.14–4.29 (m, Ala α/Ser α), 6.61 (m, Tyr φ), 6.90 (m, Tyr φ), 7.03–7.12 (br, Phe φ). Peaks corresponding to the substituent protons in the carboxy-SF: 3.7 ppm (-$\underline{CH_2}$-COOH). Peaks corresponding to the substituent protons in the hydroxy-SF: 1.8 ppm (–CH_2-$\underline{CH_2}$-CH_2-OH), 3.55 ppm (–CH_2-CH_2-$\underline{CH_2}$-OH) and 3.6 ppm (–$\underline{CH_2}$-CH_2-CH_2-OH). Peaks corresponding to the substituent protons in the methyl-SF: 0.75 ppm (–CH_2-CH_2-CH_2-$\underline{CH_3}$), 1.4 ppm (–CH_2-CH_2-$\underline{CH_2}$-CH_3) and 3.4 ppm (–$\underline{CH_2}$-CH_2-CH_2-CH_3), (the fourth peak expected at 1.67 ppm (–CH_2-$\underline{CH_2}$-CH_2-CH_3) is not clearly detectable because of overlap with the SF backbone signals).

10.8.2 Amino acid composition analysis

Amino acid composition analysis was performed using the ninhydrin method. The hydrolyzed amino acids were characterized using High-Speed Amino Acid Analyzers, L-8900 and L8500A (Hitachi-HighTech, Tokyo, Japan). In addition to the natural amino acids, DOPA and 2,4,5-trihydroxyphenylalanine (6-hydroxy-DOPA, TOPA; Sigma-Aldrich) were used to calibrate the analyzer, and accordingly, the peaks were assigned as DOPA and TOPA.

10.8.3 Synchrotron wide-angle and small-angle X-ray scattering (WAXS and SAXS) measurements

Synchrotron WAXS and SAXS measurements were performed at the at SPring-8 (Harima, Japan) using an X-ray energy of 12.4 keV (wavelength = 0.1 nm) and a beam diameter of 45 μm. All the WAXS patterns were recorded with a flat-panel detector (FPD, C9728DK-10; Hamamatsu Photonics, Hamamatsu, Japan) [35]. The sample-to-detector distances during the WAXS and SAXS measurements were 50.2 and 2036 mm, respectively. The exposure time for each scattering pattern was 10 s. The obtained two-dimensional scattering patterns were converted to one-dimensional profiles using Fit2D software [36]. Corrections were made for background scattering and the detector geometry [35] (Figs. 10.3 and 10.4).

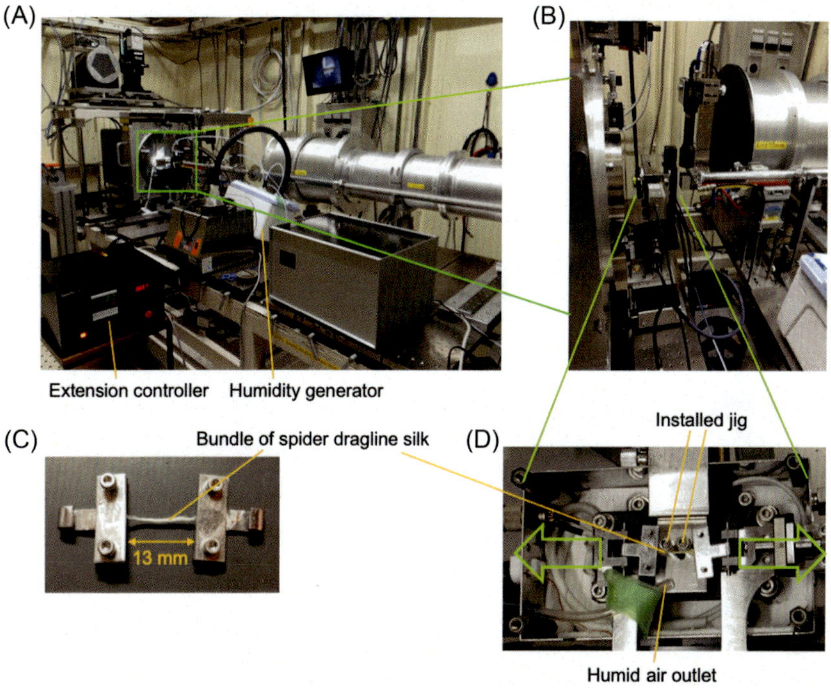

Figure 10.3 A typical synchrotron X-ray measurement setting at SPring-8, Japan. Simultaneous WAXS-tensile tests under RH control: (A) overview of the devices equipped with extension controller and humidity generator, (B) the X-ray beam radiation part, (C) the initial bundle length of silk fibers set to 13 mm, and (D) the two jigs installed to prevent vibration of the silk bundles during extension. The humid air outlet was located near the bundle of silk fibers.

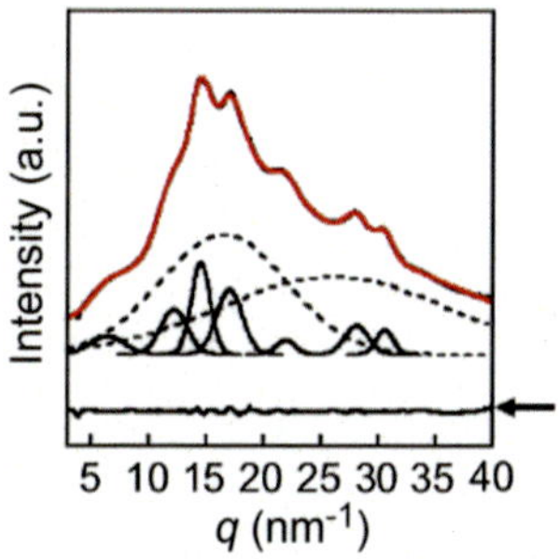

Figure 10.4 Deconvolution of one-dimensional radial integration WAXS profiles using Gaussian functions for spider dragline silk fibers. The red lines (gray color in print version) are fitted curves, the dashed lines represent the amorphous halos, the solid lines represent the crystal peaks, and the arrows show the residual between the fitted curves and the measured curves.

10.8.4 ATR-FT-IR measurements

Attenuated total reflection Fourier transform infrared (ATR-FT-IR) was measured on an IRPrestigae-21 FT-IR spectrophotometer (Shimadzu Corporation, Kyoto, Japan) with a MIRacle A single reflection ATR unit using a Ge prism. Fifty microliter of DOPA-SF solutions (15 g/L) at different pH values (4.2, 7.0, 8.0, 10, 12) were dropped on each surface and dried at room temperature for 16 h to prepare the samples. The measurements were conducted from 3800 to 600 cm^{-1}. The background spectra obtained in the same conditions were subtracted from the scan for each sample. SF with beta-sheet structure was induced by methanol treatment for 24 h, while SF with random structure was obtained from as-casted sample.

10.8.5 Scanning electron microscopy (SEM) observations

The sample morphology was analyzed by SEM. Silk samples were sliced into small pieces approximately $2 \times 2 \times 1$ mm. The sliced samples were mounted on an aluminum stub, sputter-coated with gold, and imaged by SEM (JSM6330F; JEOL Ltd., Tokyo, Japan) at an accelerating voltage of 5 kV. The cross section of the samples was prepared using a microtome (RM2265; Leica Microsystems GmbH, Wetzlar, Germany) with a diamond blade.

10.8.6 Birefringence

Retardation of the silk fiber was measured with a Birefringence Measurement System WPA-100 (Photonic Lattice Inc., Miyagi, Japan) and was characterized by the WPA-VIEW (version 1.05) software. The Birefringence Analyzing System WPA-100 can measure samples that have several thousand-nanometer phase difference values by using lasers of wavelengths 523, 543, and 575 nm. The birefringence of the silk fiber was calculated from the retardation and diameter values of the silk fibers (determined by SEM).

10.8.7 Atomic force microscopy observation

A single fiber was either placed directly or mounted with super glue (Loctite 460, Henkel, Germany) on the sample stage, and the cross section of the sample was prepared using a microtome (RM2265; Leica Microsystems GmbH, Wetzlar, Germany) with a diamond blade. The cross section of the fiber sample was observed using AFM. Images were obtained in tapping mode at 40% RH and 25°C via scanning probe microscopy (AFM5300E; Hitachi High-Technologies Corporation, Tokyo, Japan) and an RH controller (Hitachi High-Technologies Corporation). The major ampullate gland and tapering duct were mounted on mica substrates, and the skin of each organ was sliced and removed using a scalpel. The wet samples were observed using AFM under humid conditions (RH 80%). The cantilever had a spring constant of 20 N/m, and the scan frequency was 0.5 Hz. Images were obtained and analyzed with an AFM5000II controller (Hitachi High-Technologies Corporation). Under ambient conditions in air, the granules on mica exhibited a flat morphology, as reported previously for another protein [37]. The microfibril width (distance between each nanofibril) was measured in at least 20 areas in one image, and five images were acquired for each sample to ensure that the data were representative. The granule width was determined by taking into consideration the deconvolution effect of the cantilever tip [37,38]. Briefly, calibration of the cantilever tip-convolution effect was carried out to obtain the true dimensions of objects according to previously reported methods [37–39]. When the hemispherical cantilever tip interacts with the silk nanoparticle having a hemispherical cross section on the substrate, the true width of the particle (W_{true}) can be expressed as

$$W_{\text{true}} = W_{\text{AFM}} - 2w_{\text{a}} \tag{10.1}$$

where W_{AFM} is the width of the particle measured by AFM and w_{a} is the width of artifacts. The spherical part of the tip is expected to be in contact with the spherical surface of the particle during scanning, since the thickness of particle (H) may be less than the apex radius of the spherical tip (R) in this study. Therefore the value of w_{a} is given as:

$$w_{\text{a}} = \left[(R+H)^2 - R^2\right]^{1/2} - H \tag{10.2}$$

The combination of Eqs. (10.1) and (10.2) gives the true width of the particle (W_{true}) in this study. On the basis of the geometry of a cantilever tip whose radius is about 10 nm, calibration was performed to calculate the true diameters of the particles.

10.8.8 Dynamic light scattering

The silk nanoparticles were characterized by zeta potentialmeter (Zetasizer Nano-ZS; Malvern Instruments, Ltd., Worcestershire, United Kingdom). Hydrodynamic diameters of the samples were determined using the dynamic light scattering (DLS) mode of the zeta potentialmeter. Zeta potential and zeta deviation of the samples were measured three times by zeta potentialmeter, and the average data were obtained using Dispersion Technology Software version 5.03 (Malvern Instruments, Ltd).

10.8.9 Far-UV circular dichroism (CD)

CD spectra were measured at 25°C using a JASCO J820 instrument. Dilute silk protein at a final concentration of 3.7 μM was prepared in 10 mM sodium phosphate buffer (400 μL final volume), and spectra were recorded in 1 nm increments from 250–190 nm at a 100 nm min^{-1} rate with a 1 s response time in a Starna 21Q quartz cuvette. Experiments were performed three times, with identical results.

10.8.10 Raman spectroscopy

Confocal laser Raman spectra were recorded using a JASCO NRS-4100 equipped with a DU420-OE CCD detector (Andor). Samples were placed on a glass slide with a customized chamber made from vinyl adhesive tape upon which a coverslip was placed and sealed with nail polish. A 532 nm excitation source was used, with the beam focused using a 100 × Plan Achromat objective lens (oil), a grating of 900 gr/mm, and a 100 × 8000 μm slit size. Spectra were acquired from 500–2500 cm^{-1} at 10 mW beam intensity with typical exposures of 3 × 30 s. No sign of sample deterioration was observed under these conditions. All fiber samples were immersed in 0.5 M potassium phosphate, pH 5, to ensure a consistent background environment. The collected spectra were calibrated internally to the peak maximum at 1453 cm^{-1}, corresponding to deformation vibrations of CH_2/CH_3 groups. At least eight spectra were collected for each sample.

10.9 Biological assays

10.9.1 Sodium dodecyl sulfate–polyacrylamide gel electrophoresis

SDS-PAGE was applied for the silk films to investigate the change of molecular weight and its distribution. The precast 4%–15% gradient gel (BIO-RAD, CA, United States) was used. The constant voltage of 100 V was applied for 60 min. The resultant gel was stained with Coomassie Brilliant Blue G-250 (BIO-RAD, CA, United States).

10.9.2 Biodegradation test

The biochemical oxygen demand (BOD) test was performed to determine the biodegradability of the silk samples in activated sludge (Chemicals Evaluation and Research Institute, Tokyo, Japan) with an Oxitop IS-6 (WTW GmbH, Weilheim in Oberbayern, Germany) according to a previous procedure [40,41]. A sample film (approximately 10 mg) was immersed in 100 mL of activated sludge at 25°C for 30 days. The activated sludge was replaced with fresh sludge every 5 days. Before and after the biodegradation test, the morphologies and mechanical properties of the samples were characterized by the methods explained above. The biodegradability was calculated from the theoretical oxygen demand (ThOD) and BOD as follows:

$$\text{Biodegradability } (\%) = \text{BOD}/\text{ThOD} \times 100$$

10.9.3 Environmental toxicity test

The environmental toxicity of the samples was tested according to the Organisation for Economic Co-operation and Development (OECD) test guideline 202, acute immobilization test (OECD), with slight modification [42,43]. Briefly, 30 *Daphnia magna*, which were not more than 24 h old at the beginning of the test, were used for each test sample and control. For each treatment, the animals were divided into six groups of five animals, and each group was kept in 10 mL of synthetic M4 medium containing one of the test samples (or no additives as a control) for one week at a water temperature of 20 ± 1°C and a light–dark cycle of 16:8 h. The final concentration of the samples was set at 100 mg/L, which was the higher limit provided by the test guidelines. The powder samples were dispersed well by sonication just before use. Survival curves of each treatment were produced based on daily observations during the test period. The effects of the samples on their survival were evaluated by comparing the Kaplan–Meier survival curves using the log-rank test followed by Bonferroni correction.

According to the OECD test No. 211, the effect of chemicals (polymers and degradation products) on the reproductive output of *D. magna* can be assessed by the following procedures (Fig. 10.5). Young female *Daphnia* are exposed to the test substance added to water at a range of concentrations (at least five kinds). For semistatic tests, at least 10 animals need to be evaluated at each test concentration. The test duration is 21 days. If there is no

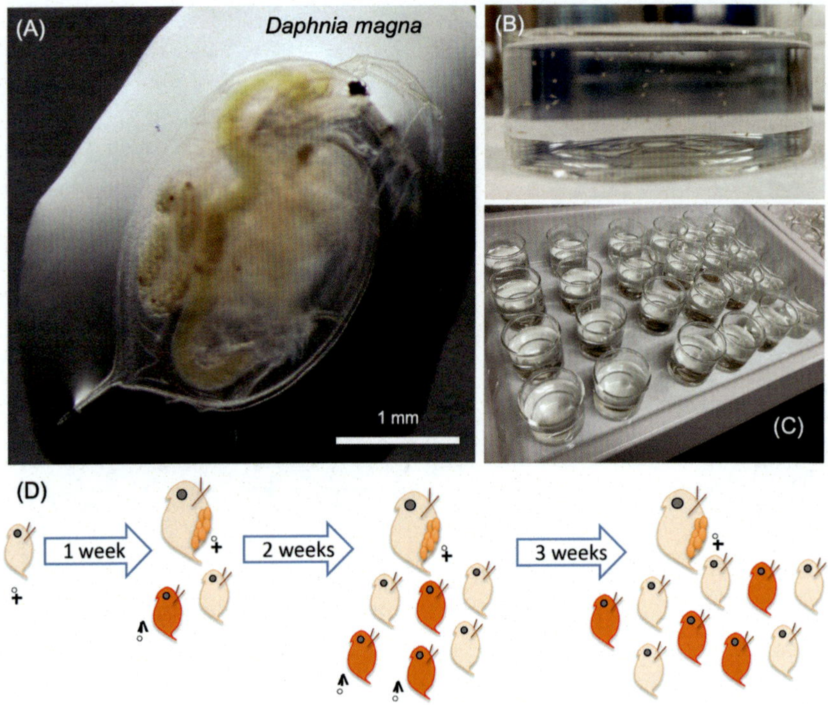

Figure 10.5 The environmental toxicity test using *Daphnia magna*. (A) *D. magna*. (B) The *D. magna* born within 24 hours. (C) Individual culture test. (D) The schematic illustrations represent the amorphous halos, the solid lines represent the crystal peaks, and the arrows show the residual between the fitted curves and the measured curves.

effect on the young female *Daphnia*, only female *Daphnia* are reproduced. However, the test substance has some effects on the reproduction, and male *Daphnia* is produced during the cultivations. Based on the individual numbers of males and females of *Daphnia*, the effect of the test substance is evaluated.

References

[1] G.B. Fields, R.L. Noble, Solid phase peptide synthesis utilizing 9-fluorenylmethoxycarbonyl amino acids, Int. J. Peptide Protein Res. 35 (1990) 161–214.

[2] R.B. Merrifield, Solid-phase peptide synthesis, Adv. Enzymol. Relat. Areas Mol. Biol. 32 (1969) 221–296.

[3] S. Fujita, K. Matsuura, Inclusion of zinc oxide nanoparticles into virus-like peptide nanocapsules self-assembled from viral beta-annulus peptide, Nanomaterials (Basel) 4 (2014) 778–791.

[4] S.K. Maji, R. Banerjee, D. Velmurugan, A. Razak, H.K. Fun, A. Banerjee, Peptide design using omega-amino acids: unusual turn structures nucleated by an n-terminal single gamma-aminobutyric acid residue in short model peptides, J. Org. Chem. 67 (2002) 633–639.
[5] T.J. Deming, Synthetic polypeptides for biomedical applications, Prog. Polym. Sci. 32 (2007) 858–875.
[6] M. Goodman, J. Hutchison, Mechanisms of polymerization of N-unsubstituted N-carboxyanhdrides, J. Am. Chem. Soc., 88, 1966, p. 3627.
[7] T. Satoh, Y. Higuchi, S. Kawakami, M. Hashida, H. Kagechika, K. Shudo, et al., Encapsulation of the synthetic retinoids Am80 and LE540 into polymeric micelles and the retinoids' release control, J. Control. Release 136 (2009) 187–195.
[8] W.H. Daly, D. Poche, The preparation of N-carboxyanhydrides of alpha-amino-acids using bis(Trichloromethyl)carbonate, Tetrahedron Lett. 29 (1988) 5859–5862.
[9] J.M. Ageitos, P.J. Baker, M. Sugahara, K. Numata, Proteinase K-catalyzed synthesis of linear and star oligo(L-phenylalanine) conjugates, Biomacromolecules 14 (2013) 3635–3642.
[10] Y.A. Ma, R. Sato, Z.B. Li, K. Numata, Chemoenzymatic synthesis of Oligo(L-cysteine) for use as a thermostable bio-based material, Macromol. Biosci. 16 (2016) 151–159.
[11] K. Numata, P.J. Baker, Synthesis of adhesive peptides similar to those found in blue mussel (Mytilus edulis) using papain and tyrosinase, Biomacromolecules 15 (2014) 3206–3212.
[12] F. Garciacarmona, J. Cabanes, F. Garciacanovas, Enzymatic oxidation by frog epidermis tyrosinase of 4-methylcatechol and para-cresol—influence of L-serine, Biochim. Biophys. Acta 914 (1987) 198–204.
[13] H.W. Duckwort, J.E. Coleman, Physicochemical and kinetic properties of mushroom tyrosinase, J. Biol. Chem. 245 (1970) 1613.
[14] K. Tsuchiya, K. Numata, Chemoenzymatic synthesis of polypeptides containing the unnatural amino acid 2-aminoisobutyric acid, Chem. Commun. 53 (2017) 7318–7321.
[15] K. Numata, B. Subramanian, H.A. Currie, D.L. Kaplan, Bioengineered silk protein-based gene delivery systems, Biomaterials 30 (2009) 5775–5784.
[16] M. Higuchi-Takeuchi, K. Morisaki, K. Toyooka, K. Numata, Synthesis of high-molecular-weight polyhydroxyalkanoates by marine photosynthetic purple bacteria, PLoS One 11 (2016) e0160981.
[17] C.P. Foong, M. Higuchi-Takeuchi, A.D. Malay, N.A. Oktaviani, C. Thagun, K. Numata, A marine photosynthetic microbial cell factory as a platform for spider silk production, Commun. Biol. 3 (2020) 357.
[18] H.J. Jin, D.L. Kaplan, Mechanism of silk processing in insects and spiders, Nature 424 (2003) 1057–1061.
[19] D.N. Rockwood, R.C. Preda, T. Yucel, X. Wang, M.L. Lovett, D.L. Kaplan, Materials fabrication from Bombyx mori silk fibroin, Nat. Protoc. 6 (2011) 1612–1631.
[20] K. Numata, S. Yamazaki, T. Katashima, J.-A. Chuah, N. Naga, T. Sakai, Silk-pectin hydrogel with superior mechanical properties, biodegradability, and biocompatibility, Macromol. Biosci. 14 (2014) 799–806.
[21] B.P. Partlow, C.W. Hanna, J. Rnjak-Kovacina, J.E. Moreau, M.B. Applegate, K.A. Burke, et al., Highly tunable elastomeric silk biomaterials, Adv. Funct. Mater. 24 (2014) 4615–4624.
[22] L.B. Rockland, Saturated salt solutions for static control of relative humidity between 5°C and 40°C, Anal. Chem. 32 (1960) 1375–1376.

[23] K. Numata, S. Yamazaki, N. Naga, Biocompatible and biodegradable dual-drug release system based on silk hydrogel containing silk nanoparticles, Biomacromolecules 13 (2012) 1383–1389.
[24] K. Numata, How to define and study structural proteins as biopolymer materials, Polym. J. 32 (2020) 1375–1376.
[25] B. Starcher, A ninhydrin-based assay to quantitate the total protein content of tissue samples, Anal. Biochem. 292 (2001) 125–129.
[26] J.M. Brewer, C.W. Roberts, W.H. Stimson, J. Alexander, Accurate determination of adjuvant-associated protein or peptide by ninhydrin assay, Vaccine 13 (1995) 1441–1444.
[27] E. Doi, D. Shibata, T. Matoba, Modified colorimetric ninhydrin methods for peptidase assay, Anal. Biochem. 118 (1981) 173–184.
[28] A.D. Malay, R. Sato, K. Yazawa, H. Watanabe, N. Ifuku, H. Masunaga, et al., Relationships between physical properties and sequence in silkworm silks, Sci. Rep. 6 (2016) 27573.
[29] J. Schindelin, I. Arganda-Carreras, E. Frise, V. Kaynig, M. Longair, T. Pietzsch, et al., Fiji: an open-source platform for biological-image analysis, Nat. Methods 9 (7) (2012) 676–682. Available from: https://doi.org/10.1038/nmeth.2019.
[30] D.N. Perkins, D.J.C. Pappin, D.M. Creasy, J.S. Cottrell, Probability-based protein identification by searching sequence databases using mass spectrometry data, Electrophoresis 20 (1999) 3551–3567.
[31] K. Numata, R.K. Srivastava, A. Finne-Wistrand, A.-C. Albertsson, Y. Doi, H. Abe, Branched poly(lactide) synthesized by enzymatic polymerization: effects of molecular branches and stereochemistry on enzymatic degradation and alkaline hydrolysis, Biomacromolecules 8 (2007) 3115–3125.
[32] Q. Yang, H. Fukuzumi, T. Saito, A. Isogai, L. Zhang, Transparent cellulose films with high gas barrier properties fabricated from aqueous alkali/urea solutions, Biomacromolecules 12 (2011) 2766–2771.
[33] P. Cebe, X. Hu, D.L. Kaplan, E. Zhuravlev, A. Wurm, D. Arbeiter, et al., Beating the heat—fast scanning melts silk beta sheet crystals, Sci. Rep. 3 (2013) 1130.
[34] P. Cebe, B.P. Partlow, D.L. Kaplan, A. Wurm, E. Zhuravlev, C. Schick, Using flash DSC for determining the liquid state heat capacity of silk fibroin, Thermochim. Acta 615 (2015) 8–14.
[35] K. Numata, H. Masunaga, T. Hikima, S. Sasaki, K. Sekiyama, M. Takata, Use of extension-deformation-based crystallisation of silk fibres to differentiate their functions in nature, Soft Matter 11 (2015) 6335–6342.
[36] Hammersley A.P., FIT2D: an introduction and overview. European Synchrotron Radiation Facility Internal Report ESRF97HA02T, 1997.
[37] K. Numata, Y. Kikkawa, T. Tsuge, T. Iwata, Y. Doi, H. Abe, Adsorption of biopolyester depolymerase on silicon wafer and poly[(R)-3-hydroxybutyric acid] single crystal revealed by real-time AFM, Macromol. Biosci. 6 (2006) 41–50.
[38] K. Numata, Y. Kikkawa, T. Tsuge, T. Iwata, Y. Doi, H. Abe, Enzymatic degradation processes of poly[(R)-3-hydroxybutyric acid] and poly[(R)-3-hydroxybutyric acid-co-(R)-3-hydroxyvaleric acid] single crystals revealed by atomic force microscopy: effects of molecular weight and second-monomer composition on erosion rates, Biomacromolecules 6 (2005) 2008–2016.
[39] K. Numata, T. Hirota, Y. Kikkawa, T. Tsuge, T. Iwata, H. Abe, et al., Enzymatic degradation processes of lamellar crystals in thin films for poly[(R)-3-hydroxybutyric acid] and its copolymers revealed by real-time atomic force microscopy, Biomacromolecules 5 (2004) 2186–2194.
[40] K. Tachibana, Y. Urano, K. Numata, Biodegradability of nylon 4 film in a marine environment, Polym. Degrad. Stabil. 98 (2013) 1847–1851.

[41] K. Numata, N. Ifuku, H. Masunaga, T. Hikima, T. Sakai, Silk resin with hydrated dual chemical-physical cross-links achieves high strength and toughness, Biomacromolecules 18 (2017) 1937–1946.

[42] R. Abe, K. Toyota, H. Miyakawa, H. Watanabe, T. Oka, S. Miyagawa, et al., Diofenolan induces male offspring production through binding to the juvenile hormone receptor in Daphnia magna, Aquat. Toxicol. 159 (2015) 44–51.

[43] B.P. Elendt, W.R. Bias, Trace nutrient deficiency in Daphnia-magna cultured in standard medium for toxicity testing—effects of the optimization of culture conditions on life-history parameters of Daphnia-Magna, Water Res. 24 (1990) 1157–1167.

Index

Note: Page numbers followed by "*f*" and "*t*" refer to figures and tables, respectively.

C

D

G

H

I

K

L

M

N

O

P

R

Printed in the United States
by Baker & Taylor Publisher Services